Lecture Notes in Computational Science and Engineering

5

Springer
Berlin
Heidelberg
New York
Barcelona
Hong Kong
London
Milan
Paris
Singapore
Tokyo

Dietmar Kröner Mario Ohlberger
Christian Rohde (Eds.)

An Introduction to Recent Developments in Theory and Numerics for Conservation Laws

Proceedings of the
International School on Theory and Numerics
for Conservation Laws,
Freiburg/Littenweiler, October 20–24, 1997

Springer

Editors

Dietmar Kröner
Mario Ohlberger
Christian Rohde

Institut für Angewandte Mathematik
Universität Freiburg
Hermann-Herder-Straße 10
D-79104 Freiburg, Germany

e-mail:
dietmar@mathematik.uni-freiburg.de
mario@mathematik.uni-freiburg.de
chris@mathematik.uni-freiburg.de

The picture on the cover shows the time-evolved density distribution of a magneto-hydrodynamic Kelvin-Helmholts instability. The colors indicate an increasing density from blue to red.
Numerics: M. Wesenberg (Freiburg/Germany)
Visualization Software: GRAPE

Cataloging-in-Publication Data applied for

Die Deutsche Bibliothek – CIP-Einheitsaufnahme

An introduction to recent developments in theory and numerics for conservation laws; proceedings of the International School on Theory and Numerics for Conservation Laws, Freiburg/Littenweiler, October 20–24, 1997 / Dietmar Kröner... (ed.). – Berlin; Heidelberg; New York; Barcelona; Hong Kong; London; Milan; Paris; Singapore; Tokyo: Springer, 1999
(Lecture notes in computational science and engineering; 5)
ISBN 3-540-65081-4

Mathematics Subject Classification (1991): primary: 35L65, 76N15, 65M12
secondary: 76N10, 65M15, 76M10

ISBN 3-540-65081-4 Springer-Verlag Berlin Heidelberg New York

Cover Design: Friedhelm Steinen-Broo, Estudio Calamar, Spain
Cover production: *design & production* GmbH, Heidelberg
Typeset by the editors using a Springer TEX macro package

SPIN 10653041 46/3143 – 5 4 3 2 1 0 – Printed on acid-free paper

Preface

The importance of hyperbolic conservation laws for scientific and industrial applications has led to a growing amount of research activity in this field. A variety of physical phenomena in fluid mechanics, astrophysics, groundwater flow, meteorology, reactive flow and several other areas can be effectively modeled by systems of conservation laws. In order to focus recent trends in theory and numerics in this research area and to stimulate further research in this field, we decided to organize the "International School on Theory and Numerics for Conservation Laws", which took place in Freiburg/Littenweiler, Germany, from 20 to 24 October 1997.
The school was sponsored by the DFG-Graduiertenkolleg "Nichtlineare Differentialgleichungen, Modellierung, Theorie, Numerik, Visualisierung" at the University of Freiburg. It was attended by about 60 young international researchers.

This volume contains the contributions of the five main lecturers of the school. Each article covers five hours of lectures on a specific new research area in the field of theory and numerics for conservation laws. Reviews of recent developments are given, accompanied by new research results of the authors.

The topics include a kinetic approach to conservation laws by Benoit Perthame, which can be used for the construction of approximate Riemann solvers for the full system of gas dynamics. Several ideas related to the stability and entropy analysis are discussed.
An introduction to the theory of non-classical shock waves is given by Philippe G. LeFloch. Non-classical shocks violate the classical entropy condition introduced by Lax and Liu, but they are sensitive to regularization mechanisms such as diffusion and dispersion. A theoretical approach to such shocks is given and a numerical approximation via finite difference schemes is investigated.
In the contribution of Athanasios E. Tzavaras the approximation of hyperbolic systems of conservation laws via viscosity and relaxation methods is discussed. In the context of zero-viscosity limits recent results on obtaining uniform BV estimates are presented. Furthermore the problem of constructing entropy weak solutions for hyperbolic conservation laws via relaxation is considered.
A major task for numerical approaches to hyperbolic problems is the development of *a posteriori* estimates and resulting adaptive strategies. Endre Süli presents an overview of recent developements in this area for finite element approximations of hyperbolic problems. The *a posteriori* analysis is based on the systematic use of hyperbolic duality arguments. The question

of computational implementation of the *a posteriori* bounds into adaptive finite element algorithms is discussed as well.
Finally Timothy J. Barth considers discretization and preconditioning algorithms for the Euler and Navier-Stokes equations on unstructured meshes. Some new results concerning congruence relationships for left or right symmetrized equations are presented. These results suggest new variants of existing finite volume, discontinuous Galerkin, Galerkin least–squares and fluctuation splitting schemes. The resulting schemes are computationally more efficient while retaining the pleasant theoretical properties achieved by entropy symmetrization.

The book is intended for students who work in the field of conservation laws and researchers who want to open up a new scope of work. Knowledge of modern methods in partial differential equations, specifically with basic concepts for hyperbolic equations, is a prerequisite.

Last but not least we thank M. Wesenberg for the support in the organization of the school and R. Axthelm and M. Werner for typesetting this book.

September 1998

Dietmar Kröner
Mario Ohlberger
Christian Rohde

Table of Contents

An Introduction to Kinetic Schemes for Gas Dynamics

Benoit Perthame

INRIA - Rocquencourt, Domaine de Voluceau, BP 105, 78153 Le Chesney Cédex, and Ecole Normale Supérieure, 45, rue d'Ulm, 75230 Paris Cédex 05, France

Abstract. In these notes we present an introduction to the theory of kinetic schemes for gas dynamics. Several ideas developed in recent years on stability and entropy analysis are reviewed. This compendium is intended to be self-contained and self-consistent even though some results are not entirely proved and a preliminary knowledge of elementary hyperbolic theory is supposed (cf Serre [28], Smoller [29], Lax [15]).

Several subjects are not treated: the modifications proposed by Prendergast & Xu [22], [31] and Deshpande [7] for improving accuracy, the early works by Brenier [1] and Giga & Miyakawa [9] on scalar equations, those of Kaniel [12] on gas dynamics with entropy conservation, discrete velocity schemes and the related subject of relaxation schemes.

Contents

1 Introduction

In these notes, we present a survey of the theory of kinetic schemes for gas dynamics. We introduce several ideas related to *a priori* bounds, entropy and stability analysis developed in recent years. The content is mainly an introductory presentation of the results proved in [13], [20], [21].

Although the idea of using the kinetic formalism to develop numerical schemes for gas dynamics is rather old (it goes back to the 70's; see Sanders and Prendergast [27], Pullin [23], Reitz [25], Harten, Lax and Van Leer [11]), a simple idea has been introduced in [20]. Namely, the classical Maxwellian plays no specific role in solving the Euler Equations (they are relevant for the Navier-Stokes Equations only; see the recent work by LeTallec, Perlat, Perthame [14] for instance). As long as one is only interested in Euler Equations, one can use simplified equilibria developed in [20], [13] which enjoy far better theoretical properties (compare [20], [13] with Estivalezes and Villedieu [8] who use a very tricky method to prove the positivity of the scheme based on the Maxwellian Equilibrium). Other advantages of these simplified equilibria are their reduced computational cost and their simplicity of implementation, while giving the same numerical results.

Several subjects are not treated in these notes. The modifications proposed by Prendergast and Xu [22], [31], and those by Raghurama Rao and Deshpande [24] for improving accuracy are not presented here, even though this lack of accuracy on steady contact discontinuities is the essential drawback of kinetic schemes. Also, the early works for scalar conservation laws by Brenier [1], Giga and Miyakawa [9], those of Kaniel [12] for gas dynamics written with entropy conservation, the discrete velocity schemes, the related subject of relaxation schemes and extended moments systems, are not presented here.

These notes have been written in order to be self–contained and self–consistent, and all the properties used for the theory of numerical schemes are introduced. Nevertheless, a preliminary knowledge of elementary hyperbolic theory is supposed (shock waves, contact discontinuities, weak solutions, entropy property). We refer to P.D. Lax [15], D. Serre [28], J. Smoller [29] for this theory.

The outline of these notes is as follows. We present the Euler Equations of compressible polytropic gases in Section 2, together with generalities on finite volume schemes and kinetic solvers. The entropy condition is discussed in Section 3: we introduce general entropic equilibrium. We also show that a specific equilibrium has the property to recover the "maximum principle on entropy". Relations to the so-called symmetric or entropic variables are also presented.

2 Kinetic schemes for gas dynamics

In this Section we give a general description of kinetic solvers for gas dynamics. We only consider consistency properties. The main other property, the entropy condition, is treated in Section 3 below.

2.1 Gas dynamics equations and entropies

The Euler system for compressible gas dynamics is formed by the complete equations for conservation of mass, momentum and energy

$$\begin{aligned} \frac{\partial \varrho}{\partial t} + \frac{\partial}{\partial x}(\varrho u) = 0\,, \quad t \geq 0, x \in \mathbb{R}\,, \\ \frac{\partial \varrho u}{\partial t} + \frac{\partial}{\partial x}(\varrho u^2 + p) = 0\,, \\ \frac{\partial}{\partial t}E + \frac{\partial}{\partial x}[(E+p)u] = 0\,, \end{aligned} \tag{1}$$

where the pressure law $p(\varrho, \mathrm{e})$ is given (for a polytropic gas, the only class we consider in this Section) by

$$p = (\gamma - 1)\varrho \mathrm{e} \qquad 1 < \gamma \leq 3\,, \tag{2}$$

and

$$E = \frac{1}{2}\varrho u^2 + \varrho e, \tag{3}$$

is the total energy (kinetic energy + internal energy).

This is a non-linear hyperbolic system whose structure has been widely studied (see the references in the introduction). In condensed form, we can write it

$$\frac{\partial}{\partial t}U + \frac{\partial}{\partial x}F(U) = 0\,, \tag{4}$$

or also

$$\frac{\partial}{\partial t}U + A(U)\partial_x U = 0\,, \qquad A = DF\,, \tag{5}$$

with

$$U = \begin{pmatrix} \varrho \\ \varrho u \\ E \end{pmatrix}, \tag{6}$$

$$F(U) = \begin{pmatrix} \varrho u \\ \varrho u^2 + p \\ (E+p)u \end{pmatrix}. \tag{7}$$

One readily checks that the matrix A has three eigenvalues

$$u - c,\ u,\ u + c, \tag{8}$$

where

$$c = \sqrt{\gamma \frac{p}{\varrho}} \tag{9}$$

is the speed of sound. These eigenvalues mean that the solution cannot propagate faster than $|u|+c$. They are easier to compute from (which is equivalent to (1) for smooth solutions)

$$\begin{aligned} \frac{\partial}{\partial t}\varrho + \frac{\partial}{\partial x}(\varrho u) &= 0\,, \\ \frac{\partial}{\partial t}u + u\frac{\partial}{\partial x}u + \frac{1}{\varrho}\frac{\partial}{\partial x}p &= 0\,, \\ \frac{\partial \mathrm{e}}{\partial t} + u\frac{\partial \mathrm{e}}{\partial x} + (\gamma-1)\mathrm{e}\frac{\partial u}{\partial x} &= 0\,, \end{aligned} \tag{10}$$

which also indicates that the densities ϱ and internal energy e should remain non–negative. From (10) we also obtain the formal relation

$$\frac{\partial}{\partial t}\left(\frac{\mathrm{e}}{\varrho^{\gamma-1}}\right) + u\frac{\partial}{\partial x}\left(\frac{\mathrm{e}}{\varrho^{\gamma-1}}\right) = 0$$

which induces the other formal relation, for all functions $F(\cdot)$,

$$\frac{\partial}{\partial t}\varrho F\left(\frac{\mathrm{e}}{\varrho^{\gamma-1}}\right) + \frac{\partial}{\partial x}\varrho u F\left(\frac{\mathrm{e}}{\varrho^{\gamma-1}}\right) = 0\,.$$

These equalities are not valid for weak solutions (ones with shocks for instance).
Nevertheless, we may ask the *entropy inequalities* to hold for weak solutions

$$\frac{\partial}{\partial t}\varrho s(\varrho,\mathrm{e}) + \frac{\partial}{\partial x}\varrho u s(\varrho,\mathrm{e}) \le 0 \tag{11}$$

for all specific entropies $s(\varrho,\mathrm{e}) = F(\mathrm{e}\varrho^{1-\gamma})$ such that $\varrho s(\varrho,\mathrm{e})$ is convex in U. There are several motivations for imposing (11): it is known that it selects uniquely shock waves, it holds true for the full diffusion approximation of (4)

$$\frac{\partial}{\partial t}U + \frac{\partial}{\partial x}F(U) = \epsilon \Delta U$$

with $\epsilon\Delta\varrho s$ in the righthand side of (11), it gives *a priori* bounds – see (13). These convex entropies are characterized by the following classical result.

Lemma 1. *The following are equivalent*

1. $\varrho s(\varrho, \mathrm{e}) = \varrho F(\mathrm{e}\varrho^{1-\gamma})$ *is convex in* U,
2. $s(\varrho, \mathrm{e}) = F(\mathrm{e}\varrho^{1-\gamma})$ *is convex in* $(\mathrm{e}, \varrho^{-1})$,
3. $F' \leq 0$, $F'' \geq 0$ *on* $(0, +\infty)$.

The *physical entropy* is

$$\mathrm{S} = \varrho \ln \frac{\varrho^{\gamma-1}}{\mathrm{e}}, \tag{12}$$

and corresponds to $F(r) = -\ln r$, and it is convex. But we also deduce from (11)

$$\frac{\varrho^{\gamma-1}}{\mathrm{e}}(t,x) \leq \max_{y\in\mathbb{R}} \frac{\varrho^{\gamma-1}}{\mathrm{e}}(s,y), \qquad \forall t \geq s,\ x \in \mathbb{R}. \tag{13}$$

Proof of (13). To prove this, it is enough to set

$$M = \min_{y\in\mathbb{R}} \mathrm{e}\varrho^{1-\gamma}(s,y),$$

and to choose

$$F(r) = (M - r)_+^2 .$$

Then

$$\int_{\mathbb{R}} \varrho F(\mathrm{e}\varrho^{1-\gamma})(s,x)\, dx = 0 .$$

Since F is convex non–increasing, the entropy inequality (11) holds true. Therefore

$$\int_{\mathbb{R}} \varrho F(\mathrm{e}\varrho^{1-\gamma})(t,x)\, dx \leq \int_{\mathbb{R}} \varrho F(\mathrm{e}\varrho^{1-\gamma})(s,x)\, dx = 0,$$

and thus, for all $x \in \mathbb{R}$

$$\mathrm{e}\varrho^{1-\gamma}(t,x) \geq M,$$

which proves (13). □

More on the entropy inequalities can be found in Serre [28], Smoller [29], Godlewski and Raviart [10]. The above result was proved in Tadmor [30].

2.2 Finite volumes approximation

An interesting class of numerical methods for the hyperbolic systems (4) consists in the *finite volume* approximation ($i \in \mathbb{Z}$, $n \in \mathbb{N}$).

$$U_i^{n+1} - U_i^n + \sigma(F_{i+\frac{1}{2}}^n - F_{i-\frac{1}{2}}^n) = 0, \tag{14}$$

$$\sigma = \frac{\Delta t}{\Delta x}. \tag{15}$$

Here n refers to time $t_n = n\Delta t$, i refers to the cell centers $x_i = i\Delta x$ of the cell $(x_{i-\frac{1}{2}}, x_{i+\frac{1}{2}})$. The equation (14) is deduced from (4) with

$$U_i^n = \int_{x_{i-\frac{1}{2}}}^{x_{i+\frac{1}{2}}} U(t_n, x)\, dx\,,$$

$$F_{i+\frac{1}{2}}^n = \frac{1}{\Delta t} \int_{t_n}^{t_{n+1}} F(U(s, x_{i+\frac{1}{2}}))\, ds\,. \tag{16}$$

The three points finite volume schemes consist in approximating the last equality by

$$F_{i+\frac{1}{2}}^n \simeq \mathcal{F}(U_i^n, U_{i+1}^n)\,. \tag{17}$$

The reason to consider this structure (14) is that it preserves the conservation law

$$\sum_{i\in\mathbb{Z}} U_i^n = \sum_{i\in\mathbb{Z}} U_i^0\,, \qquad \forall n \geq 0$$

whenever the F_i^n vanish as i goes to infinity. In view of the entropy considerations we may add two different types of entropy conditions. Either we ask for the inequality

$$S_i^{n+1} - S_i^n + \sigma\left(\eta_{i+\frac{1}{2}}^n - \eta_{i-\frac{1}{2}}^n\right) \leq 0 \;\; \text{(weak entropy condition)}, \tag{18}$$

for some $S_i^n = \varrho_i^n F\left(\mathrm{e}_i^n (\varrho_i^n)^{1-\gamma}\right)$, satisfying the property 3 from Lemma 1 and

$$\eta_{i+\frac{1}{2}}^n = \eta\left(U_i^n, U_{i+1}^n\right)\,, \tag{19}$$

or we ask for all the convex entropy inequalities, and then the maximum principle holds. Then, we require

$$\frac{\left(\varrho_i^{n+1}\right)^{\gamma-1}}{\mathrm{e}_i^{n+1}} \leq \max_{j=i-1,i,i+1} \frac{\left(\varrho_j^n\right)^{\gamma-1}}{\mathrm{e}_j^n} \;\; \text{(strong entropy condition)}. \tag{20}$$

Many theories have been devoted to the construction of appropriate numerical fluxes $\mathcal{F}(\cdot,\cdot)$. We refer to Godlewski & Raviart [10] for a recent account of this theory, Harten, Lax & Van Leer [11], LeVeque [16].

It remains that we still do not have a "perfect" numerical flux in the sense that the resulting scheme would satisfy

Property 2.

1. $\varrho_i^{n+1} \geq 0$, $\mathrm{e}_i^{n+1} \geq 0$ *whenever* $\varrho_i^n \geq 0$, $\mathrm{e}_i^n \geq 0$, $\forall i \in \mathbb{Z}$ *(weak stability)*,
2. $\mathcal{F}(U,U) = F(U)$ *(dynamical consistency)*,

3. $\mathcal{F}(U,V) = (0,\, p,\, 0)^T$ *for* $U = (\varrho_L,\, 0,\, \varrho_L \mathrm{e}_L)^T$, $V = (\varrho_R,\, 0,\, \varrho_R \mathrm{e}_R)^T$ *with* $p(\varrho_L, \mathrm{e}_L) = p(\varrho_R, \mathrm{e}_R) = p$ *(steady consistency),*
4. in cell entropy inequality (18), (19) hold,
5. maximum principle on entropy (20) holds.

In fact, only the Godunov scheme satisfies all these properties to the expense of a fixed point which makes it too slow for several practical use.

Lax–Friedrichs or kinetic schemes satisfy all of them except 3., which makes them too dissipative for applications (even in high order versions).

Therefore a compromise has been adopted in several codes which consists in using Roe scheme [26] which does not satisfy 1., 4., 5. (but almost does), and is more precise especially because it is exact on stationary contact discontinuities (property 3. above).

Let us comment on these requirements. The first condition 1. implies the l^1 bounds

$$\sum_{i\in\mathbb{Z}} \varrho_i^n \le \sum_{i\in\mathbb{Z}} \varrho_i^0 ,$$

$$\sum_{i\in\mathbb{Z}} \varrho_i^n (u_i^n)^2 + \varrho_i^n \mathrm{e}_i^n \le \sum_{i\in\mathbb{Z}} \varrho_i^0 (u_i^0)^2 + \varrho_i^0 \mathrm{e}_i^0$$

and therefore

$$\sum_{i\in\mathbb{Z}} \varrho_i^n |u_i^n| \le \Big(\sum_{i\in\mathbb{Z}} \varrho_i^n\Big)^{\frac{1}{2}} \Big(\sum_{i\in\mathbb{Z}} \varrho_i^n (u_i^n)^2\Big)^{\frac{1}{2}} \le C(\varrho^0, E^0) .$$

These are the L^1–stability of the scheme.
The consistency condition 2. is classical and implies the scheme is at least first order (see LeVeque [16], Godlewski & Raviart [10] and the references therein).

It is also known that an entropy condition 4. has to be imposed to ensure the convergence to the right physical solution. Otherwise rarefaction shocks can be generated which are unphysical. It is also a stability condition because it gives an *a priori* bound.

The condition 5. is a stronger stability condition compared to 4., it gives for instance

$$\begin{aligned}\sum_{i\in\mathbb{Z}} (\varrho_i^n)^\gamma &\le \sum_{i\in\mathbb{Z}} \varrho_i^n \mathrm{e}_i^n \sup_j \frac{\left(\varrho_j^n\right)^{\gamma-1}}{\mathrm{e}_j^n} \\ &\le \sum_{i\in\mathbb{Z}} E_i^0 \sup_j \frac{\left(\varrho_j^0\right)^{\gamma-1}}{\mathrm{e}_j^0} \le C_1(U^0) .\end{aligned}$$

Notice that the condition 3. means that the scheme is exact on steady contact discontinuities. These are important in practice because they represent shear layers or boundary layers in higher dimensions. It is essential to impose it because for long times small errors on steady contacts could add up and destroy the whole computation. Actually it could be imposed as an accuracy requirement for higher order schemes. But, in practice, it turns out that we do not know how to recover it when it is not satisfied at first order.

2.3 Boltzmann's formalism

Kinetic physics represents a flow by the density $f(t,x,\xi)$ of particles at time t, position x, with velocity ξ. Then, the main equation for the evolution of f is the Boltzmann equation

$$\frac{\partial}{\partial t} f + \xi \cdot \nabla_x f = \frac{1}{\epsilon} Q(f) \,, \qquad t \geq 0,\ x, \xi \in \mathbb{R}^3 \,. \tag{21}$$

Here $Q(f)$ is a quadratic operator whose main properties are

$$\begin{aligned} &\textstyle\int \left(1, \xi, \frac{1}{2}|\xi|^2\right) Q(f)\, d\xi = (0,0,0) \,, \qquad \forall f \\ &Q(f) = 0 \Leftrightarrow f = \mathcal{M}(\varrho, T; \xi - u) \,, \end{aligned} \tag{22}$$

the *Maxwellian* $\mathcal{M}$ being defined by

$$\mathcal{M}(\varrho, T; \xi - u) = \frac{\varrho}{(2\pi T)^{3/2}} \exp\left(-\frac{|\xi - u|^2}{2T}\right) . \tag{23}$$

The Euler equations of gas dynamics are recovered setting (formally) ϵ to 0. Then $Q(f) \to 0$, which forces f to be a Maxwellian.

Therefore the limiting equation for (21) becomes, with $\mathcal{M}$ given by (23) for some $\varrho(t,x)$, $u(t,x)$, $T(t,x)$, (we change notation for Q)

$$\frac{\partial}{\partial t}\mathcal{M} + \xi \cdot \nabla_x \mathcal{M} = Q \,.$$

Using (22) and integrating this equation against $(1, \xi, \frac{|\xi|^2}{2})$ gives

$$\frac{\partial}{\partial t} \int \begin{pmatrix} 1 \\ \xi \\ \frac{|\xi|^2}{2} \end{pmatrix} \mathcal{M}\, d\xi + \sum_{i=1}^{3} \frac{\partial}{\partial x_i} \int \xi_i \begin{pmatrix} 1 \\ \xi \\ \frac{|\xi|^2}{2} \end{pmatrix} \mathcal{M}\, d\xi = 0$$

which, in terms of ϱ, u, T, can also be written

$$\partial_t \varrho + \mathrm{div} \varrho \mathrm{u} = 0 \,,$$

$$\partial_t \varrho u + \mathrm{div} \varrho \mathrm{u} \otimes \mathrm{u} + \nabla(\varrho \mathrm{T}) = 0 \,, \tag{24}$$

$$\partial_t E + \mathrm{div}[(\mathrm{E} + \mathrm{p})\mathrm{u}] = 0 \,, \tag{25}$$

where $E = \frac{1}{2}\varrho|u|^2 + \varrho \mathrm{e}$. These are just the 3–dimensional Euler equations for $\gamma = \frac{5}{3}$ (monatomic gas), because here $\varrho \mathrm{e} = \frac{3}{2}\varrho T$.

Notice that this system should be completed with the entropy relation

$$\frac{\partial}{\partial t}\mathrm{S}(\varrho,\mathrm{e})+\frac{\partial}{\partial x}\mathrm{S}(\varrho,\mathrm{e})\leq 0\,, \tag{26}$$

for entropies

$$\mathrm{S}(\varrho,\mathrm{e})=\varrho F\left(\frac{\mathrm{e}}{\varrho^{\gamma-1}}\right),$$

which are convex in $(\varrho,\varrho u,E)$, see Section 2.1.

For the theory of Boltzmann–Equations we refer to Cercignani [2], Cercignani, Illner & Pulvirenti [3], Lifshitz & Pitaevskii [17].

The main restriction of this formalism is to monatomic gases ($\gamma=\frac{5}{3}$ in 3–d, $\gamma=\frac{d+2}{d}$ in general). Therefore a more complicated structure has been introduced which takes into account the internal degrees of freedom of molecules. This structure will be developed in Section 3.

2.4 Kinetic schemes (generalities)

The reason why the Boltzmann equation generates the right fluid limit is simple. It can be generalized as follows. We introduce two functions defined from $\mathbb{R}$ into $\mathbb{R}^+$,

$$\chi(\cdot)\geq 0\,,\quad \chi(w)=\chi(-w)\,,\int_{\mathbb{R}}\chi(w)\;dw=1\,,\int_{\mathbb{R}}w^2\chi(w)\;dw=1 \tag{27}$$

$$\zeta(\cdot)\geq 0\,,\quad \zeta(w)=\zeta(-w)\,,\quad \int_{\mathbb{R}}\zeta(w)\;dw=\frac{3-\gamma}{2(\gamma-1)}\,. \tag{28}$$

Notice, we only consider $1<\gamma\leq 3$ here, which is the main restriction of kinetic schemes. Then we have by an easy computation

$$\begin{aligned}
\varrho&=\int_{\mathbb{R}}\frac{\varrho}{\sqrt{T}}\chi\left(\frac{\xi-u}{\sqrt{T}}\right)\;d\xi\,,\\
\varrho u&=\int_{\mathbb{R}}\frac{\varrho}{\sqrt{T}}\chi\left(\frac{\xi-u}{\sqrt{T}}\right)\xi\;d\xi\,,\\
E=\frac{1}{2}\varrho u^2+\varrho\mathrm{e}&=\int_{\mathbb{R}}\frac{\varrho}{\sqrt{T}}\chi\left(\frac{\xi-u}{\sqrt{T}}\right)\frac{\xi^2}{2}\;d\xi+\int_{\mathbb{R}}\varrho\sqrt{T}\zeta\left(\frac{\xi-u}{\sqrt{T}}\right)\;d\xi
\end{aligned}$$

with $\varrho\mathrm{e}=\frac{1}{2}\varrho T+\frac{3-\gamma}{2(\gamma-1)}\varrho T=\frac{1}{\gamma-1}\varrho T$, and also

$$\begin{aligned}
\varrho u^2+\varrho T&=\int_{\mathbb{R}}\frac{\varrho}{\sqrt{T}}\chi\left(\frac{\xi-u}{\sqrt{T}}\right)\xi^2\;d\xi\,,\\
(E+\varrho T)u&=\int_{\mathbb{R}}\frac{\varrho}{\sqrt{T}}\chi\left(\frac{\xi-u}{\sqrt{T}}\right)\xi^3\;d\xi\,.
\end{aligned}$$

In other words, setting $p = \varrho T$, we have, with the notations of Section 2.1

$$\begin{aligned} U \quad &= \int_{\mathbb{R}} \begin{pmatrix} 1 \\ \xi \\ \frac{\xi^2}{2} \end{pmatrix} \frac{\varrho}{\sqrt{T}} \chi\left(\frac{\xi-u}{\sqrt{T}}\right) \, d\xi \\ &+ \int_{\mathbb{R}} \begin{pmatrix} 0 \\ 0 \\ 1 \end{pmatrix} \varrho\sqrt{T} \zeta\left(\frac{\xi-u}{\sqrt{T}}\right) \, d\xi \,, \\ F(U) \quad &= \int_{\mathbb{R}} \xi \begin{pmatrix} 1 \\ \xi \\ \frac{\xi^2}{2} \end{pmatrix} \frac{\varrho}{\sqrt{T}} \chi\left(\frac{\xi-u}{\sqrt{T}}\right) \, d\xi \\ &+ \int_{\mathbb{R}} \xi \begin{pmatrix} 0 \\ 0 \\ 1 \end{pmatrix} \varrho\sqrt{T} \zeta\left(\frac{\xi-u}{\sqrt{T}}\right) \, d\xi \,. \end{aligned} \tag{29}$$

This kinetic representation of the conservative variables and the fluxes in the Euler equation indicates the general relations between kinetic–transport equations and the Euler equation. The Euler equation (1) can be rewritten

$$\begin{aligned} \frac{\partial}{\partial t} f_{\text{equi}}(t,x,\xi) + \xi \frac{\partial}{\partial x} f_{\text{equi}}(t,x,\xi) &= Q_1 \,, \\ \partial_t g_{\text{equi}}(t,x,\xi) + \xi \frac{\partial}{\partial x} g_{\text{equi}}(t,x,\xi) &= Q_2 \,, \end{aligned} \tag{30}$$

where the *equilibria* are defined by

$$\begin{aligned} f_{\text{equi}} \quad &= \frac{\varrho(t,x)}{\sqrt{T(t,x)}} \chi\left(\frac{\xi-u(t,x)}{\sqrt{T(t,x)}}\right) \,, \\ g_{\text{equi}} &= \varrho(t,x)\sqrt{T(t,x)} \zeta\left(\frac{\xi-u(t,x)}{\sqrt{T(t,x)}}\right) \,, \end{aligned} \tag{31}$$

and Q_1, Q_2 are some unknown functions satisfying for all t, x

$$\begin{aligned} \int_{\mathbb{R}} Q_1(t,x,\xi) \, d\xi &= 0 \,, \\ \int_{\mathbb{R}} \xi Q_1(t,x,\xi) \, d\xi &= 0 \,, \\ \int_{\mathbb{R}} \left[\frac{\xi^2}{2} Q_1(t,x,\xi) + Q_2(t,x,\xi) \right] \, d\xi &= 0 \,. \end{aligned} \tag{32}$$

In other words, we have reduced the non–linear system of Euler to two simple linear transport equations (with a maximum principle !) with singular r.h.s.'s. It is an open question to characterize these r.h.s.'s for general non-smooth solutions.

This allows us to derive a finite volume scheme by discretizing the transport equations. We use a classical upwind scheme, with a classical procedure to treat the r.h.s.,

$$\begin{aligned} &f_i^{n+1^-}(\xi) - f_i^n(\xi) + \sigma\xi\left[f_{i+\frac12}^n(\xi) - f_{i-\frac12}^n(\xi)\right] = 0\,,\\ &g_i^{n+1^-}(\xi) - g_i^n(\xi) + \sigma\xi\left[g_{i+\frac12}^n(\xi) - g_{i-\frac12}^n(\xi)\right] = 0\,, \end{aligned} \tag{33}$$

with

$$h_{i+\frac12}^n(\xi) = \begin{cases} h_i^n(\xi) & \text{for } \xi \geq 0\,,\\ h_{i+1}(\xi) & \text{for } \xi \leq 0\,, \end{cases}$$

for $h = f$ or g. At this level we do not take into account the collisions, we use them in a second step, introducing a discontinuity at time t_{n+1} on f and g, and setting and replacing f_i^{n-} and g_i^{n-} by an equilibrium

$$\begin{aligned} f_i^n(\xi) &= \frac{\varrho_i^n}{\sqrt{T_i^n}}\chi\left(\frac{\xi - U_i^n}{\sqrt{T_i^n}}\right),\\ g_i^n(\xi) &= \varrho_i^n\sqrt{T_i^n}\zeta\left(\frac{\xi - U_i^n}{\sqrt{T_i^n}}\right), \end{aligned} \tag{34}$$

$$U_i^{n+1} = \int_{\mathbb{R}} \begin{pmatrix} 1\\ \xi\\ \frac{\xi^2}{2} \end{pmatrix} f_i^{n+1^-}(\xi)\,d\xi + \int_{\mathbb{R}} \begin{pmatrix} 0\\ 0\\ 1 \end{pmatrix} g_i^{n+1^-}(\xi)\,d\xi\,. \tag{35}$$

Notice that f^{n+1}, g^{n+1} are discontinuous ($f^{n+1} \neq f^{n+1^-}$, $g^{n+1} \neq g^{n+1^-}$) but not U_i^{n+1} ($U_i^{n+1} = U_i^{n+1^-}$!).

Therefore we have obtained a finite volume scheme for Euler equations by integrating the discretized equations (33) to produce the quantities (35). We obtain

$$\begin{aligned} &U_i^{n+1} - U_i^n + \sigma\left(F_{i+\frac12}^n - F_{i-\frac12}^n\right) = 0\,,\\ &F_{i+\frac12}^n = \mathcal{F}\left(U_i^n, U_{i+1}^n\right), \end{aligned}$$

and the numerical flux $\mathcal{F}$ is a vector flux-splitting

$$\begin{aligned} \mathcal{F}(U,V) &= F_+(U) + F_-(V)\,,\\ F_+(U) &= \int_{\xi\geq 0} \xi \begin{pmatrix} 1\\ \xi\\ \frac{\xi^2}{2} \end{pmatrix} \frac{\varrho}{\sqrt{T}}\chi\left(\frac{\xi-u}{\sqrt{T}}\right) d\xi\\ &\quad + \int_{\xi\geq 0} \xi \begin{pmatrix} 0\\ 0\\ 1 \end{pmatrix} \varrho\sqrt{T}\zeta\left(\frac{\xi-u}{\sqrt{T}}\right) d\xi\,, \end{aligned} \tag{36}$$

$$F_-(V) = \int\limits_{\xi\le 0} \xi \begin{pmatrix} 1 \\ \xi \\ \frac{\xi^2}{2} \end{pmatrix} \frac{\varrho}{\sqrt{T}}\chi\left(\frac{\xi-u}{\sqrt{T}}\right)\, d\xi$$
$$+ \int\limits_{\xi\le 0} \xi \begin{pmatrix} 0 \\ 0 \\ 1 \end{pmatrix} \varrho\sqrt{T}\zeta\left(\frac{\xi-u}{\sqrt{T}}\right)\, d\xi\,.$$

This class of scheme produces good schemes in the sense that we have

Theorem 3. *Assume that χ, ζ satisfy (27–28), that $\chi(w)=0$, $\zeta(w)=0$ for $|w|>\sqrt{\gamma_{\max}}$, then under the* CFL *condition*

$$\Delta t\left(\max_i |u_i^n| + \sqrt{\gamma_{\max} T_i^n}\right) \le \Delta x \tag{37}$$

the kinetic scheme (36) satisfies the positivity property 1. *and the dynamical consistency property* 2. *of Properties (2).*

Proof. We have, from (33)

$$f_i^{n+1^-}(\xi) = f_i^n(\xi)(1-\sigma|\xi|) + \sigma f_{i+1}^n(\xi)\xi_- + \sigma f_{i-1}^n(\xi)\xi_+ \tag{38}$$

with $\xi_+=\max(0,\xi)$, $\xi_-=\min(0,-\xi)$. We only have to consider the ξ's such that $f_i^n(\xi)$ or $f_{i+1}^n(\xi)$ or $f_{i-1}^n(\xi)$ do not vanish; these are given by

$$\chi\left(\frac{\xi-u_j^n}{\sqrt{T_j^n}}\right)\neq 0\,,\quad \zeta\left(\frac{\xi-u_j^n}{\sqrt{T_j^n}}\right)\neq 0\,,$$

for $j=i,i+1,i-1$ and thus by

$$|\xi-u_j^n| \le \sqrt{\gamma_{\max}T_j^n}\,.$$

For such ξ's, by the CFL condition, we have

$$\sigma|\xi|\le 1\,.$$

Hence, we have $f_i^{n+1^-}(\xi)\ge 0$ and thus

$$\varrho_i^{n+1} = \int\limits_{\mathbb{R}} f_i^{n+1^-}(\xi)\, d\xi \ge 0\,,$$

$$(\varrho T)_i^{n=1} = \int\limits_{\mathbb{R}} |\xi-u|^2 f_i^{n+1^-}(\xi)\, d\xi \ge 0\,,$$

and 1. is proved.

The consistency relation is easy

$$
\begin{aligned}
\mathcal{F}(V,V) &= F_+(V) + F_-(V) \\
&= \int_{\mathbb{R}} \xi \begin{pmatrix} 1 \\ \xi \\ \frac{\xi^2}{2} \end{pmatrix} \frac{\varrho}{\sqrt{T}} \chi\left(\frac{\xi - u_j^n}{\sqrt{T_j^n}}\right) d\xi + \int_{\mathbb{R}} \xi \begin{pmatrix} 0 \\ 0 \\ 1 \end{pmatrix} \varrho\sqrt{T}\zeta\left(\frac{\xi - u_j^n}{\sqrt{T_j^n}}\right) d\xi \\
&= F(V)
\end{aligned}
$$

by the relation (28).

□

Notice however that

(*i*) $\gamma_{\max} \geq 1$ (hint: use only the condition on χ)
(*ii*) the "usual" CFL condition is
$\Delta t \, \sup_i \left(|u_i^n| + \sqrt{\gamma e_i^n}\right) \leq \Delta x$.
Therefore good choices of χ, ζ should impose $\gamma_{\max} \geq \gamma$.
(*iii*) The property 3. cannot be satisfied, whatever the choice of χ, ζ. This motivated several authors to improve the method (see the introduction).

We show in Section 3 how to ensure the entropy conditions 4., 5..

2.5 Lax–Friedrichs as a kinetic scheme

The kinetic schemes are better than the Lax–Friedrichs scheme which can be seen as a kinetic discretisation of the kinetic transport equation (30) as (neglecting again Q and R in a first step)

$$
\begin{aligned}
f_i^{n+1^-} - f_i^n &+ \frac{\sigma}{2}(\xi + \xi_{\max}) \left[f_i^n(\xi) - f_{i-1}^n(\xi)\right] \\
&+ \frac{\sigma}{2}(\xi - \xi_{\max}) \left[f_{i+1}^n(\xi) - f_i^n(\xi)\right] = 0 \, ,
\end{aligned}
$$

(and the same discretisation for the equation on g).

Then, following the proof of Theorem 3, we can see that positivity and consistency hold true (imposing $|\xi| < \xi_{\max}$ for the ξ's for which χ and ζ do not vanish). The resulting scheme is

$$
\begin{aligned}
\mathcal{F}_{LF}(U,V) &= F_{LF+}(U) + F_{LF-}(V) \\
F_{LF+}(U) &= \int_{\mathbb{R}} (\xi + \xi_{\max}) \begin{pmatrix} 1 \\ \xi \\ \frac{\xi^2}{2} \end{pmatrix} f_{\text{equi}} \, d\xi + \int_{\mathbb{R}} (\xi + \xi_{\max}) \begin{pmatrix} 0 \\ 0 \\ 1 \end{pmatrix} g_{\text{equi}} \, d\xi \\
&= F(U) + \xi_{\max} U \, , \\
F_{LF-} &= F(U) - \xi_{\max} U \, .
\end{aligned}
$$

This is the non–staggered version of the Lax–Friedrichs scheme ($\xi_{\max}$ larger than the largest eigenvalue of $A(U) = \frac{\partial F(V)}{\partial V}$).

2.6 The simplest kinetic scheme and Van–Leer flux vector splitting

The simplest choice for the functions χ, ζ in the above theory is

$$\chi(w) = \zeta(w) = \begin{cases} \frac{1}{2\sqrt{3}} & \text{for } |w| \leq \sqrt{3} \ , \\ 0 & \text{otherwise} \ . \end{cases} \tag{39}$$

This leads to particularly simple formulas for the fluxes $F_{i+1/2} = F_+(U_i) + F_-(U_{i+1})$. We set

$$M_+ = \max(0, \frac{u}{\sqrt{T}} + \sqrt{3}) \, ,$$
$$M_- = \max(0, \frac{u}{\sqrt{T}} - \sqrt{3}) \, ,$$

then

$$F_+^{\varrho}(U) = \frac{1}{4\sqrt{3}} \varrho \sqrt{T} (M_+^2 - M_-^2) \, ,$$
$$F_+^{\varrho u}(U) = \frac{1}{6\sqrt{3}} \varrho T (M_+^3 - M_-^3) \, ,$$
$$F_+^{E}(U) = \frac{1}{16\sqrt{3}} \varrho T^{3/2} (M_+^4 - M_-^4) + \frac{3-\gamma}{2(\gamma-1)} T F_+^{\varrho}(U) \, ,$$

(recall the conventions: $E = \frac{1}{2}\varrho u^2 + \varrho e$, $p = (\gamma - 1)\varrho e = \varrho T$, $1 < \gamma \leq 3$). The F_- can be obtained from the formula $F_-(U) = F(U) - F_+(U)$, or simply replacing max by min in the above formulas.

For $\gamma = 3$, we recover the Van Leer formulas.

2.7 Sanders and Prendergast scheme

The first kinetic scheme was introduced by Sanders & Prendergast [27] in 1974.
It consists in choosing

$$\chi(w) = \frac{\gamma - 1}{\gamma} \delta(w) + \frac{1}{2\gamma} \delta(w - \sqrt{\gamma}) + \frac{1}{2\gamma} \delta(w + \sqrt{\gamma}) \, ,$$
$$\zeta(w) = \frac{3-\gamma}{2(\gamma-1)} \delta(w) \, .$$

Here δ represents the Dirac mass. It gives even simpler formulas than the scheme presented in Section 6. However results are not very good, which seems to be related to the violation of the entropy condition.

2.8 Maxwellian based scheme

The most natural choice, from a physical point of view, is to choose Maxwellians for χ, ζ. This was introduced by Reitz [25], Pullin [23], Deshpande [6], [7]. It leads to

$$\chi(w) = \frac{1}{\sqrt{2\pi}} \exp(-\frac{w^2}{2}),$$
$$\zeta(w) = \frac{3-\gamma}{2(\gamma-1)} \frac{1}{\sqrt{2\pi}} \exp(-\frac{w^2}{2}).$$

And fluxes are ($z = (\xi - u)/\sqrt{T}$, $\mu = u/\sqrt{T}$)

$$F_+^\varrho(U) = \frac{\varrho}{\sqrt{2\pi}} \int\limits_{w \geq \frac{u}{\sqrt{T}}} (z+\mu) \exp(\frac{-z^2}{2})\, dz$$

$$F_+^{\varrho u}(U) = \frac{\varrho}{\sqrt{2\pi}} \int\limits_{w \geq \frac{u}{\sqrt{T}}} (z+\mu)^2 \exp(\frac{-z^2}{2})\, dz$$

$$F_+^E(U) = \frac{\varrho}{\sqrt{2\pi}} \int\limits_{w \geq \frac{u}{\sqrt{T}}} \frac{(z+\mu)^3}{2} \exp(\frac{-z^2}{2})\, dz + \frac{3-\gamma}{2(\gamma-1)} T F_+^\varrho(U).$$

These formulas can be made more explicit. But it remains necessary to use quantities like $\int_{w \geq \alpha} \exp(\frac{-w^2}{2})\, dw$. These are more expensive than the formulas in 2.6, for similar numerical results (higher smoothness of the fluxes of the same type as (6), can be obtained using $\chi(w) = \zeta(w) = \frac{3}{4\sqrt{5}} \left(1 - \frac{w^2}{5}\right)_+$. Also, they do not satisfy the support property in Theorem 3. Nevertheless positivity was proved under a CFL condition by Estivalezes & Villedieu [8].

3 The entropy inequality for kinetic schemes

In this Section we discuss the entropy inequality for kinetic schemes. We show that the choice of the kinetic entropy gives a pair of equilibrium functions (χ, ζ) in the above construction. We begin with a simple example (at the multidimensional and continuous level). Then we generalize this theory, and finally we explain how it is possible to recover the maximum principle on the specific entropy, which turns out to be equivalent to the maximum principle at the kinetic level.

3.1 Kinetic internal energy variable or kinetic internal energy distribution: the log entropy

Following the general principles of thermodynamics the Maxwellian equilibrium minimizes the kinetic entropy. An example is the following

Proposition 4. *Let $0 < \delta < 1$. Then, for all $\varrho > 0$, $u \in \mathbb{R}^d$, $E = \frac{1}{2}\varrho|u|^2 + \varrho e$ with $e > 0$, the following minimum is achieved*

$$S(\varrho, e) = \min \Big\{ \int_{\mathbb{R}^d} f(\xi) \ln \frac{f(\xi)}{g^\delta(\xi)}\, d\xi\,; \qquad f \ge 0,\, g \ge 0$$

$$\text{and} \int_{\mathbb{R}^d} f(\xi)\, d\xi = \varrho, \int_{\mathbb{R}^d} \xi f(\xi)\, d\xi = \varrho u, \int_{\mathbb{R}^d} \Big[\frac{|\xi|^2}{2} f(\xi) + g(\xi)\Big]\, d\xi = E \Big\}.$$

The minimizer is

$$\begin{aligned} f_{\text{equi}} &= \tfrac{\varrho}{(2\pi T)^{d/2}} \exp\Big(-\tfrac{|v-u|^2}{2T}\Big)\,, \\ g_{\text{equi}} &= \tfrac{\delta}{1-\delta} T f_{\text{equi}}\,, \end{aligned} \tag{40}$$

with $\varrho T = (\gamma - 1)\varrho e$, $\frac{d+2-d\gamma}{2(\gamma-1)} = \frac{\delta}{1-\delta}$, $1 < \gamma \le 5/3$.

Proof. Writing the Euler–Lagrange equations for the above minimum gives, for $H(f, g) = f \ln \frac{f}{g^\delta}$,

$$\begin{aligned} H_f(f,g) &= \ln \frac{f}{g\delta} + 1 = \alpha_0 + \alpha_1 \xi + \alpha_2 \frac{|\xi|^2}{2}\,, \\ H_g(f,g) &= -\frac{\delta f}{g} = \alpha_2\,. \end{aligned} \tag{41}$$

We set

$$\alpha_2 = -\frac{1-\gamma}{T}\,,$$

and choosing α_0, α_1 appropriately, we obtain the formula (40). Notice that the integral constraints are indeed satisfied since

$$\begin{aligned} \int \frac{|\xi|^2}{2} f_{\text{equi}}(\xi)\, d\xi &= \frac{d}{2}\varrho T\,, \\ \int g_{\text{equi}}(\xi)\, d\xi &= \frac{\delta}{1-\delta}\varrho T\,. \end{aligned}$$

Next, since $H(\cdot,\cdot)$ is strictly convex, for $(f, g) \neq (f_{\text{equi}}, g_{\text{equi}})$ we have

$$\begin{aligned} H(f,g) > H(f_{\text{equi}}, g_{\text{equi}}) &+ H'_f(f_{\text{equi}}, g_{\text{equi}})(f - f_{\text{equi}}) \\ &+ H'_g(f_{\text{equi}}, g_{\text{equi}})(g - g_{\text{equi}})\,. \end{aligned}$$

Therefore as soon as (f, g) satisfies the constraints we obtain from (41)

$$\int_{\mathbb{R}^d} H(f(\xi), g(\xi))\, d\xi \ge \int_{\mathbb{R}^d} H(f_{\text{equi}}(\xi), g_{\text{equi}}(\xi))\, d\xi$$

and the result is proved. (Notice that the translational invariance shows that the entropy S is independent of u.)

□

A consequence of this Proposition is on the gas dynamics system (24).

Proposition 5. *The function $S(\varrho, \mathrm{e})$ is a strictly convex (in ϱ, ϱu, E) entropy for the multidimensional gas dynamics system (24), asssociated with the γ-law pressure; γ and δ are related as in Proposition 4.*

Proof. For smooth functions $\varrho(t,x)$, $u(t,x)$, $e(t,x)$, system (24) is equivalent to, using the notation (40),

$$\begin{aligned} \frac{\partial}{\partial t} f_{\text{equi}} + \xi \nabla_x f_{\text{equi}} &= Q_1(t,x,\xi)\,, \\ \frac{\partial}{\partial t} g_{\text{equi}} + \xi \nabla_x g_{\text{equi}} &= Q_2(t,x,\xi)\,, \end{aligned} \tag{42}$$

with $\int_{\mathbb{R}^d} Q_1(t,x,\xi)\ d\xi = 0$ (this gives the mass conservation equation), $\int_{\mathbb{R}^d} \xi Q_1(t,x,\xi)\ d\xi = 0$ (this gives the momentum equation with the pressure $p = \varrho T$), $\int_{\mathbb{R}^d} \left[\frac{|\xi|^2}{2} Q_1 + Q_2\right]\ d\xi = 0$ (energy equation).

Indeed, we have

$$\begin{aligned} \varrho(t,x) &= \int_{\mathbb{R}^d} f_{\text{equi}}(t,x,\xi)\ d\xi\,, \\ \varrho u(t,x) &= \int_{\mathbb{R}^d} \xi f_{\text{equi}}(t,x,\xi)\ d\xi\,, \\ E(t,x) &= \int_{\mathbb{R}^d} \left[\frac{|\xi|^2}{2} f_{\text{equi}} + g_{\text{equi}}\right]\ d\xi\,, \end{aligned} \tag{43}$$

and

$$\begin{aligned} F^{\varrho} &= \varrho u = \int_{\mathbb{R}^d} \xi f_{\text{equi}}\ d\xi\,, \\ F^{\varrho u} &= \varrho u \otimes u + pI = \int_{\mathbb{R}^d} \xi \otimes \xi f_{\text{equi}}\ d\xi\,, \\ F^{E} &= (E+p)u = \int_{\mathbb{R}^d} \xi \left[\frac{|\xi|^2}{2} f_{\text{equi}} + g_{\text{equi}}\right]\ d\xi\,, \end{aligned} \tag{44}$$

(we omit the details of this computation).

Next, we can recover the entropy condition as follows. From (43) we deduce

$$\begin{aligned} &\frac{\partial}{\partial t} \int_{\mathbb{R}^d} H(f_{\text{equi}}, g_{\text{equi}})\ d\xi + \operatorname{div} \int_{\mathbb{R}^d} \xi H(f_{\text{equi}}, g_{\text{equi}})\ d\xi \\ &= \int_{\mathbb{R}^d} \left[H'_f Q_1 + H'_g Q_2\right]\ d\xi \\ &= 0 \end{aligned}$$

by using the relations (41) and the above cancellation properties of Q_1, Q_2.

It remains to notice that

$$\int_{\mathbb{R}^d} H(f_{\text{equi}}, g_{\text{equi}})\, d\xi = S(\varrho, \mathrm{e}),$$

$$\int_{\mathbb{R}^d} \xi H(f_{\text{equi}}, g_{\text{equi}})\, d\xi = uS(\varrho, \mathrm{e}), \tag{45}$$

to obtain the entropy equality (26) for smooth solutions.

The strict convexity is a consequence of a general result: minimizing a strictly convex functional with linear constraints gives a strictly convex function of the constraints.

□

In order to conclude this Section, we would like to comment on the formalism developed above. We have used two densities f, g, which is not the usual way to do it, but which turns out to be a more general way to treat internal energy (see [14], Coquel & Perthame [4] for details).

The usual method (see Deshpande [6,7]) consists in introducing a kinetic internal energy variable I. For instance, we can obtain the same results as in the Proposition 4 for the minimization problem

$$S(\varrho, \mathrm{e}) = \min\Big\{ \int_{\mathbb{R}^d \otimes \mathbb{R}} f \ln f \, d\xi dI; \quad f(\xi, I) \geq 0 \text{ and}$$

$$\int_{\mathbb{R}^d \otimes \mathbb{R}} f(\xi, I)\, d\xi dI = \varrho, \quad \int_{\mathbb{R}^d \otimes \mathbb{R}} \xi f(\xi, I)\, d\xi dI = \varrho u,$$

$$\int_{\mathbb{R}^d \otimes \mathbb{R}} \left(\frac{|\xi|^2}{2} + I^\delta \right) f(\xi, I)\, d\xi dI = E \Big\}.$$

The minimizer is

$$f_{\text{equi}}(\xi, I) = \lambda_\delta \frac{\varrho}{\sqrt{2\pi T}^d T^{\frac{1}{\delta}}} \exp\left(-\frac{|\xi - u|^2}{2T} - \frac{I^\delta}{T} \right).$$

The constant λ_δ is determined by the mass constraint above or equivalently

$$\lambda_\delta \int_{\mathbb{R}} \mathrm{e}^{-\zeta^\delta}\, d\zeta = 1.$$

Then we have

$$\int_{\mathbb{R}^d \otimes \mathbb{R}} |\xi - u|^2 f(\xi, I)\, d\xi dI = d\varrho T$$

and, integrating by parts,

$$\lambda_\delta \int_{\mathbb{R}} \zeta^\delta \mathrm{e}^{-\zeta^\delta}\, d\zeta = \frac{\lambda_\delta}{\delta} \int \mathrm{e}^{-\zeta^\delta}\, d\zeta = \frac{1}{\delta},$$

which gives

$$E = \frac{1}{2}\varrho|u|^2 + \frac{d}{2}\varrho T + \frac{\varrho T}{\delta} = \frac{1}{2}\varrho|u|^2 + \varrho \mathrm{e},$$

therefore the γ–law is reached with the relation

$$\frac{d+2-d\gamma}{2(\gamma-1)} = \frac{1}{\delta}.$$

In this case the Euler equations are equivalent to the statement

$$\frac{\partial}{\partial t} f_{\mathrm{equi}}(t,x,\xi,I) + \xi \nabla_x f_{\mathrm{equi}}(t,x,\xi,I) = Q$$

with

$$\int_{\mathbb{R}^d \otimes \mathbb{R}} Q\, d\xi dI = 0, \quad \int_{\mathbb{R}^d \otimes \mathbb{R}} \xi Q\, d\xi dI = 0, \quad \int_{\mathbb{R}^d \otimes \mathbb{R}} \left(\frac{|\xi|^2}{2} + I^\delta\right) Q\, d\xi dI = 0.$$

And the entropy equality for smooth solutions also follows by the same argument as before.

3.2 Homogenous entropies

We restrict ourselves to one dimension for simplicity.

In order to find equilibrium densities with a bounded support, we have to change the kinetic entropy function H. We choose

$$\begin{aligned} H(f,0) &= +\infty && \text{for } f > 0,\ H(0,0) = 0, \\ H(f,g) &= f\left(f^{\gamma+1} g^{\gamma-3}\right)^{\frac{p}{\gamma-1}}, && p > 0,\ 1 < \gamma < 3. \end{aligned}$$

Then the Proposition 4 admits the following variant

Proposition 6. *The minimization problem*

$$\begin{aligned} S(\varrho,\mathrm{e}) &= \min\Big\{ \int_{\mathbb{R}} H(f(\xi),g(\xi))\, d\xi;\ f \geq 0,\ g \geq 0 \\ &\int_{\mathbb{R}} f(\xi)\, d\xi = \varrho,\ \int_{\mathbb{R}} \xi f(\xi)\, d\xi = \varrho u,\ \int_{\mathbb{R}} \left[\tfrac{|\xi|^2}{2} f(\xi) + g(\xi)\right] d\xi = E \Big\} \end{aligned} \tag{46}$$

admits a unique minimizer (whenever $\varrho > 0$, $\mathrm{e} > o$)

$$\begin{aligned} f_{\mathrm{equi}} &= \alpha_p \tfrac{\varrho}{\sqrt{T}} \left(1 - \tfrac{|\xi-u|^2}{\beta_p T}\right)_+^{\frac{1+2p\lambda}{2p}}, \\ g_{\mathrm{equi}} &= \delta_p T f_{\mathrm{equi}} \left(1 - \tfrac{|\xi-u|^2}{\beta_p T}\right)_+, \end{aligned} \tag{47}$$

with α_p, β_p, δ_p *the only constants which fulfil the constraints in (46) and*

$$\lambda = \frac{3-\gamma}{2(\gamma-1)}.$$

Proof. Again the proof consists in writing the Euler–Lagrange equations: when $f, g > 0$

$$\begin{aligned} H_f(f,g) = q\tfrac{f^{q-1}}{g^r} &= \alpha_0 + \alpha_1\xi + \alpha_2\tfrac{\xi^2}{2}\,, \\ H_g(f,g) \qquad &= -r\tfrac{f^q}{g^{r+1}} = \alpha_2\,, \end{aligned} \tag{48}$$

for $q = 1+p\frac{\gamma+1}{\gamma-1}$, $r = p\frac{3-\gamma}{\gamma-1}$. This leads to the formulas (47) for the minimizers. And again the strict convexity of H (this requires $q > 1$, $r > 0$, $q-r-1 > 0$ which are always satisfied) allows us to conclude as before:

$$\begin{aligned} H(f,g) &\geq H(f_{\text{equi}}, g_{\text{equi}}) + H'_f(f_{\text{equi}}, g_{\text{equi}})(f - f_{\text{equi}}) \\ &\quad H'_g(f_{\text{equi}}, g_{\text{equi}})(g - g_{\text{equi}}) \\ &\geq H(f_{\text{equi}}, g_{\text{equi}}) - \left[\left(\alpha_0 + \alpha_1\xi + \alpha_2\frac{\xi^2}{2}\right) f_{\text{equi}} + \alpha_2 g_{\text{equi}}\right] \\ &\quad + \left(\alpha_0 + \alpha_1\xi + \alpha_2\frac{\xi^2}{2}\right) f + \alpha_2 g\,. \end{aligned}$$

The last inequality is an equality inside the support of $(f_{\text{equi}}, g_{\text{equi}})$ and is obvious outside since $\alpha_0 + \alpha_1\xi + \alpha_2\frac{\xi^2}{2} < 0$ and $\alpha_2 < 0$ then. Therefore

$$\int_{\mathbb{R}} H(f(\xi), g(\xi))\, d\xi \geq \int_{\mathbb{R}} H(f_{\text{equi}}(\xi), g_{\text{equi}}(\xi))\, d\xi$$

and the Proposition 6 is proved.

Again one readily checks the compatibility with the γ–law Euler equations by computing

$$\begin{aligned} U &= \int_{\mathbb{R}} \begin{pmatrix} 1 \\ \xi \\ \frac{\xi^2}{2} \end{pmatrix} f_{\text{equi}}\, d\xi + \int_{\mathbb{R}} \begin{pmatrix} 0 \\ 0 \\ 1 \end{pmatrix} g_{\text{equi}}\, d\xi\,, \\ F(U) &= \int_{\mathbb{R}} \xi \begin{pmatrix} 1 \\ \xi \\ \frac{\xi^2}{2} \end{pmatrix} f_{\text{equi}}\, d\xi + \int_{\mathbb{R}} \xi \begin{pmatrix} 0 \\ 0 \\ 1 \end{pmatrix} g_{\text{equi}}\, d\xi\,. \end{aligned} \tag{49}$$

3.3 A class of entropic kinetic schemes

We can use the above theory to prove that the choices

$$\begin{aligned} \chi &= \alpha_p \left(1 - \tfrac{w^2}{\beta_p}\right)_+^{\frac{1+2p\lambda}{2p}}\,, \\ \zeta &= \delta_p \chi(w) \left(1 - \tfrac{w^2}{\beta_p}\right)_+\,, \end{aligned} \tag{50}$$

give an entropy scheme.

Theorem 7. *With the CFL condition (37) and* $\gamma_{\max} = \sqrt{\beta_p}$, *the choice (50) in the kinetic scheme (37) gives an in–cell entropy inequality (18) for*

$$S_i^n = S(\varrho_i^n, \mathrm{e}_i^n)$$

with S *given in Proposition 6, and the entropy flux is*

$$\begin{aligned}
\eta(U,V) &= \eta_+(U) + \eta_-(V) \\
\eta_+(U) &= \int\limits_{\xi \geq 0} \xi H(f_{\mathrm{equi}}, g_{\mathrm{equi}})\, d\xi\,, \\
\eta_-(U) &= \int\limits_{\xi \leq 0} \xi H(f_{\mathrm{equi}}, g_{\mathrm{equi}})\, d\xi\,,
\end{aligned}$$

(see the notations and properties of Section 3.2).

Proof. We come back to the form (38) of the kinetic scheme

$$\begin{aligned}
f_i^{n+1^-}(\xi) &= (1-\sigma|\xi|) f_i^n + \sigma\xi_+ f_{i-1}^n + \sigma\xi_- f_{i+1}^n\,, \\
g_i^{n+1^-}(\xi) &= (1-\sigma|\xi|) g_i^n + \sigma\xi_+ g_{i-1}^n + \sigma\xi_- g_{i+1}^n\,.
\end{aligned} \tag{51}$$

These are convex combinations, and H is convex, therefore

$$\begin{aligned}
H(f_i^{n+1^-}(\xi), g_i^{n+1^-}(\xi)) \leq{} & (1-\sigma|\xi|) H(f_i^n(\xi), g_i^n(\xi)) \\
& + \sigma\xi_+ H(f_{i-1}^n(\xi), g_{i-1}^n(\xi)) \\
& + \sigma\xi_- H(f_{i+1}^n(\xi), g_{i+1}^n(\xi))\,.
\end{aligned} \tag{52}$$

Now, from the moments equality which defines U^{n+1} from f^{n+1^-}

$$U_i^{n+1} = \int\limits_{\mathbb{R}} \begin{pmatrix} 1 \\ \xi \\ \frac{\xi^2}{2} \end{pmatrix} f_i^{n+1^-}(\xi)\, d\xi + \int\limits_{\mathbb{R}} \begin{pmatrix} 0 \\ 0 \\ 1 \end{pmatrix} g_i^{n+1^-}(\xi)\, d\xi$$

and from the construction of the equilibrium (f_i^n, g_i^n) (in (34) and Proposition 6) we have

$$S_i^{n+1} = \int\limits_{\mathbb{R}} H(f_i^{n+1}(\xi), g_i^{n+1}(\xi))\, d\xi \leq \int\limits_{\mathbb{R}} H(f_i^{n+1^-}(\xi), g_i^{n+1^-}(\xi))\, d\xi\,.$$

Combined with (52) this inequality yields

$$\begin{aligned}
S_i^{n+1} \leq S_i^n &+ \sigma \int\limits_{\mathbb{R}} \xi_+ \left[H(f_{i-1}^n, g_{i-1}^n) - H(f_i^n, g_i^n) \right] d\xi \\
&- \sigma \int\limits_{\mathbb{R}} \xi_- \left[H(f_i^n, g_i^n) - H(f_{i+1}^n, g_{i+1}^n) \right] d\xi\,,
\end{aligned}$$

which is the in–cell entropy inequality (18) with the fluxes of Theorem 7.

□

Remark 8. From a computational point of view these functions χ, ζ in (50) are not very good when $\frac{1+2p\lambda}{2p}$ is not integer. It is however always possible to choose $p \in (\frac{1}{2}, +\infty]$ which realizes this constraint.

3.4 A kinetic scheme with the maximum principle on the specific entropy $\gamma = 3$

Among the class of equilibria (and schemes) developed in Section 3.3, a very special one plays a central role in the sense that it gives the strong entropy inequality of Section 1: the maximum principle on entropy.
We begin with the very simple case $\gamma = 3$. We need a preliminary remark.

Lemma 9. *The dual minimisation problems*

$$\Sigma = \min \{ \|f\|_{L^\infty};\ f(\xi) \geq 0,\ \int_{\mathbb{R}} f(\xi)\ d\xi = \varrho\,, \tag{53}$$

$$\int_{\mathbb{R}} \xi f(\xi)\ d\xi = 0,\ \int_{\mathbb{R}} \xi^2 f(\xi)\ d\xi = \varrho T \}$$

$$\varrho T = \min \{ \int_{\mathbb{R}} \xi^2 f(\xi)\ d\xi;\ f(\xi) \geq 0,\ \int_{\mathbb{R}} f(\xi)\ d\xi = \varrho\,, \tag{54}$$

$$\int_{\mathbb{R}} \xi f(\xi)\ d\xi = 0,\ \|f(\xi)\|_{L^\infty} = \Sigma \}\,,$$

are achieved, denoting $\Sigma = \frac{\varrho}{2\sqrt{3T}}$, *by*

$$f_{\text{equi}} = \begin{cases} \Sigma & \textit{for } |\xi| \leq \sqrt{3T}\ , \\ 0 & \textit{otherwise}\ . \end{cases} \tag{55}$$

We do not prove this easy Lemma. We just note that it can be considered as a limiting case of the problem in the Proposition 6 as $p \to +\infty$.

When choosing this equlibrium χ in the kinetic scheme (for $\gamma = 3$, there is no need of ζ) we obtain

$$f_i^{n+1^-}(\xi) = f_i^n(\xi)(1 - \sigma|\xi|) + \sigma\xi_+ f_{i-1}^n(\xi) + \sigma\xi_- f_{i+1}^n(\xi)\,,$$

therefore

$$\|f_i^{n+1^-}(\xi)\|_{L^\infty(\mathbb{R}_\xi)} \leq \sup_{j=i,i-1,i+1} \|f_j^n(\xi)\|_{L^\infty(\mathbb{R}_\xi)}\,,$$

and by Lemma 9, the equilibrium $f_i^{n+1}(\xi)$ satisfying the constraints, we obtain

$$\|f_i^{n+1}(\xi)\|_{L^\infty(\mathbb{R}_\xi)} \leq \sup_{j=i,i-1,i+1} \|f_j^n(\xi)\|_{L^\infty(\mathbb{R}_\xi)}\,.$$

Since

$$\|f_i^n(\xi)\|_{L^\infty} = \frac{\varrho_i^n}{2\sqrt{3T_i}}$$

is the specific entropy - up to multiplication by a constant - we have proved the strong inequality (20).

3.5 A kinetic scheme with the maximum principle on the specific entropy $\gamma < 3$

The extension of the above argument to general values of γ relies on the following variant of Lemma 9.

Lemma 10. *For $1 < \gamma < 3$, the dual minimisation problems*

$$\Sigma = \min\Big\{\|f^{\frac{\gamma+1}{2}} g^{\frac{\gamma-3}{2}}\|_{L^\infty},\ f(\xi) \geq 0,\ g(\xi) \geq 0, \tag{56}$$

$$\int_{\mathbb{R}} f(\xi)\, d\xi = \varrho,\ \int_{\mathbb{R}} \xi f(\xi)\, d\xi = 0,\ \int_{\mathbb{R}} \left[\frac{\xi^2}{2} f(\xi) + g(\xi)\right]\, d\xi = \varrho e\Big\}$$

$$\varrho e = \min\Big\{\int_{\mathbb{R}} \left[\frac{\xi^2}{2} f(\xi) + g(\xi)\right]\, d\xi;\ f(\xi) \geq 0,\ g(\xi) \geq 0, \tag{57}$$

$$\int_{\mathbb{R}} f(\xi)\, d\xi = \varrho,\ \int_{\mathbb{R}} \xi f(\xi)\, d\xi = 0,\ f^{\frac{\gamma+1}{2}} g^{\frac{\gamma-3}{2}} \leq \Sigma\Big\}$$

are achieved for $\Sigma = \alpha^{\frac{\gamma+1}{2}} \delta^{\frac{\gamma-3}{2}} \frac{\varrho^{\gamma-1}}{T}$, $\varrho T = (\gamma - 1)\varrho e$ by

$$\begin{aligned} f_{\text{equi}}(\xi) &= \alpha \tfrac{\varrho}{\sqrt{T}} \left(1 - \tfrac{\xi^2}{\beta T}\right)_+^{\lambda}, \\ g_{\text{equi}}(\xi) &= \delta T f_{\text{equi}}(\xi) \left(1 - \tfrac{\xi^2}{\beta T}\right). \end{aligned} \tag{58}$$

And α, β, δ are the only constants satisfying the constraints of the minimisation problem.

Again this is a limiting case of the Proposition 6 as $p \to +\infty$. A proof can be found in Khobalatte & Perthame [13]. A consequence of this is

Theorem 11. *The kinetic scheme obtained, for $1 < \gamma < 3$ with*

$$\chi(w) = \alpha \left(1 - \frac{w^2}{\beta}\right)_+^{\lambda},\ \lambda = \frac{3-\gamma}{2(\gamma-1)},$$

$$S(w) = \delta \chi(w) \left(1 - \frac{w^2}{\beta}\right)_+,$$

satisfies the maximum principle on specific entropy (20), under the CFL condition (37) with $\gamma_{\max} = \beta$.

Proof. We sketch it because it is an easy extension of the previous one.

Since $H(f,g) = f^{\frac{\gamma+1}{2}} g^{\frac{\gamma-3}{2}}$ is a convex function we have, using that $f_i^n(\xi), g_i^n(\xi)$ are the equilibrium in (58),

$$\begin{aligned} H(f_i^{n+1^-}(\xi), g_i^{n+1^-}(\xi)) &\leq H(f_i^n(\xi), g_i^n(\xi))(1-\sigma|\xi|) \\ &\quad +\sigma\xi_+ H(f_{i-1}^n(\xi), g_{i-1}^n(\xi)) + \sigma\xi_- H(f_{i+1}^n(\xi), g_{i+1}^n(\xi)) \\ &\leq \max_{j=i,i-1,i+1} (\Sigma_j^n), \end{aligned}$$

and again the relation (56) gives

$$\Sigma_i^{n+1} \leq \|H(f_i^{n+1^-}, g_i^{n+1^-})\|_{L^\infty} \leq \max_{j=i,i-1,i+1} (\Sigma_j^n)$$

and the inequality (20) is proved.

Associated to this principle, a singular entropy inequality holds which can be found in Khobalatte & Perthame [13].

3.6 Symmetric variables

Here we would like to present the kinetic symmetrisation of Euler equations. The ideas have been developed by Croisille & Delorme [5]. We again refer to Serre [28] for this notion and further references.

We come back to the frameworks of Proposition 4 and Proposition 6. The general Euler–Lagrange equations for the minimizers are

$$\nabla H(f_{\text{equi}}, g_{\text{equi}}) = (\alpha_0 + \alpha_1\xi + \alpha_2\frac{\xi^2}{2}, \alpha_2),$$

or in other words, using the Legendre transform $H^\star$ of H

$$(f_{\text{equi}}, g_{\text{equi}}) = \nabla H^\star(\alpha_0 + \alpha_1\xi + \alpha_2\frac{\xi^2}{2}, \alpha_2),$$

where the Lagrange multipliers $\alpha_0(U)$, $\alpha_1(U)$, $\alpha_2(U)$ are given so as to satisfy the mass, momentum and energy constraints. It can also be rewritten

$$\nabla H^\star\left(\alpha_3 - \frac{|\xi-u|^2}{2T}, -\frac{1}{2T}\right)$$

which allows us to see more explicitly the evenness property which generates Euler equation through the transport equations on f and g.

Proposition 12. *The variables $\alpha(U) = (\alpha_0, \alpha_1, \alpha_2)$ are the symmetric variables associated to the entropy S, which is also characterised by its Legendre transform*

$$S^\star(\alpha) = \int_{\mathbb{R}} H^\star\left(\alpha_0 + \alpha_1\xi + \alpha_2\frac{\xi^2}{2}, \alpha_2\right)\, d\xi\,. \tag{59}$$

Proof. We denote by $K(\xi)$ the 3×2–matrix

$$K(\xi) = \begin{pmatrix} 1 & 0 \\ \xi & 0 \\ \xi^2/2 & 1 \end{pmatrix}.$$

Then the Euler equations are written

$$\frac{\partial}{\partial t} K \cdot \nabla H^\star(\alpha \cdot K) + \xi \frac{\partial}{\partial x} K \cdot \nabla H^\star(\alpha \cdot K) = Q(t, x, \xi), \tag{60}$$

where Q is a vector with three entries and 0 integral. This is also written

$$K \cdot D^2 H^\star \cdot K^t \frac{\partial \alpha}{\partial t} + \xi K \cdot D^2 H^\star \cdot K^t \frac{\partial \alpha}{\partial x} = Q,$$

and one notices that $K \cdot D^2 H^\star \cdot K^t$ is a symmetric nonnegative 3×3–matrix. Integrating in ξ gives the structure of the symmetric system

$$A \cdot \frac{\partial \alpha}{\partial t} + B \cdot \frac{\partial \alpha}{\partial x} = 0, \tag{61}$$

where A is a symmetric positive definite 3×3–matrix and B is a symmetric 3×3–matrix. Indeed

$$A(\alpha) = \int_{\mathbb{R}} K(\xi) \cdot D^2 H^\star(\cdots) \cdot K^t(\xi)\, d\xi,$$

$$B(\alpha) = \int_{\mathbb{R}} \xi K(\xi) \cdot D^2 H^\star(\cdots) \cdot K^t(\xi)\, d\xi.$$

The relation (59) on $S^\star(\alpha)$ is a simple application of a classical computation. We have

$$\begin{aligned} S^\star(\alpha) &= \max_U (\alpha \cdot U - S(U)) \\ &= \sup_{U, f, g \in \mathcal{A}} \left\{ \alpha \cdot U - \int H(f, g)\, d\xi \right\} \\ &= \sup_{f \geq 0,\, g \geq 0} \left\{ \int_{\mathbb{R}} \left[(\alpha_0 + \alpha_1 \xi + \alpha_2 \frac{\xi^2}{2}) f + \alpha_2 g - H(f, g) \right] d\xi \right\} \\ &= \int_{\mathbb{R}} H^\star(\alpha_0 + \alpha_1 \xi + \alpha_2 \frac{\xi^2}{2}, \alpha_2)\, d\xi, \end{aligned}$$

where $\mathcal{A}$ is the admissible set of the minimisation problem in Proposition 4 and Proposition 6. The last equality requires some arguments which can be found in Coquel & Perthame [4].

□

Notice that a numerical application of this theory is to build upwind linearised schemes setting

$$A_+ = \int_{\xi \geq 0} K \cdot D^2 H^\star \cdot K^t \, d\xi \,,$$

$$B_+ = \int_{\xi \geq 0} \xi K \cdot D^2 H^\star \cdot K^t \, d\xi \,.$$

These are positive definite symmetric matrices.

References

1. Brenier, Y.: Résolution d'équations d'évolution quasilinéaire en dimension N d'espace à l'aide d'équations linéaires en dimension N+1. J. diff. Eq. **50** (1983) 375–390
2. Cercignani, C.: The Boltzmann Equation and its applications. Springer–Verlag, Berlin, New–York (1994)
3. Cercignani, C., Illner, R., Pulvirenti, M.: The Mathematical Theory of Dilute Gases. Springer–Verlag, Berlin, New–York (1994)
4. Coquel, F., Perthame, B.: A kinetic formalism for pressure laws of real gases. Work in preparation.
5. Croisille, J.P., Delorme, P.: Kinetic symmetrisation and pressure laws for the Euler equations. Phys. **D 57** (1992) N. 3–4, 395–426
6. Deshpande, S.: Kinetic Theory based new upwind methods for inviscid compressible flows. AIAA paper **96–0275** (1986)
7. Deshpande, S.: A second order accurate, kinetic theory based method for inviscid compressible flows. NASA Technical paper **N 2613** (1986)
8. Estivalezes, J.L., Villedieu, P.: Higher order positivity preserving kinetic schemes for the compressible Euler equations. SIAM J. Num. Anal. **33** (1996) N. 5, 2050–2067
9. Giga, Y., Miyakawa, T.: A kinetic construction of global solutions of first-order quasilinear equations. Duke Math. J. **50** (1983) 505–515
10. Godlewski, E., Raviart, P.–A.: Numerical Approximation of Hyperbolic Systems of Conservation Laws. Appl. Math. **118**. Springer Verlag (1996)
11. Harten, A., Lax, P.D., Van Leer, B.: On upstream differencing and Godunov–type schemes for hyperbolic conservation laws. SIAM Rev. **25** (1983) 35–61
12. Kaniel, S.: A kinetic model for the compressible flow equations. Ind. Univ. Math. J. (1988)**N.3** 537–563
13. Khobalatte, B., Perthame, B.: Maximum principle on the entropy and second order kinetic schemes. Math. Comp. **62 (205)** (1994) 119–135
14. LeTallec, P., Perlat, J.P., Perthame, B.: The Gaussian BGK model of Boltzmann equation with small Prandtl number. Work in preparation.
15. Lax, P.D.: Hyperbolic systems of conservation laws and the mathematical theory of shock waves. Conf. Board Math. Sc. **11** SIAM (1973)

16. LeVeque, R.J.: Numerical Methods for Conservation Laws. Lectures in Mathematics, ETH Zürrich, Birkhäuser (1992)
17. Lifshitz, E.M., Pitaevskii, L.P.: Course in theoretical physics **10**, Physical kinetics. Pergamon Press (1981)
18. Lions, P.–L.: Compactness in Boltzmann's equation via Fourier integral operators and applications. Parts I to III. J. of Math. of Kyoto Univ. **34 N 2** (1994) 391–461
19. Mandal, J.C., Deshpande, S.M.: Kinetic Flux Vector Splitting for Euler Equations. Computers and fluids **23 No. 2** 447.
20. Perthame, B.: Boltzmann type schemes and the entropy condition. SIAM J. on Num. Anal. **27,6** (1990), 1405–1421
21. Perthame, B.: Second Order Boltzmann Schemes for compressible Euler Equations in one and two space dimensions. SIAM J. on Num. Anal. **29,1** (1992), 1–19
22. Prendergast, K.H., Xu, K.: Numerical Hydrodynamics from gas kinetic theory. J. Comp. Phys. **109** (1993) 53
23. Pullin, D.I.: Direct simulation method for compressible inviscid ideal–gas–flow. J. Comput. Phys. **34** (1980) 231–244
24. Raghurama Rao, S.V., Deshpande, S.M.: Peculiar velocity based upwind method for inviscid compressible flows, Comp. Fluid Dynamics J., Japan **3(4)** (1994), 415–432.
25. Reitz, R.D.: One–dimensional compressible gas dynamics calculations using the Boltzmann equations. J. Comput. Phys. **42** (1981) 108–123
26. Roe, P.L.: Approximate Riemann Solvers, Parameter Vectors and Difference Schemes. J. Comput. Phys. **43** 357.
27. Sanders, R., Prendergast, K.H.: the possible relation of the three kiloparsec arm to explosions in the galactic nucleus. Astrophysical Journal **188** (1974)
28. Serre, D.: Systèmes hyperboliques de lois de conservation, Parties **I** et **II**. Diderot, Paris (1996)
29. Smoller, J.: Shock waves and reaction-diffusion equations. Springer, Berlin (1982)
30. Tadmor, E.: A minimum entropy principle in the gas dynamics equations. Appl. Num. Math. **2** (1986) 211–219
31. Xu, K., Prendergast, K.H.: Multidimensional hydrocode from kinetic theory. J. Comp. Physics, **114** (1994) 9–17

An Introduction to Nonclassical Shocks of Systems of Conservation Laws

Philippe G. LeFloch

Centre de Mathématiques Appliquées & Centre National de la Recherche Scientifique, UA 756, Ecole Polytechnique, 91128 Palaiseau Cedex, France

Abstract. We review a recent activity on *nonclassical* shock waves of strictly hyperbolic systems of conservation laws, generated by balanced diffusion and dispersion effects. These shocks do not satisfy the standard Lax and Liu entropy criteria, and in fact are *undercompressive* and satisfy a *single entropy inequality.* The selection of admissible nonclassical shocks requires a strengthened version of the entropy inequality, called a *kinetic relation*, which constrains the entropy dissipation. The kinetic function is determined from traveling wave solutions to a system of equations augmented with diffusion and dispersion.

For nonconvex scalar conservation laws and non-genuinely nonlinear, strictly hyperbolic systems, the existence and uniqueness of nonclassical shocks is investigated using successively the traveling wave analysis, the front tracking algorithm and the compensated compactness method. Nonclassical shocks may also be generated by finite difference schemes.

The kinetic relation provides a useful tool to study the properties of nonclassical shocks and, in particular, their sensitivity to regularization parameters.

Contents

1 Introduction

We are interested in discontinuous solutions of nonlinear hyperbolic systems of conservation laws. We review a recent activity on shock waves that are sensitive to regularization mechanisms such as diffusion and dispersion, and violate the classical entropy conditions introduced by Lax and Liu. The material surveyed here comes primarily from a series of papers by Hayes and the author [15–17] and from [4] and was initially motivated by a result of Jacobs, McKinney and Shearer [25].

This research is driven by problems in continuum mechanics and physics. The importance of capillarity on the dynamics of phase transitions in fluids and solids is well recognized and is studied in particular by Abeyaratne and Knowles [1,2], Slemrod [42] (also [11]) and Truskinovsky [46,47]. The Hall term in magnetohydrodynamics may have a driving effect in certain important applications; see Wu [48] and Kennel and Wu [49]. From a mathematical standpoint, one has to understand how dispersion terms in balance with diffusion terms drive the dynamics of shock waves.

Consider a system of conservation laws of the general form

$$\partial_t u + \partial_x f(u) = 0, \qquad u(x,t) \in \mathbb{R}^N,\ x \in \mathbb{R},\ t > 0, \tag{1}$$

endowed with (at least) one entropy-entropy flux pair $(U, F) : \mathbb{R}^N \to \mathbb{R}^2$, i.e. a pair of functions satisfying the compatibility condition $DF = DU \cdot Df$, which moreover is strictly convex at each point u where the matrix $Df(u)$ is hyperbolic. Here $f : \mathbb{R}^N \to \mathbb{R}^N$ is a given smooth mapping, called the flux-function of (1). Discontinuous solutions defined in the weak sense of distributions are not uniquely determined by their initial data. It is therefore necessary to specify a selection mechanism of the solutions of interest (see Dafermos [10] for a review.) As it is natural, at least for the examples derived from physical modeling, we shall always constrain the solutions to satisfy the *entropy inequality*

$$\partial_t U(u) + \partial_x F(u) \le 0, \tag{2}$$

that is, the mathematical entropy dissipation measure is non-positive. In presence of diffusion and dispersion, only a *single* entropy inequality can generally be employed, as we will show in Section 2 below. When (1) is genuinely nonlinear – a convexity-like condition on the flux f – the solutions of the Riemann problem determined by (1)-(2) and

$$u(x,0) = \begin{cases} u_l & \text{for } x < 0, \\ u_r & \text{for } x > 0, \end{cases} \tag{3}$$

are *unique* for u_l, $u_r \in \mathbb{R}^N$ with $|u_r - u_l| << 1$. This holds, say, in the class of solutions consisting of a combination of elementary waves (shock waves and rarefaction waves). On the other hand, the entropy inequality (2) is *not* sufficiently discriminating when one (or several) characteristic field fails to be genuinely nonlinear. To deal with this difficulty, Liu [34] introduced a *more stringent* entropy condition which still selects a unique, stable solution to the Riemann problem. This solution however does not coincide with the one obtained by a vanishing diffusion-dispersion approximation, and a generalization of Liu's construction is necessary to deal with the models of, for instance, phase dynamics and magnetohydrodynamics.

One of our objectives will be to investigate (2) for strictly hyperbolic systems and non-genuinely nonlinear fields, and assess its role in reducing the class of admissible solutions (Section 4). We shall show that uniqueness for the Riemann problem can be recovered (Section 5) by completing (2) with a strengthened admissibility condition, called a *kinetic relation*, which constrains the *value* of the entropy dissipation measure, rather than just its *sign*.

On the other hand, we are especially interested in solutions generated by diffusive-dispersive regularizations of the general form

$$\partial_t u^\varepsilon + \partial_x f(u^\varepsilon) = \partial_x R\big(\varepsilon\, u_x^\varepsilon, \varepsilon^2\, u_{xx}^\varepsilon, \cdots\big). \tag{4}$$

Under some assumption on R, the limiting solutions $\bar{u} = \lim_{\varepsilon \to 0} u^\varepsilon$ of (4) are compatible with the single entropy inequality (2). On the other hand, by an analysis of traveling wave solutions of (4), we shall show that these solutions may include *nonclassical shocks*, i.e., waves violating the classical Lax and Liu entropy criteria (Section 2). We shall see that a nonclassical shock is *undercompressive*, in the sense that the number of characteristics impinging on the discontinuity is less than what is required for the linearized stability analysis and Lax shock admissibility inequalities fail. Interestingly, for certain choices of regularization R, these shocks may have an *arbitrarily small* amplitude. An analysis of the traveling wave solutions of (4) yields the associated kinetic relation and leads, in several instances, to a unique solution of the Riemann problem. This will be demonstrated in Section 3 with the conservation law with cubic flux

$$\partial_t u + \partial_x u^3 = 0, \qquad u(x,t) \in \mathbb{R}, \tag{5}$$

which is an important model to understand nonclassical behavior.

The Cauchy problem in the class of solutions containing nonclassical shocks is treated in Section 6 using successively the wave front tracking algorithm and the compensated compactness method. The numerical approximation of nonclassical shocks via finite difference schemes is dealt with in Section 7. The concept of a kinetic relation provides an efficient tool for the selection of undercompressive shocks and for the study of their sensitive dependence on critical parameters of regularization.

An extensive literature is available on undercompressive as well as overcompressive shocks for non-strictly hyperbolic systems or hyperbolic-elliptic systems, from both the standpoint of their existence and their properties [5,6,12,23,24,37,39,41] and the standpoint of nonlinear stability [13,14,36]; see also the references cited therein. Regularization-sensitive shock waves having a similar behavior to the nonclassical shocks studied here arise in nonconservative hyperbolic systems

$$\partial_t u + A(u)\, \partial_x u = 0, \qquad u(x,t) \in \mathbb{R}^N, \tag{6}$$

$A(u)$ not being a Jacobian matrix, and in nonlinear boundary conditions (Joseph and LeFloch [27]).

2 Entropy Stability and Examples

In this section, we emphasize the role played by the entropy inequality (2) in the selection of nonclassical solutions. The main focus is on systems that are strictly hyperbolic but not genuinely nonlinear. However the discussion also applies to hyperbolic-elliptic systems. In both cases, the Riemann problem (1)–(3) does not possess a unique solution.

We start by considering a scalar conservation law,

$$\partial_t u^\varepsilon + \partial_x f(u^\varepsilon) = \varepsilon\, u^\varepsilon_{xx} + \delta(\varepsilon)\, u^\varepsilon_{xxx}, \qquad u^\varepsilon(x,t) \in \mathbb{R} \tag{7}$$

with $\varepsilon, \delta(\varepsilon) \to 0$. Let $U : \mathbb{R} \to \mathbb{R}$ be convex and define $F : \mathbb{R} \to \mathbb{R}$ by $F' := U'\, f'$. From the balance law

$$\begin{aligned}\partial_t U(u^\varepsilon) + \partial_x F(u^\varepsilon) &= \varepsilon\, U(u^\varepsilon)_{xx} - \varepsilon\, U''(u^\varepsilon)\big(u^\varepsilon_x\big)^2 \\ &\quad + \delta(\varepsilon)\, \big(U'(u^\varepsilon) u^\varepsilon_{xx} - \tfrac{1}{2} U''(u^\varepsilon)(u^\varepsilon_x)^2\big)_x + \tfrac{\delta(\varepsilon)}{2}\, U'''(u^\varepsilon)\big(u^\varepsilon_x\big)^3\end{aligned}$$

and since the conservative terms $\varepsilon\, \partial_x(\ldots)$ and $\delta(\varepsilon)\partial_x(\ldots)$ are expected to vanish as $\varepsilon \to 0$, we deduce at least formally that

$$\partial_t U(\bar u) + \partial_x F(\bar u) = -\lim_{\varepsilon\to 0} \left(\varepsilon\, U''(u^\varepsilon)\big(u^\varepsilon_x\big)^2 + \lim_{\varepsilon \to 0} \frac{\delta(\varepsilon)}{2}\, U'''(u^\varepsilon)\big(u^\varepsilon_x\big)^3 \right)$$

for the limiting function

$$\bar u := \lim_{\varepsilon \to 0} u^\varepsilon.$$

In the right hand side, the first term is non-positive since $U'' \geq 0$, while the second one has *no definite sign* in general, except if $U''' \equiv 0$. Therefore we call the entropy

$$U(u) = u^2$$

compatible with the diffusive-dispersive regularization (7). This entropy is unique, up to a multiplicative constant and up to a linear function. The function $\bar u$ thus satisfies the constraint

$$\partial_t u^2 + \partial_x \left(\int^u 2\, v f'(v)\, dv \right) \leq 0. \tag{8}$$

It can be checked that the entropy inequalities (2) for the *other* entropies U are generally violated by $\bar u$.

When

$$\delta(\varepsilon) << \varepsilon^2,$$

the dispersion in (7) is negligable and the diffusion dominates. The functions u^ε converge in a strong topology (Section 6) to the classical entropy solutions $\bar u$ of the scalar conservation law

$$\partial_t u + \partial_x f(u) = 0, \tag{9}$$

selected by the infinite set of inequalities (2) for *all* U. Equivalently, the limiting solutions can be characterized with the *Oleinik entropy criterion.* A shock wave from u_l to u_r is admissible iff – considering for instance the case $u_l > u_r$ – the following criterion is fulfilled:

Oleinik entropy criterion: The line connecting the points with coordinates $(u_l, f(u_l))$ and $(u_r, f(u_r))$ is above the graph of the flux f in the interval $[u_r, u_l]$. (10)

On the other hand, when

$$\delta(\varepsilon) >> \varepsilon^2,$$

the dispersion dominates. The functions u^ε become highly oscillating as $\varepsilon \to 0$ and fail to converge in a strong topology. This regime is investigated by Lax and Levermore in [30].

We shall focus on the "intermediate" regime,

$$\delta(\varepsilon) = \gamma\,\varepsilon^2, \qquad \gamma \text{ fixed.}$$

The solutions u^ε converge in a strong topology (Section 6) but, in general, $\bar{u} := \lim_{\varepsilon\to 0} u^\varepsilon$ is *not* the classical entropy solution. Oscillations having small amplitude are observed near the shocks but vanish as $\varepsilon \to 0$ (Section 3). However, when the flux f is *convex* and for piecewise smooth solutions at least, the single entropy inequality (8) suffices to select a *unique* solution. For regularizations that are compatible with a convex entropy, we conclude that

Any diffusive-dispersive limit of a scalar conservation law with convex flux is a classical entropy solution.

In the present work, we are primarily interested in *nonconvex fluxes*.

Definition 1. Consider the scalar conservation law (9) with a nonconvex flux. Let (U, F) be a fixed, strictly convex entropy pair.

A shock wave solution of (9) is said to be *nonclassical* iff it satisfies the *single* entropy inequality (2) *but not* the Oleinik entropy criterion (10).

In Section 3 we will return to the discussion of nonclassical shocks for scalar equations. We now turn to systems of conservation laws. First of all, we rely on the assumption that a strictly convex entropy U exists. System (1) is hyperbolic and can be symmetrized via the change of variable

$$v = \hat{v}(u) := \nabla_u U(u),$$

the so-called entropy variable. The mapping $u \mapsto v$ is one-to-one and (1) may be rewritten as

$$\partial_t \tilde{u}(v) + \partial_x \tilde{f}(v) = 0, \qquad \tilde{u}(v) := u, \quad \tilde{f}(v) := f(u). \tag{11}$$

Define also $\tilde{U}(v) := U(u)$,... From $DF = DU \cdot Df$, we deduce that $\nabla_u^2 U \cdot D_u f$ is a symmetric matrix and, then, that the matrices $D_v \tilde{u}(v)$ and $D_v \tilde{f}(v)$ are symmetric.

Consider the following regularization of the symmetric form (11) of (1):

$$\partial_t \tilde{u}(v^\varepsilon) + \partial_x \tilde{f}(v^\varepsilon) = \varepsilon\, v^\varepsilon_{xx} + \gamma\, \varepsilon^2\, v^\varepsilon_{xxx}. \tag{12}$$

We are adding diffusion and dispersion terms that are *linear* in the *entropy variable.* The parameter γ represents the ratio of the dispersion to the diffusion coefficients. In agreement with the discussion made with scalar equations, the diffusion and the dispersion are kept in balance.

The formal limit $\bar{v} := \lim_{\varepsilon \to 0} v^\varepsilon$ is a solution of (11). Multiplying (12) by v^ε, we observe that the left hand side of (12) takes the *conservative* form

$$\partial_t \tilde{U}(v^\varepsilon) + \partial_x \tilde{F}(v^\varepsilon).$$

On the other hand, in the right hand side of (12), the diffusion is the sum of a *conservative* term and a *dissipative* one,

$$\varepsilon\, v^\varepsilon \cdot v^\varepsilon_{xx} = \frac{\varepsilon}{2} \left(|v^\varepsilon|^2\right)_{xx} - \varepsilon \left|v^\varepsilon_x\right|^2,$$

while the dispersion is fully *conservative*

$$\gamma\, \varepsilon^2 v^\varepsilon \cdot v^\varepsilon_{xxx} = \gamma\, \varepsilon^2 \Big(v^\varepsilon \cdot v^\varepsilon_{xx} - \frac{1}{2}|v^\varepsilon_x|^2\Big)_x.$$

It follows that the function $\bar{v}$ satisfies

$$\partial_t \tilde{U}(v) + \partial_x \tilde{F}(v) \leq 0, \tag{13}$$

which is equivalent to the entropy inequality in the conservative variable, (2), provided we define a solution $\bar{u}$ of (1) from the solution $\bar{v}$ of (11) by $\bar{u} = \big(\nabla_u U\big)^{-1}(\bar{v})$.

The approximation scheme (12) represents only one instance of regularization compatible with a given entropy. Variants of this general structure seem to always exist for hyperbolic systems derived from physical modeling in fluid dynamics, MHD and phase dynamics. For instance the Hall terms in MHD have a conservative form and do not contribute to the limiting entropy dissipation, similarly as we demonstrated for the dispersive terms in (7) and (12).

The regularization (12) can also be regarded in the conservative variables. Consider a hyperbolic-elliptic system and its *nonconvex* entropy function U. Consider the following regularization of (1)

$$\partial_t u^\varepsilon + \partial_x f(u^\varepsilon) = \varepsilon\, \hat{v}(u^\varepsilon)_{xx} + \gamma\, \varepsilon^2\, \hat{v}(u^\varepsilon)_{xxx}. \tag{14}$$

This is identical with (12) whenever the entropy variable is a change of variable. The proof given above shows that the formal limit $\bar{u} := \lim_{\varepsilon \to 0} u^\varepsilon$

satisfies the conservation law (1) and the entropy inequality (2). Once again the diffusion $\varepsilon\,\hat{v}(u^\varepsilon)_{xx}$ is dissipative for the entropy U, while the dispersion $\gamma\,\varepsilon^2\,\hat{v}(u^\varepsilon)_{xxx}$ is conservative for the entropy U.

We end this section with a fundamental example, the system of elasticity for a nonlinear material

$$\begin{aligned} \partial_t v - \partial_x \sigma(w) &= \varepsilon\, v_{xx} - \gamma\,\varepsilon^2\, w_{xxx}, \\ \partial_t w - \partial_x v &= 0, \end{aligned} \tag{15}$$

where $v(x,t)$ and $w(x,t)$ are the velocity and the deformation gradient of the material at the point (x,t), respectively. The strain-stress relation $w \mapsto \sigma(w)$ depends on the material under consideration. A typical (nonconvex) constitutive equation of interest is

$$\sigma(w) = w^3 + a\,w, \tag{16}$$

a being a real parameter. In the right hand side of (15), the coefficients ε and $\gamma\,\varepsilon^2$ (with $\gamma > 0$) represent the diffusion and capillarity of the material respectively.

Consider the system (15) with $\varepsilon = 0$,

$$\begin{aligned} \partial_t v - \partial_x \sigma(w) &= 0, \\ \partial_t w - \partial_x v &= 0, \end{aligned} \tag{17}$$

When $a > 0$, it is strictly hyperbolic and admits two real and distinct wave speeds, $\pm c(w) := \pm\sqrt{3\,w^2 + a}$. When $a = 0$, it is strictly hyperbolic except on the line $\{w = 0\}$. When $a < 0$, it is strictly hyperbolic in the range $\{3\,w^2 > |a|\}$, but elliptic in $\{3\,w^2 < |a|\}$. The system fails to be genuinely nonlinear at $w = 0$, that is the point where σ fails to be convex or concave. The total energy associated with (15) with $\varepsilon = 0$ can be used as a mathematical entropy pair

$$U(v,w) = \frac{v^2}{2} + \frac{w^4}{4} + \frac{a\,w^2}{2}, \qquad F(v,w) := -v\,\left(w^3 + a\,w\right). \tag{18}$$

The analysis presented in this section for general systems applies to (15) only if the dispersion w_{xxx} was replaced by $\big(\sigma(w)\big)_{xxx}$, which is *conservative* for the entropy (17). The dispersion w_{xxx} in (15) requires a variant of our general argument given now. Observe that

$$\begin{aligned} &\partial_t \left(U(v,w) + \gamma\,\varepsilon^2 \tfrac{w_x^2}{2}\right) + \partial_x F(v,w) \\ &= \varepsilon\,\left(v\,v_x\right)_x - \varepsilon\,v_x^2 - \gamma\,\varepsilon^2\,\left(-v\,w_{xx} + w_t\,w_{xx}\right)_x. \end{aligned}$$

Thus, formally (see Section 6 for a rigorous statement), we recover the entropy inequality

$$\partial_t \left(\frac{v^2}{2} + \frac{w^4}{4} + a\,\frac{w^2}{2}\right) - \partial_x\,\left(v\,(w^3 + a\,w)\right) \le 0. \tag{19}$$

Note that (19) holds for either $a > 0$ or $a < 0$ and the entropy (18) is strictly convex iff $a > 0$.

3 Traveling Wave Analysis and the Riemann Problem

For scalar conservation laws with nonconvex flux, the Oleinik entropy criterion (10) selects a unique solution to the Riemann problem, depending continuously upon its initial states. Focusing on a model example, the scalar conservation law with cubic flux, we show that (nonclassical) shocks *violating* the Oleinik entropy criterion can be limits of diffusive-dispersive approximations. They are associated with traveling waves connecting two *saddle equilibria*. (A classical traveling wave connects a saddle and a node.)

Nonclassical shocks therefore should be retained when solving the Riemann problem if shock waves are driven by balanced diffusion and dispersion effects. A general Riemann solver allowing nonclassical shocks is presented at the end of this section. An alternative to Oleiniks admissibility condition is needed to characterize the nonclassical solutions among all weak solutions of the Riemann problem (see Sections 4 and 5).

Consider the conservation law with cubic flux,

$$\partial_t u + \partial_x u^3 = 0, \qquad u(x,t) \in \mathbb{R}, \tag{20}$$

and augment it with a *nonlinear* diffusion and a *linear* dispersion [15]:

$$\partial_t u^\varepsilon + \partial_x u_\varepsilon^3 = \varepsilon\, \partial_x \left(|u_x^\varepsilon| u_x^\varepsilon\right) + \delta(\varepsilon)\, u_{xxx}^\varepsilon \tag{21}$$

with $\varepsilon, \delta(\varepsilon) > 0$. We search for traveling wave solutions

$$u^\varepsilon(x,t) = v(\xi), \qquad \xi = \frac{x - st}{\sqrt{\delta(\varepsilon)}},$$

where the wave speed s has to be determined. The solution should satisfy the ordinary differential equation

$$-s\, v' + \left(v^3\right)' = \frac{\varepsilon}{\delta(\varepsilon)} \left(|v'| v'\right)' + v''', \tag{22}$$

which we supplement with the boundary conditions

$$\lim_{-\infty} v = u_l, \quad \lim_{+\infty} v = u_r, \quad \lim_{\pm\infty} v' = \lim_{\pm\infty} v'' = 0, \tag{23}$$

where u_l, u_r are given states. From now on, we assume that

$$\gamma = \frac{\delta(\varepsilon)}{\varepsilon} \qquad \text{is a constant}$$

and we study the problem (22)-(23). We will determine the set of u_r that can be reached from a given u_l. This will provide us with the shock curve (or rather, the shock set) associated with u_l. Before continuing, we want to point out that the *linear* diffusion $\varepsilon\, u_{xx}^\varepsilon$ was first treated by Jacobs, McKinney and

Shearer [25]. Recently their result was extended by Hayes and Shearer [19] to general fluxes.

For a traveling wave connecting $u_l > 0$ (for definiteness) to u_r, the Rankine-Hugoniot condition imposes that

$$s = \frac{u_l^3 - u_r^3}{u_l - u_r} = u_l^2 + u_l\, u_r + u_r^2.$$

Integrating (22) once yields

$$v'' + \frac{1}{\gamma}|v'|v' = M_1 - s\,v + v^3, \tag{24}$$

where $M_2 = s\,u_l - u_l^3 = u_l u_r (u_l + u_r)$. We rewrite (24) as a first-order system

$$\begin{aligned} v' &= w, \\ w' &= p(v) - \tfrac{1}{\gamma}\,|w|\,w, \end{aligned} \tag{25}$$

where $p(v) \equiv C - s\,v + v^3$.

The *equilibrium points* are $(v, w) = (\bar{v}, 0)$ where $p(\bar{v}) = 0$. For simplicity, we restrict our attention to the (interesting) case that the cubic function $p(v)$ has three distinct roots denoted by $v_l > v_m > v_r$ with $v_l = u_l > 0$. From the cubic form of p, it is checked that

$$v_l + v_m + v_r = 0, \tag{26}$$

$v_r < 0$ and $v_m \in (\frac{-u_l}{2}, u_l)$. We may fix $u_l > 0$ and view s as a parameter varying in the interval $(\frac{3}{4}u_l^2, 3u_l^2)$.

The linearization of (25) about the equilibria leads to considering the eigenvalues

$$\lambda_\pm(\bar{v}) = \pm\sqrt{p'(\bar{v})} = \pm\sqrt{3\,\bar{v}^2 - s}. \tag{27}$$

The outer equilibria, v_l and v_r, are *saddle points*, corresponding to two real eigenvalues with distinct signs. The middle equilibrium v_m is an *elliptic* point, corresponding to purely imaginary eigenvalues $\lambda_\pm(v_m)$.

For a trajectory leaving from u_l and reaching some point u_r, we have

– either $u_r = v_m$ and the connection is associated with a *classical* shock: the line connecting the points with coordinates $(v_l, f(v_l)$ and $(v_m, f(v_m))$ is above the graph of f in the interval $[v_m, v_l]$.

– or $u_r = v_r$ and we obtain a *nonclassical* shock, violating the Oleinik entropy criterion.

It can be checked that whenever $v_m \geq 0$, which corresponds to the range of speed $s \in [u_l^2, 3u_l^2]$, the traveling wave always connects $u_l = v_l$ to v_m. The trajectory spirals into the elliptic point $(v_m, 0)$ and the profile $\xi \mapsto v(\xi)$ contains oscillations about the right-end state u_r having fast decay at $\xi = +\infty$.

Consider now the case $v_m < 0$, thus $s \in (\frac{3}{4}u_l^2, u_l^2)$. The state v_r is restricted by (26) and the fact that $v_r < v_m$, so that $-u_l < v_r < -\frac{u_l}{2}$. Both cases $u_r = v_r$ or $u_r = v_m$ are possible now. In fact it can be checked that there is a *single* value v_r in the range $-u_l < v_r < -\frac{u_l}{2}$, which can be reached from u_l. To illustrate this, we integrate the system (25) as follows.

For solutions having $w < 0$ (i.e., $v' < 0$, which is the case for an orbit leaving from u_l and connecting to $v_r < u_l$), the second equation in (25) becomes

$$\frac{dw}{d\xi} - \frac{1}{\gamma}w^2 = p(v).$$

It can be explicitly integrated if we regard w as a function of v:

$$w(v)^2 = M_2 e^{2v/\gamma} - \gamma v^3 - \frac{3\gamma^2}{2}v^2 + \left(s - \frac{3\gamma^2}{2}\right)\gamma v + \frac{s}{\gamma^2} - M_1\gamma - \frac{3\gamma^4}{4}. \quad (28)$$

The constant M_2 is determined by the condition $\lim_{-\infty} v' = 0$, that is $w(u_l) = 0$, so

$$M_2 = \frac{\gamma^2}{2}e^{-2v_l/\gamma}\left(3v_l^2 - s + 3v_l\gamma + \frac{3}{2}\gamma^2\right).$$

The trajectory leaves u_l and decreases in v (since $v' = w < 0$). We look for the particular trajectory joining u_l to the other saddle point v_r. This imposes one more condition, $w(v_r) = 0$, to be plugged in the equation (28). We end up with an equation connecting v_l, v_r and γ

$$e^{2(v_r - v_l)/\gamma}\left(3v_l^2 - s + 3v_l\gamma + \frac{3}{2}\gamma^2\right) = 3v_r^2 - s + 3v_r\gamma + \frac{3}{2}\gamma^2, \quad (29)$$

where $s = v_r^2 + v_r v_l + v_l^2$.

For $v_l > 0$ and $\gamma > 0$, the equation (29) is solved by a unique value $v_r \in (-v_l, -v_l/2)$. Interestingly, the value scales with $u\gamma$. If we introduce the quantities

$$\psi = \frac{v_r}{v_l} \qquad \text{and} \qquad \Gamma = \frac{v_l}{\gamma},$$

we can rewrite (29) as

$$e^{2\Gamma(\psi-1)}\left((2 - \psi - \psi^2)\Gamma^2 + 3\Gamma + \frac{3}{2}\right) = (2\psi^2 - \psi - 1)\Gamma^2 + 3\psi\Gamma + \frac{3}{2}, \quad (30)$$

so that ψ depends only on Γ, say $\psi = \psi(\Gamma)$.

The above analysis shows that, given $u_l > 0$, there exists a *single* connection leaving from u_l and violating the Oleinik criterion. It connects to a state

$$u_r = u_l\,\psi(\frac{u_l}{\gamma}) \in (-u_l, -u_l/2).$$

On the other hand it can be proved that only those points

$$u_r = v_m \in \left(-u_l(1 + \psi(\frac{u_l}{\gamma})), u_l\right)$$

can be realized by a traveling wave leaving from u_l. Thus we conclude:

Theorem 2. *Given any $u_l > 0$, the set $S(u_l)$ consisting of all states u_r that can be achieved through a nonlinear diffusive-dispersive traveling wave of* (21)*, taking the values u_l and u_r at the left and the right ends, respectively, is*

$$S(u_l) = \{u_l\,\psi(\frac{u_l}{\gamma})\} \cup [-u_l(1+\psi(\frac{u_l}{\gamma}), u_l),$$

where $\gamma = \frac{\delta(\varepsilon)}{\varepsilon}$ is constant and the function $\psi(\Gamma)$ is given by the implicit relation (30)*.*

Combining the *admissible shocks* found in Theorem 2 with rarefaction waves, the Riemann problem can be solved *uniquely.*

Theorem 3. *Consider the conservation law with cubic flux* (20)*. Given any $u_l > 0$, the solution of the Riemann problem with initial data u_l and u_r is*

(i) a rarefaction wave if $u_r \geq u_l$;
(ii) a classical shock wave if $u_r \in [-u_l(1+\psi(\frac{u_l}{\gamma})), u_l)$;
(iii) two shock waves, if $u_r \in (u_l\,\psi(\frac{u_l}{\gamma}), -u_l(1+\psi(\frac{u_l}{\gamma})))$; i.e., a nonclassical shock from u_l to $u_l\,\psi(\frac{u_l}{\gamma})$ followed by a classical shock connecting to u_r;
(iv) a shock wave and a rarefaction wave, if $u_r \leq u_l\,\psi(\frac{u_l}{\gamma})$; that is: a nonclassical shock wave from u_l to $u_l\,\psi(\frac{u_l}{\gamma})$, followed by a rarefaction wave connecting to u_r.

The construction in Theorem 2 can be extended by replacing $u_l\,\psi(\frac{u_l}{\gamma})$ by a general function $\varphi(u_l)$. This yields a general nonclassical Riemann solver for the cubic conservation law, which allows us to encompass *all limits* of dispersive-diffusive approximations. For definiteness, we choose the entropy-entropy flux pair

$$U(u) := u^2, \qquad F(u) := \frac{3u^4}{2}. \tag{31}$$

This choice is consistent with (7) and (21) as the inequality (8) holds for the corresponding limiting solutions. Given a left state u_l, a shock connecting u_l to a right state, u_r, satisfies (2)-(31) iff $u_r \in [-|u_l|, |u_l|]$. When $u_l > 0$, the shock wave

– either satisfies the Oleinik condition if $u_r \in [-\frac{u_l}{2}, u_l)$,
– or violates this condition if $u_r \in [-u_l, -\frac{u_l}{2})$.

To have uniqueness for the Riemann problem, for every left state u one selects a unique right state $\varphi(u)$ that can be connected to u with a nonclassical shock. The function $\varphi : \mathbb{R} \mapsto \mathbb{R}$ is called a *kinetic function.* We suppose that φ is Lipschitz continuous and

$$-u < \varphi(u) \leq -\frac{u}{2} \quad \text{for } u > 0, \qquad -\frac{u}{2} \leq \varphi(u) < -u \quad \text{for } u < 0. \tag{32}$$

Set $\alpha(u) = -u - \varphi(u)$ and observe that (32) is equivalent to

$$-\frac{u}{2} \le \alpha(u) < 0 \quad \text{for } u > 0, \qquad 0 < \alpha(u) \le -\frac{u}{2} \quad \text{for } u < 0.$$

Note that $\varphi(0) = \alpha(0) = 0$ and

$$\frac{u^3 - \varphi(u)^3}{u - \varphi(u)} = \frac{u^3 - \alpha(u)^3}{u - \alpha(u)}.$$

Thus the line connecting $\big(u, f(u)\big)$ and $\big(\varphi(u), f(\varphi(u)\big)$ intersects the graph of f exactly at the point $\big(\alpha(u), f(\alpha(u)\big)$.

The following definition is a natural extension of Theorem 3. It can be checked that it represents the *unique* class of solutions of the Riemann problem that can be constructed when one entropy inequality is enforced. This will follow from the results in Sections 4 and 5.

Definition 4. Let $\varphi : \mathbb{R} \to \mathbb{R}$ be a function satisfying (32). Consider the Riemann problem for the conservation law with cubic flux (20) and with the initial data $u_l > 0$ and $u_r \in \mathbb{R}$. The *admissible nonclassical solution* u based on the *kinetic function* φ is defined as follows:

(*i*) If $u_r \ge u_l$, u is an (increasing) rarefaction wave connecting u_l to u_r.
(*ii*) If $u_r \in \big[\alpha(u_l), u_l\big)$, u is a classical shock wave connecting u_l to u_r.
(*iii*) If $u_r \in \big(\varphi(u_l), \alpha(u_l)\big)$, u contains a (slow and decreasing) nonclassical shock connecting u_l to $\varphi(u_l)$ followed by a (fast and increasing) classical shock connecting to u_r.
(*iv*) If $u_r \le \varphi(u_l)$, u contains a nonclassical shock connecting u_l to $\varphi(u_l)$ followed by a (non-attached) rarefaction connecting to u_r.

The four cases are depicted in Figure 1.

The Riemann problem with left data $u_l < 0$ is solved in a completely similar fashion, using the value $\varphi(u_l) > 0$. For $u_l = 0$, the Riemann problem is a single rarefaction wave, connecting monotonically u_l to u_r. Oleinik's solution is recovered by choosing

$$\varphi(u) = -\frac{u}{2} \qquad \text{for all } u, \tag{33}$$

which also implies $\alpha(u) = -\frac{u}{2}$.

At the critical value $u_r = \alpha(u_l)$, one switches from a monotone increasing, single wave pattern (Case ii) to a *non-monotone* two-wave pattern (Case iii): the wave speeds in the Riemann solution change continuously, but the intermediate state $\varphi(u_l)$ does not approach the left or right states of the shock, u_l and u_r. The (local) total variation of $x \mapsto u(x,t)$ regarded as a function of u_r is also discontinuous at $u_r = \alpha(u_l)$. This is a typical feature of the nonclassical solution, not encountered when the Oleinik entropy condition is enforced.

More importantly observe that the nonclassical shock connecting u_l to $\varphi(u_l)$ is *undercompressive*, in the sense that

$$\min\big(f'(\varphi(u_l)), f'(u_l)\big) \geq \frac{f(u_l) - f(\varphi(u_l))}{u_l - \varphi(u_l)}. \tag{34}$$

The characteristics are passing through the shock instead of impinging on it from both sides. In the following section, undercompressive shocks are also found to exist for strictly hyperbolic systems.

The solutions can also be computed numerically as we show in Figure 2 and 3 below, in the cases (ii) and (iv) of Definition 4, respectively. The numerical results were obtained with the Beam-Warming scheme [15].

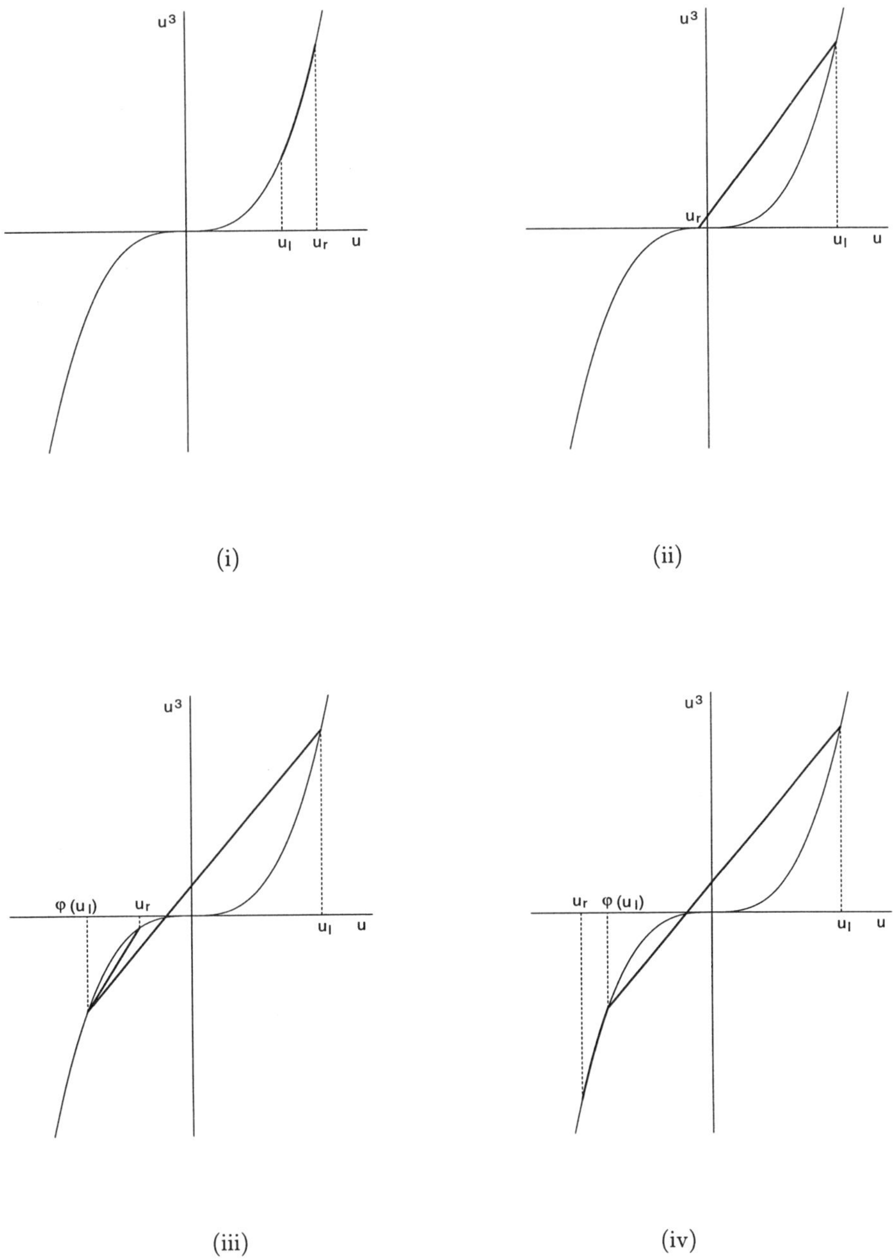

Fig. 1. Nonclassical Riemann solution

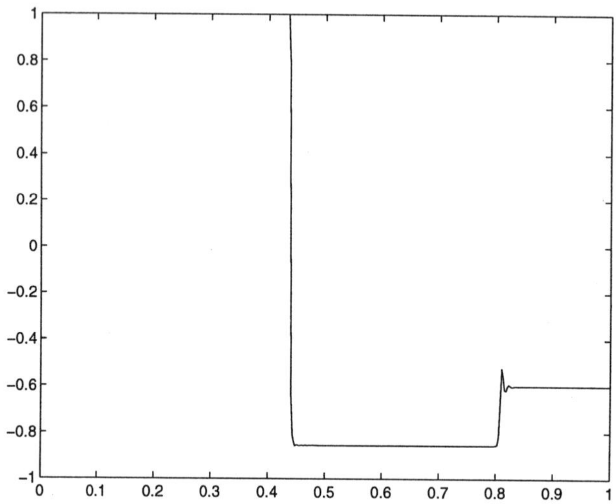

Fig. 2. A nonclassical shock and a classical shock.

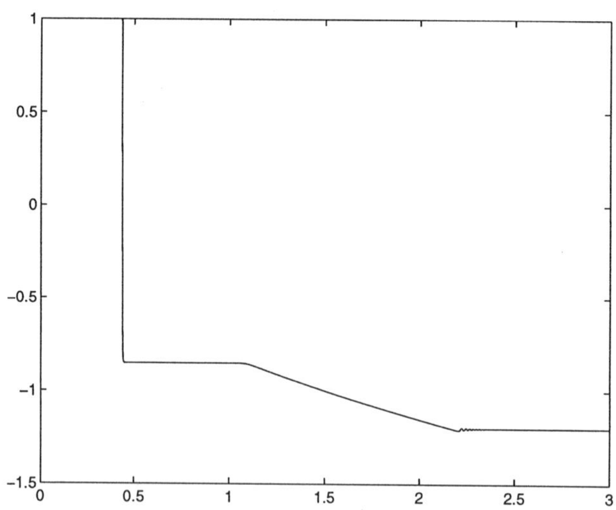

Fig. 3. A nonclassical shock plus a rarefaction.

4 Multi-Parameter Family of Solutions

The analysis in Section 2 motivates us to constrain the solutions of (1) with the single entropy inequality (2). Not surprisingly, when one (or several) characteristic field of the system is not genuinely nonlinear, the entropy inequality does not guarantee uniqueness for the Riemann problem. This is investigated here where we construct a *multi-parameter* family of solutions of (1)–(3). Liu's construction arises as a special selection in our family of Riemann solutions. Liu's criterion is adapted to the regularization (4) with $R(\varepsilon\, u_x^\varepsilon, \cdots) = \varepsilon\, u_x^\varepsilon$, which is compatible with *any* convex entropy associated with (1). Our construction encompasses all possible limits of diffusive-dispersive approximations compatible with a given entropy pair.

The following extends Definition 1 to systems.

Definition 5. A shock wave is called *nonclassical* when it satisfies the single entropy inequality (2) but does *not* fulfill the Liu entropy criterion (see, below, (39)).

We shall see that, for strictly hyperbolic systems,

(*i*) nonclassical shocks are *undercompressive*, in the sense (40)-(41) given below,
(*ii*) and nonclassical shocks with *arbitrarily small* strength may arise.

Note that, in independent work, Azevedo, Marchesin, Plohr and Zumbrun [6] also found undercompressive shocks of small amplitude for *viscous* pertubations of strictly hyperbolic systems. The existence and stability of undercompressive waves in *several* space dimensions are also studied in Freistühler [13]. In both papers, there is no consideration of the entropy inequality (2).

We restrict our attention to $\mathcal{U} := B(u_*, \delta)$, the ball centered at u_* and of small radius $\delta > 0$. The matrix $A(u) := Df(u)$ admits N real and distinct eigenvalues $\lambda_1(u) < \lambda_2(u) < \cdots < \lambda_N(u)$. Choose also a basis of right-eigenvectors $r_j(u)$ and a basis of left-eigenvectors $l_j(u)$ satisfying $l_j(u) \cdot r_j(u) = \delta_{ij}$, the Kronecker symbol. For $\delta << 1$ and $u, u' \in \mathcal{U}$, the same properties remain valid for the averaged matrix

$$\bar{A}(u, u') := \int_0^1 Df(\theta\, u + (1-\theta)\, u')\, d\theta,$$

together with $\bar{\lambda}_j(u, u')$, $\bar{r}_j(u, u')$ and $\bar{l}_j(u, u')$, defined in the obvious way.

Assume that there is a subset with P elements, $\mathbf{P} \subset \{1, 2, \cdots, N\}$ such that, for $j \notin \mathbf{P}$, $\nabla\lambda_j(u) \cdot r_j(u) > 0$ for all u and for $j \in \mathbf{P}$, the set

$$\mathcal{M}_j = \{u \in \mathcal{U} \mid \nabla\lambda_j(u) \cdot r_j(u) = 0\}$$

is a smooth affine manifold with dimension $N - 1$ containing the point u_*. Let $u \mapsto \bar{\mu}_j(u)$ be a scalar-valued function satisfying $\nabla\bar{\mu}_j \cdot r_j > 0$. If the

j-field is genuinely nonlinear, choose $\bar{\mu}_j(u) = \lambda_j(u)$. This function will serve as a parameter along the wave curves. Assume also that

$$\bar{\mu}_j(u) = 0 \qquad \text{iff} \quad \nabla\lambda_j(u) \cdot r_j(u) = 0,$$

and for definiteness

$$\bar{\mu}_j(u) \quad \text{and} \quad \nabla\lambda_j(u) \cdot r_j(u) \quad \text{have the same sign.} \tag{35}$$

The case that

$$\bar{\mu}_j(u) \quad \text{and} \quad \nabla\lambda_j(u) \cdot r_j(u) \quad \text{have the opposite sign} \tag{36}$$

is treated similarly [16]. Observe that, under the condition (35), $\nabla\lambda_j \cdot r_j$ changes sign across $\mathcal{M}_j$ and the point $\bar{\mu}_j(u) = 0$ corresponds to a *minimum* of the wave speed, as it the case for the cubic scalar flux $f(u) = u^3$.

We now define the one-parameter families of shock and rarefaction waves, to be used as building blocks for solving the Riemann problem (1)–(3).

Given $u_0 \in \mathcal{U}$ and $j = 1, \cdots, N$, let $\mathcal{O}_j(u_0) = \{v_j(\mu_j; u_0) \in \mathcal{U}\}$ be the integral curve of the vector field r_j issued from u_0. Note that $r_j(u_0)$ is the tangent vector of $\mathcal{O}_j(u_0)$ at u_0. We adopt $\mu_j = \bar{\mu}_j(v_j(\mu_j; u_0))$ (defined here implicitly) as a parameter along the rarefaction curve.

On the other hand, consider the Hugoniot locus $\{w \,|\, -s\,(w - u_0) + f(w) - f(u_0) = 0\}$. The Hugoniot set decomposes into N Hugoniot curves $\mathcal{H}_j(u_0) = \{w_j(\mu_j; u_0) \in \mathcal{U}\}$ passing through u_0 and having the tangent vector $r_j(u_0)$ at u_0. Again we normalize the parameter via the condition $\mu_j = \bar{\mu}_j(w_j(\mu_j; u_0))$. Along the j-shock curve, the shock speed $s = \bar{\lambda}_j(u_0, w_j)$ satisfies

$$\bar{\lambda}_j(u_0, w_j) = \lambda_j(u_0) + \frac{\mu_j}{2}\,\nabla\lambda_j(u_0) \cdot r_j(u_0) + O(\mu_j^2). \tag{37}$$

We assume that, for $j \in \mathbf{P}$, these elementary curves are *transverse* to the manifold $\mathcal{M}_j$: each Hugoniot curve and each integral curve intersect the manifold at exactly one point. Observe that for δ small enough it is sufficient to assume that r_j is transverse to the manifold $\mathcal{M}_j$ at the point u_*, that is $\nabla(\nabla\lambda_j \cdot r_j) \cdot r_j\big|_{u=u_*} \neq 0$. The tranversality assumption implies that, for $j \in \mathbf{P}$, the wave speed $\mu_j \mapsto \lambda_j(v_j(\mu_j; u_0))$ has exactly one critical point along each integral curve, while the shock speed $\mu_j \mapsto \bar{\lambda}_j(u_0, w_j(\mu_j; u_0))$ also admits (at most) one critical point along the Hugoniot curve (Proposition 6, below).

Recall that, in famous works, Lax and then Liu constructed a unique, stable solution of the Riemann problem for strictly hyperbolic systems. A j-shock connecting u_0 to u_1 with speed $\bar{\lambda}_j(u_0, u_1)$ is admissible in the sense of Lax [28,29] iff

$$\text{Lax shock inequalities:} \qquad \lambda_j(u_0) \geq \bar{\lambda}_j(u_0, u_1) \geq \lambda_j(u_1). \tag{38}$$

When all characteristic fields are genuinely nonlinear, (38) leads to uniquely defined wave curves and to a unique solution of the Riemann problem. Each wave curve contains two distinct parts, half of the Hugoniot curve and half of the integral characteristic curve.

For non-genuinely nonlinear characteristic fields, Liu requires that

$$\text{Liu entropy criterion:} \qquad \bar{\lambda}_j(u_0, w_j(\mu_j; u_0)) \geq \bar{\lambda}_j(u_0, u_1) \tag{39}$$

along the Hugoniot curve $\mathcal{H}_j(u_0)$ for all μ_j between $\mu_j(u_0)$ and $\mu_j(u_1)$. This means that the shock speed for μ_j in the above range achieves its *minimum* at the point u_1. Liu [35] constructs a unique wave curve based on (39). The wave curves may be composed of more than two pieces, and the Riemann solution contains composite waves mixing shocks and rarefactions.

An arbitrary j-shock, connecting u_0 to u_1, can be either a *Lax shock*, in which case (38) holds, an *undercompressive shock* satisfying either

$$\bar{\lambda}_j(u_0, u_1) \leq \min\big(\lambda_j(u_0), \lambda_j(u_1)\big), \tag{40}$$

or

$$\bar{\lambda}_j(u_0, u_1) \geq \max\big(\lambda_j(u_0), \lambda_j(u_1)\big), \tag{41}$$

or *a rarefaction shock*:

$$\lambda_j(u_0) < \bar{\lambda}_j(u_0, u_1) < \lambda_j(u_1). \tag{42}$$

The monotonicity of $\lambda_j(u_0)$ and $\bar{\lambda}_j(u_0, u_1)$ and the shock admissibility conditions along a wave curve are investigated in Propositions 6 and 7 below. See also Figure 4.

Proposition 6. *For $j = 1, \cdots, N$, consider the Hugoniot curve $\mathcal{H}_j(u_0)$ issued from a state u_0 satisfying $\bar{\mu}_j(u_0) > 0$.*

The wave speed $\mu_j \mapsto g(\mu_j; u_0) := \lambda_j(w_j(\mu_j; u_0))$ is decreasing for $\mu_j < 0$ and increasing for $\mu_j > 0$ and achieves its minimum at $\mu_j = 0$.

There exists a value $\mu_j^\star(u_0) \leq 0$ such that the shock speed $\mu_j \mapsto h(\mu_j; u_0) := \bar{\lambda}_j(u_0, w_j(\mu_j; u_0))$ is decreasing for $\mu_j < \mu_j^\star(u_0)$ and increasing for $\mu_j > \mu_j^\star(u_0)$ and achieves its minimum at $\mu_j^\star(u_0)$.

At the critical value of the shock speed, both speeds coincide

$$g(\mu_j^\star(u_0); u_0) = h(\mu_j^\star(u_0); u_0). \tag{43}$$

Moreover we have

$$\begin{aligned} &h(\mu_j; u_0) - g(\mu_j; u_0) > 0 \qquad &&\text{for } \mu_j \in \big(\mu_j^\star(u_0), \mu_j(u_0)\big), \\ &h(\mu_j; u_0) - g(\mu_j; u_0) < 0 \qquad &&\text{for } \mu_j < \mu_j^\star(u_0) \ \text{ or } \ \mu_j > \mu_j(u_0). \end{aligned} \tag{44}$$

Proposition 6 is due to Liu and is the key to Liu's construction for the Riemann problem for which we refer the reader to [34,35]. In view of Proposition 6, we can introduce $\mu_j^{\star\star}(u_0)$ as the value $\mu_j < \mu_j^\star(u_0)$ such that

$$h(\mu_j^{\star\star}(u_0); u_0) = h(\mu_j(u_0); u_0). \tag{45}$$

Proposition 7. *For $j = 1, \cdots, N$, consider the Hugoniot curve $\mathcal{H}_j(u_0)$ issued from a state u_0 satisfying $\bar{\mu}_j(u_0) > 0$.*

A shock connecting u_0 to $u_1 = w_j(\mu_j(u_1); u_0)$ is

- *a rarefaction shock (see (42)) if $\mu_j(u_1) > \mu_j(u_0)$ or $\mu_j(u_1) < \mu_j^{\star\star}(u_0)$,*
- *a Lax shock (see (38)) if $\mu_j(u_1) \in \big[\mu_j^\star(u_0), \mu_j(u_0)\big]$,*
- *an undercompressive shock (see (40)) if $\mu_j(u_1) \in \big[\mu_j^{\star\star}(u_0), \mu_j^\star(u_0)\big)$.*

In the second case the shock also satisfies the Liu criterion (39).

The entropy dissipation

$$D(u_0, u_1) := -\bar{\lambda}_j(u_0, u_1)\big(U(u_1) - U(u_0)\big) + F(u_1) - F(u_0)$$

is dealt with now. See also Figure 5.

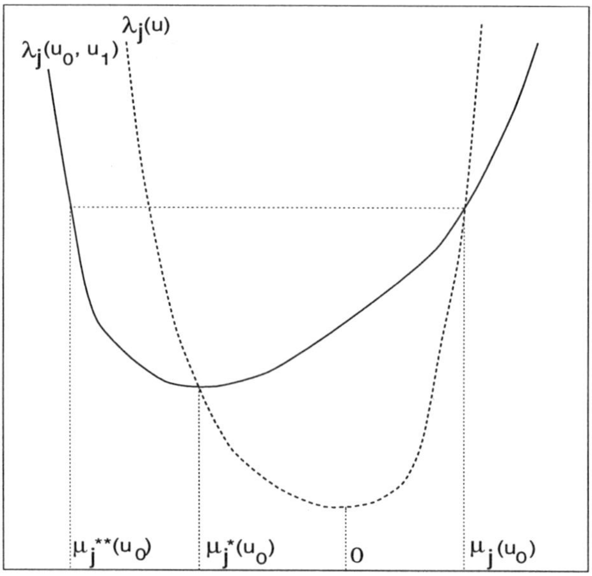

Fig. 4. Wave speed and shock speed.

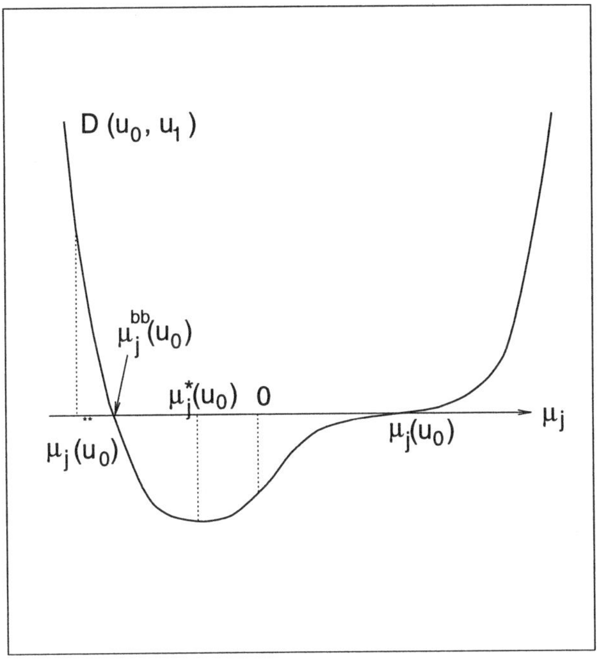

Fig. 5. Entropy dissipation.

Proposition 8. *For $j = 1, \cdots, N$, consider the Hugoniot curve $\mathcal{H}_j(u_0)$ issued from a state u_0 satisfying $\bar{\mu}_j(u_0) > 0$.*

1. *The entropy dissipation function $\mu_j \mapsto D(u_0, w_j(\mu_j; u_0))$ vanishes at the point $\mu_j(u_0)$ and at a point $\mu_j^{bb}(u_0)$ in the interval $\big(\mu_j^{\star\star}(u_0), \mu_j^\star(u_0)\big)$. The entropy dissipation is decreasing for $\mu_j < \mu_j^\star(u_0)$, increasing for $\mu_j > \mu_j^\star(u_0)$, and achieves a negative maximum value at the critical point of the wave speed, that is $\mu_j^\star(u_0)$.*
2. *A shock satisfying the entropy inequality,*
$$D(u_0, w_j(\mu_j; u_0)) \leq 0, \tag{46}$$
cannot be a rarefaction shock. As a corollary, a nonclassical shock is undercompressive and satisfies $\mu_j \in \big(\mu_j^{bb}(u_0), \mu_j^\star(u_0)\big)$.
3. *Any shock satisfying the Liu criterion* (39) *also satisfies the entropy inequality* (46)*.*

Proposition 8 shows that, under our assumption (a "single point" where genuine nonlinearity is lost), the Lax inequalities and the Liu criterion are equivalent! Proposition 8 shows that the (larger) part of the Hugoniot $\mathcal{H}_j(u_0)$ corresponding to $\mu_j \in [\mu_j^{bb}(u_0), \bar{\mu}_j(u_0)]$ is admissible, according to (46). It is the key to our construction of a multi-parameter family of solutions.

For u_l and u_r given in $\mathcal{U}$, the Riemann problem (1)–(3) admits up to a P-parameter family of solutions containing N separated wave fans, each of them being composed of (at most) two waves. Specifically we obtain the following description of the classical and nonclassical waves.

Consider a j-wave fan with left-hand state u_0 and right-hand state u, with for instance $\bar{\mu}_j(u_0) > 0$. For $j \notin \mathbf{P}$, the wave fan is either a rarefaction wave if $\bar{\mu}_j(u) > \bar{\mu}_j(u_0)$, or a classical shock if $\bar{\mu}_j(u) < \bar{\mu}_j(u_0)$.

On the other hand, for $j \in \mathbf{P}$, the j-wave fan using only classical waves contains

- either a rarefaction from u_0 to $u \in \mathcal{O}_j(u_0)$ if $\bar{\mu}_j(u) \geq \bar{\mu}_j(u_0)$,
- or a classical shock from u_0 to $u \in \mathcal{H}_j(u_0)$ if $\bar{\mu}_j(u) \in \big(\mu_j^\star(u_0), \bar{\mu}_j(u_0)\big)$,
- or a classical shock from u_0 to $u^\star := w_j\big(\mu_j^\star(u_0); u_0\big)$ followed by an attached rarefaction connecting to $u \in \mathcal{O}_j(u^\star)$ if $\bar{\mu}_j(u) \leq \mu_j^\star(u_0)$.

This completes the description of the classical wave curve $\mathcal{W}_j^c(u_0)$ given by Liu.

However other solutions also exist:

Theorem 9. *The j-wave fan may also contain a (slow) nonclassical j-shock connecting u_0 to* any state $u^\flat \in \mathcal{H}_j(u_0)$ *with $\bar{\mu}_j(u^\flat) \in \big(\mu_j^{\flat\flat}(u_0), \mu_j^\star(u_0)\big)$ followed by*

- *either a* non-attached *rarefaction connecting $u^\flat$ to $u \in \mathcal{O}_j(u^\flat)$ if $\bar{\mu}_j(u) < \bar{\mu}_j(u^\flat)$,*
- *or by a (fast) classical shock connecting $u^\flat$ to $u \in \mathcal{H}_j(u^\flat)$ if $\bar{\mu}_j(u) > \bar{\mu}_j(u^\flat)$.*

This defines a two-parameter family of u that can be reached from u_0 by nonclassical solutions.

For a given $u^\flat$, the classical shock with largest strength and connecting $u^\flat$ to some $u = u^\sharp \in \mathcal{H}_j(u^\flat)$ is characterized by the condition $\bar{\lambda}_j(u^\flat, u^\sharp) = \bar{\lambda}_j(u_0, u^\flat)$ and, in that situation, one also has $u^\sharp \in \mathcal{H}_j(u_0)$. In particular the nonclassical shock with largest possible strength connects the point $u^{\flat\flat} := w_j(\mu_j^{\flat\flat}(u_0); u_0)$ to the point $u^{\sharp\sharp} := w_j(\mu^{\sharp\sharp}(u_0); u^{\flat\flat})$, where $\mu^{\sharp\sharp}(u_0)$ is defined by $u^{\sharp\sharp} \in H_j(u_0)$. Moreover one has

$$\mu_j^{\star\star}(u_0) \leq \mu_j^{\flat\flat}(u_0) \leq \mu_j^{\flat}(u_0) \leq \mu_j^\star(u_0) \leq \mu_j^\sharp(u_0) \leq \mu_j^{\sharp\sharp}(u_0) \leq \bar{\mu}_j(u_0).$$

Based on these results, we introduce the following terminology. Given u_0, the set of all states that can be reached using only j-waves is called the *j-wave set* issuing from u_0 and is denoted by $\mathcal{S}_j(u_0)$, by analogy with the notion of j-wave curve known for classical solutions. The wave set is represented in Figure 6.

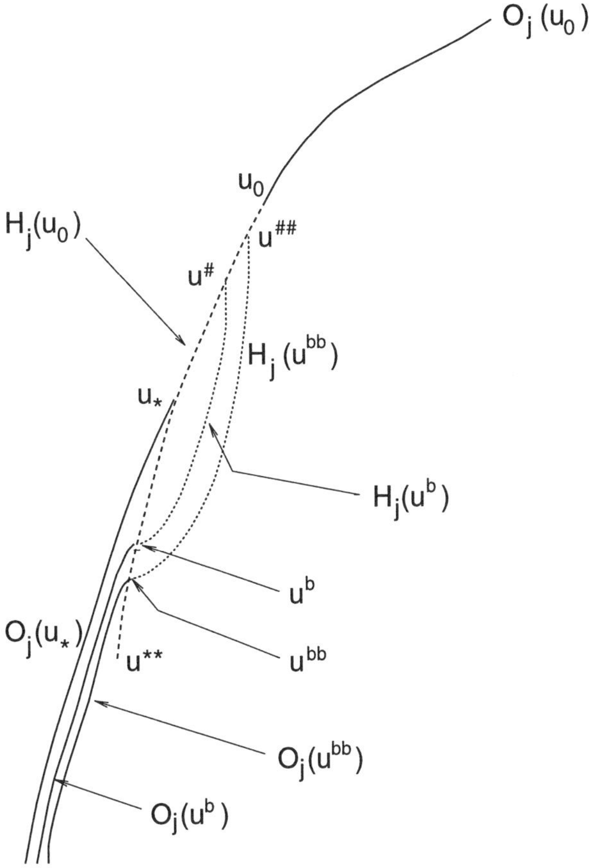

Fig. 6. Wave set $\mathcal{S}_j(u_0)$ issuing from u_0.

5 Selection by Kinetic Relations

In Section 4 we constructed all of the solutions to the Riemann problem constrained by a single entropy inequality. We found that one free-parameter should be determined for every non-genuinely nonlinear wave family. Indeed, by Theorem 9, the wave set $\mathcal{S}_j(u_0)$ issued from a state u_0 is two-dimensional when $j \in \mathbf{P}$. To select a nonclassical wave curve $\mathcal{W}_j^{nc}(u_0)$ in the wave set, we now postulate that, for each state u_0 and each index $j \in \mathbf{P}$, there exists a *single* right state u_1 that can be reached by a nonclassical shock. Precisely, we stipulate that the *entropy dissipation*

$$D(u_0, u_1) = -s\,(U(u_1) - U(u_0)) + F(u_1) - F(u_0) \tag{47}$$

of a nonclassical shock connecting u_0 to u_1 and having the speed $s = \bar{\lambda}_j(u_0, u_1)$– is a given "constitutive" function, called a *kinetic relation*, which represents certain small-scale properties that have been neglected at the hyperbolic level of modeling. When defining nonclassical shocks as limits of a particular diffusive-dispersive regularization, the kinetic relation is to be determined from traveling wave analysis, as we explain at the end of this section.

In some physical systems, the entropy dissipation is related to the mechanical energy, and may be interpreted as *a force driving the propagation* of the shock. The kinetic relation provides a one-to-one relationship between the propagation speed and the driving force. This standpoint was developed by Abeyaratne and Knowles [1] and Truskinovsky [46,47] for propagating phase boundaries in solids undergoing phase transformations. A mathematical interpretation of the kinetic relation was proposed by the author in [31].

For simplicity, we restrict attention to piecewise Lipschitz continuous solutions and to the Riemann problem.

Definition 10. Let $\phi_j : \mathcal{U} \to \mathbb{R}_-$ be a given function for each $j \in \mathbf{P}$.

A solution $u(x,t)$ of the system (1) is called an *admissible nonclassical solution* if it satisfies the entropy inequality (2) and, furthermore, the entropy dissipation of any nonclassical j-shock in u connecting two states u_0 to u_1 satisfies the *kinetic relation*

$$D(u_0, u_1) = \phi_j(u_0). \tag{48}$$

The kinetic function ϕ_j can also be regarded a function of the right state u_1 or as a function of a variable "symmetric" in u_0 and u_1, which is more realistic from the physical standpoint. For instance one may use the shock speed or – for problem in fluid dynamics and material science – the mass flux across the discontinuity. For simplicity we will here restrict attention to kinetic functions depending solely on the shock speed s, i.e.,

$$D(u_0, u_1) = \varphi(s). \tag{49}$$

The function φ need be defined only on the union of intervals $\Lambda = \bigcup_{j \in \mathbf{P}} [\lambda_j^{\min}, \lambda_j^{\max}]$ where $\lambda_j^{\min} \leq \lambda_j(u) \leq \lambda_j^{\max}$ for all u in consideration. For scalar conservation laws and for the system (15), the kinetic function can always be expressed as a function of the shock speed.

Denote by $D_j^\star(u_0)$ the *maximal negative value* of the entropy dissipation $D(u_0, u_1)$ along the Hugoniot curve $\mathcal{H}_j(u_0)$:

$$D_j^\star(u_0) = \min_{u_1 \in \mathcal{H}_j(u_0)} D(u_0, u_1). \tag{50}$$

The maximum is achieved at the critical value $\mu_j^\star(u_0)$ for the shock speed. The entropy dissipation can also be regarded as a function of the shock speed:

$$d^\star(s) = \max \big\{ D_j^\star(u_0) \,|\, u_0 \in \mathcal{U},\, j \in \mathbf{P}, \quad \lambda_j\big(u_0, w_j(\mu_j^\star(u_0); u_0)\big) = s \big\}. \tag{51}$$

Knowing the entropy dissipation of the admissible nonclassical shocks determines a unique solution of the Riemann problem.

Theorem 11. *Let $\varphi : \Lambda \to \mathbb{R}_-$ be a Lipschitz continuous function satisfying*

$$d^\star(s) \leq \varphi(s) \leq 0, \quad \frac{d\varphi}{ds}(s) \leq 0, \qquad \textit{for all } s \in \Lambda. \tag{52}$$

Given any $u_0 \in \mathcal{U}$ and $j \in \mathbf{P}$, there exists a unique wave curve $\mathcal{W}_j^{nc}(u_0)$ selected from the wave set $\mathcal{S}_j(u_0)$, using nonclassical shocks satisfying the kinetic relation (49)*. There exist $\mu_j^\flat(u_0)$ and $\mu_j^\sharp(u_0)$ satisfying*

$$\mu_j^{\star\star}(u_0) \leq \mu_j^{\flat\flat}(u_0) \leq \mu_j^\flat(u_0) \leq \mu_j^\star(u_0) \leq \mu_j^\sharp(u_0) \leq \mu_j^{\sharp\sharp}(u_0) \leq \mu_j(u_0) \tag{53}$$

and such that

$$\mathcal{W}_j^{nc}(u_0) = \begin{cases} \mathcal{O}_j(u_0) & \textit{for } \mu_j \geq \mu_j(u_0), \\ \mathcal{H}_j(u_0) & \textit{for } \mu_j^\sharp(u_0) \leq \mu_j \leq \mu_j(u_0), \\ \mathcal{H}_j(u^\flat) & \textit{for } \mu_j^\flat(u_0) \leq \mu_j < \mu_j^\sharp(u_0), \\ \mathcal{O}_j(u^\flat) & \textit{for } \mu_j \leq \mu_j^\flat(u_0), \end{cases} \tag{54}$$

with

$$u^\flat := w_j\big(\mu_j^\flat(u_0); u_0\big).$$

The curve $\mathcal{W}_j^{nc}(u_0)$ is continuous and monotone in the parameter μ_j. It is of class C^2 except at $\mu_j = \mu_j^\sharp(u_0)$ where it is only Lipschitz continuous.

Theorem 12. *For any u_l and u_r in $\mathcal{U}$, the Riemann problem* (1)-(3) *admits a unique solution in the class of admissible nonclassical solutions. Furthermore it depends continuously in the L^1 norm upon its end states.*

Theorems 11 and 12 follow from Theorem 9 and Proposition 8. We can recover the classical curve $\mathcal{W}_j^c(u_0)$ with the (maximal) choice

$$\phi_j(u_0) = D_j^\star(u_0), \tag{55}$$

using a kinetic function of the general form (48). As a matter of fact the classical wave curve and the corresponding classical Riemann solution are always available and provide a possible alternative to the nonclassical construction.

The kinetic function in (48)-(49) can be determined from a regularization of (1) in the form (4). Suppose that the solutions u^ε of (4) remain bounded in the total variation norm and converge almost everywhere to a limit $\bar{u}$. Suppose also that (4) is compatible with the entropy pair (U, F) in the sense that

$$\bar{\mu}_U := - \lim_{\varepsilon \to 0} \partial_x \nabla U(u^\varepsilon) \cdot R\big(\varepsilon\, u_x^\varepsilon, \varepsilon^2\, u_{xx}^\varepsilon, \cdots\big) \le 0, \tag{56}$$

which implies that $\bar{u}$ satisfies (2). Since the entropy inequality is too lax to guarantee uniqueness for the Riemann problem and an additional Rankine-Hugoniot relation is necessary for nonclassical shocks, we go beyond the entropy *inequality* and, instead, write down the entropy *equality* (see [31]):

$$\partial_t U(\bar{u}) + \partial_x F(\bar{u}) = \bar{\mu}_U \le 0. \tag{57}$$

The bounded, non-positive Borel measure $\bar{\mu}_U$ incorporates some information on the small-scale effects generated by the sequence u^ε. It is supported in the set of points of discontinuity of $\bar{u}$. The mass of the measure along a shock curve of $\bar{u}$ is precisely the entropy dissipation $D(.,.)$ defined in (47).

For each shock we deduce from (1)-(2) the Rankine-Hugoniot relation

$$-\bar{\lambda}_j(u_0, u_1)\,\big(u_1 - u_0\big) + f(u_1) - f(u_0) = 0,$$

and the entropy inequality

$$-\bar{\lambda}_j(u_0, u_1)\ \big(U(u_1) - U(u_0)\big) + F(u_1) - F(u_0) \le 0.$$

When solving the Riemann problem, they uniquely determine the propagation of a *classical* shocks. For nonclassical shocks, we need the entropy dissipation measure $\bar{\mu}_U$ and derive it from the traveling wave solutions of (4), as we explain now.

A traveling wave solution $u_\varepsilon(x,t) = w((x - s\,t)/\varepsilon)$ associated with (4) is determined by solving the following ordinary differential equation in the variable $\xi = (x - s\,t)/\varepsilon$:

$$-s\,w' + f(w)' = R(w', w'', \cdots)'. \tag{58}$$

It should satisfy the boundary conditions

$$\begin{aligned} &\lim_{\xi \to -\infty} w(\xi) = u_0, \qquad \lim_{\xi \to \infty} w(\xi) = u_1. \\ &\lim_{\pm\infty} w' = \lim_{\pm\infty} w'' = \cdots = 0. \end{aligned} \tag{59}$$

Integrating (58) once and using (59), we obtain

$$-s\,(w - u_0) + f(w) - f(u_0) = R(w', w'', \cdots). \tag{60}$$

The internal structure of the shock is described by the corresponding trajectory $\xi \mapsto w(\xi)$. The entropy dissipation measure at the hyperbolic level is computed as follows:

$$\begin{aligned} D(u_0, u_1) &= -\bar{\lambda}_j(u_0, u_1)\,\big(U(u_1) - U(u_0)\big) - F(u_1) + F(u_0) \\ &= \int_{\mathbb{R}} \nabla U(w) \cdot \big(-\bar{\lambda}_j(u_0, u_1) + Df(w)\big)\, w'\, d\xi \\ &= -\int_{\mathbb{R}} w' \cdot \nabla^2 U(w) \cdot \big(-\bar{\lambda}_j(u_0, u_1)\,(w - u_0) + f(w) - f(u_0)\big)\, d\xi, \end{aligned}$$

thus

$$D(u_0, u_1) = -\int_{\mathbb{R}} (w')^T \nabla^2 U(w) R(w', w'', \cdots)\, d\xi \le 0. \tag{61}$$

In view of (61), the entropy dissipation should in general depend on the small-scale effects induced by the specific regularization in consideration. In several examples of interest, the right-hand side of (61) can be computed explicitly and expressed as a function of the state u_0 or the shock speed s.

Consider the system of nonlinear elasticity. A similar discussion holds for the scalar conservation laws. We discuss here the derivation of the kinetic relation associated with the regularization (15). A traveling wave solution $\xi \mapsto (v, w)$ connecting a state (v_0, w_0) to a state (v_1, w_1) and with the speed s must satisfy the following O.D.E.

$$\begin{aligned} s\, w' + v' &= 0, \\ s\, v' + \sigma(w)' &= -v'' + \gamma\, w''', \end{aligned} \tag{62}$$

with

$$\begin{aligned} &v(\xi) \to v_0, \quad w(\xi) \to w_0 \qquad \text{as } \xi \to -\infty, \\ &v(\xi) \to v_1, \quad w(\xi) \to w_1 \qquad \text{as } \xi \to +\infty. \end{aligned} \tag{63}$$

Eliminating the variable v, we obtain a *single equation* for the scalar-valued function w:

$$-s^2\, w' + \sigma(w)' = s\, w'' + \gamma\, w''',$$

and integrating once,

$$-s^2\,\big(w - w_0\big) + \sigma(w) - \sigma(w_0) = s\, w' + \gamma\, w''. \tag{64}$$

For each shock speed s, there are at most three equilibrium points w, i.e.:

$$-s^2\,\big(w - w_0\big) + \sigma(w) - \sigma(w_0) = 0. \tag{65}$$

Namely w_0 itself, and (at most) two additional points w_1 and w_2 such that $w_0 + w_1 + w_2 = 0$. For instance suppose that $w_0 > 0$ and $w_2 < w_1 < 0$ which holds in a certain range of values s.

For 2-waves propagating with $s > 0$, it follows from the Liu criterion applied to the system of elasticity that a traveling wave connecting w_0 to w_1 represents a classical shock, while a connection from w_0 to w_2 is nonclassical. In fact, for the scalar equation there exists a critical value for the slope $s^\sharp$ such that a traveling wave trajectory connecting to w_1 exists for speeds $s > s^\sharp$ and there exists a connection to w_2 when $s = s^\sharp$.

In particular, given the left state w_0, there is a *unique* state w_2 and a *unique* speed such that w_0 and w_2 can be connected by a nonclassical shock. Henceforth, an analysis of traveling waves yields the explicit relation

$$w_2 = \tilde{\varphi}(w_0) \qquad \text{and} \qquad s = \tilde{s}(w_0). \tag{66}$$

The entropy dissipation of the nonclassical shock is then computed as a function of the left-state of the shock and this precisely gives us the kinetic function:

$$\phi(w_0) := D(w_0, w_2) = D\big(w_0, \tilde{\varphi}(w_0)\big). \tag{67}$$

Provided the relation $s = \tilde{s}(w_0)$ is one-to-one, we also obtain the kinetic function as a function of the propagation speed, that is:

$$\varphi(s) := \phi(w_0) \qquad \text{with } s^2 = w_0^2 + \tilde{\varphi}(w_0)^2 + w_0\, \tilde{\varphi}(w_0) + a. \tag{68}$$

We want to point out that the possibility of writing the kinetic function as a function of a single variable (here w), and hence as a function of the speed s, is a special property of the system of elasticity *and* the regularization (15). Other regularizations for which a scalar equation like (64) could not be derived, may require a kinetic function of the general form $\phi(v_0, w_0)$.

Finally Figures 7, 8 and 9, 10 show the structure of the Riemann solution for the system of nonlinear elasticity. We consider the hyperbolic regime where $a > 0$ and various sets of data for the Riemann problem. The dash line represents the solution when the capillarity is turned off, $\beta = 0$. The continuous line is the classical solution. For instance in Figure 7 the solution contains, in the second wave family, a nonclassical shock followed by a classical one. and in the first wave family a rarefaction. Figure 8 has one shock in the first family and one shock followed by a constant state and then a rarefaction in the second family. Figures 9 and 10 show that the same behavior may arise in both wave families.

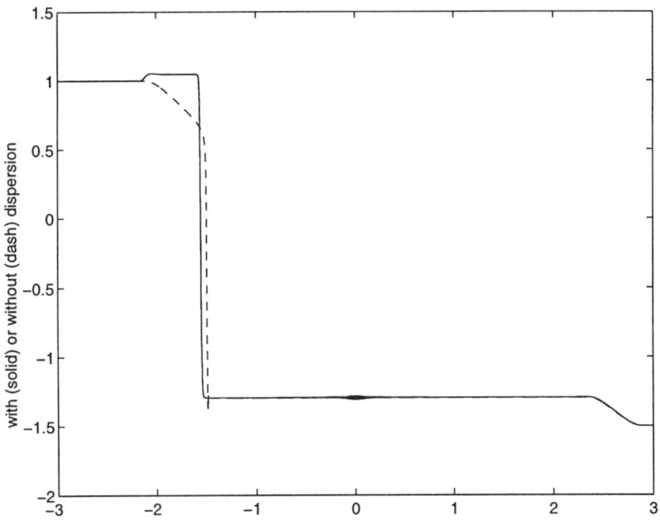

Fig. 7. Nonclassical solution with nonclassical 1-shock

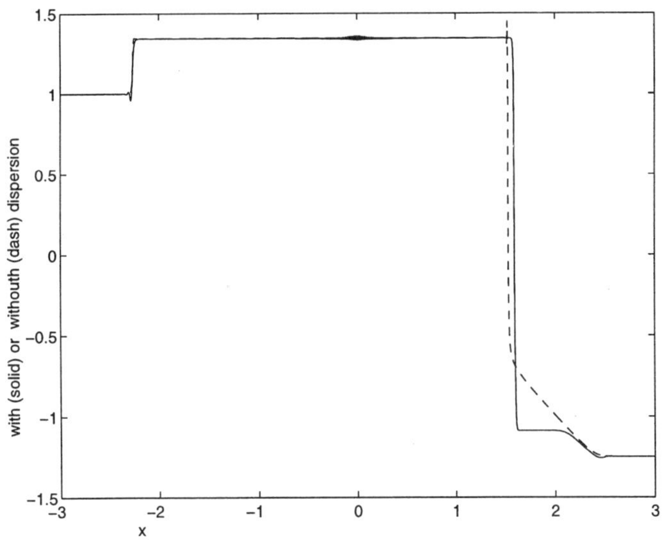

Fig. 8. Nonclassical solution with nonclassical 2-shock.

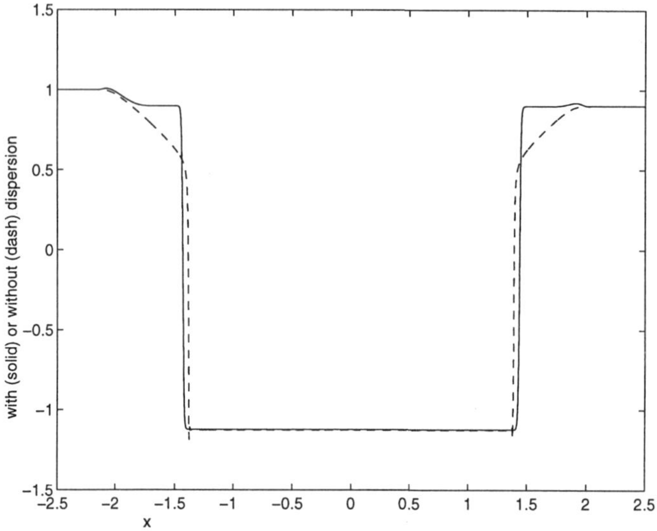

Fig. 9. Nonclassical shocks in both characteristic families: w-component.

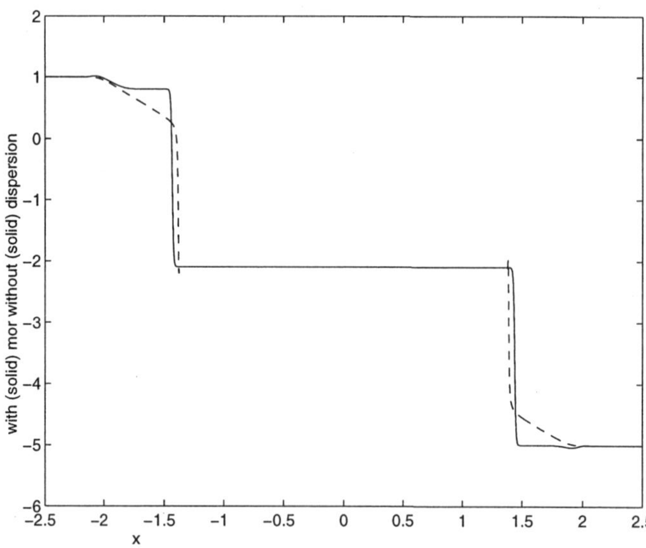

Fig. 10. Nonclassical shocks in both characteristic families: v-component.

6 Existence for the Cauchy Problem

The Cauchy problem for equations or systems may be solved in the class of admissible nonclassical solutions by applying the wave front tracking algorithm with the nonclassical Riemann solver built from a traveling wave analysis and a kinetic relation. Another approach is to consider directly the Cauchy problem for equations or systems augmented with vanishing diffusion and dispersion terms and rely on the method of compensated compactness in cases where it applies ($N \leq 2$).

We first apply the wave front tracking method and establish the existence of nonclassical solutions. We treat the scalar conservation law with cubic flux [4] and refer to [7] for more generality. The notations introduced at the end of Section 3 are used. Two assumptions will be needed, that is

$$\text{the function } \varphi \text{ is decreasing} \tag{69}$$

$$\text{the function } \alpha = -u - \varphi \text{ is decreasing,} \tag{70}$$

but not necessarily strictly decreasing. Both (4) and (9) yield kinetic functions that are consistent with these assumptions.

Given an initial data $u_0 : \mathbb{R} \to \mathbb{R}$ having bounded total variation, we use the wave front tracking algorithm and for $\nu = 1, 2, \cdots$ we define an approximate solution $u^\nu : \mathbb{R} \times \mathbb{R}_+ \to \mathbb{R}$ to the corresponding Cauchy problem. First of all, choose a piecewise constant approximation $u_0^\nu : \mathbb{R} \to \mathbb{R}$ with *finitely many* jumps such that as $\nu \to 0$

$$u_0^\nu \to u_0 \quad \text{in the } L^1 \text{ norm}, \qquad TV(u_0^\nu) \to TV(u_0). \tag{71}$$

Solve the Riemann problem at each jump, using the nonclassical Riemann solver from Definition 4. Each shock is propagated with the standard speed given by the Rankine-Hugoniot relation. Each rarefaction at time $t = 0+$ is approximated by several (small) entropy-violating jumps having strength $< 1/\nu$, each of them being propagated with the characteristic speed of their right state (say). The function u^ν is well-defined until two wave fronts interact. At each interaction one solves a Riemann problem again by using the nonclassical solver. If a rarefaction wave is generated at a time $t > 0$ in one of the Riemann solutions, it is propagated as a single jump, again with the characteristic speed of its left state, but it is not decomposed further.

It can be checked that:

Proposition 13. *The admissible nonclassical solution u of the Riemann problem, as given in Definition 4, depends continuously in the L^1-norm upon its initial states and satisfies the maximum principle:*

$$\|u\|_\infty \leq \max\big(|u_l|, |u_r|\big). \tag{72}$$

Therefore the wave front tracking scheme satisfies, for every $t > 0$ and $\nu \geq 1$,

$$\big\|u_\nu(t)\big\|_\infty \leq \big\|u_\nu(0)\big\|_\infty \leq C. \tag{73}$$

By Proposition 13, the approximate solutions u^ν are uniformly bounded in L^∞. The main difficulty in getting the convergence of u^ν is the derivation of a uniform bound for the total variation $TV(u^\nu)$. To this end we study the interactions of two wave fronts: certain types of interaction do *increase* the total variation norm. Namely, the interaction of a classical shock with an (arbitrarily small!) rarefaction may produce a (large) nonclassical shock plus a classical shock. This causes a significant increase of the total variation. A nonclassical solution does not reduce the total variation.

We shall use a *modified total variation* functional that decreases along approximate solutions. For simplicity we present this functional in the case that

$$\varphi(u) = -\frac{u}{2} \qquad \text{if } |u| \le \beta. \tag{74}$$

The general kinetic functions are treated in [7]. Let $u : \mathbb{R} \mapsto \mathbb{R}$ be a piecewise constant function and let x_α, $\alpha = 1, \dots, N$, be its points of discontinuity. Set

$$\mathbf{V}(u) := \sum_{\alpha=1}^{N} \sigma\big(u(x_\alpha-), u(x_\alpha+)\big), \tag{75}$$

where $\sigma(u_l, u_r)$ is a measure of the strength of the wave connecting u_l to u_r. If $\sigma(u_l, u_r)$ were equal to $|u_r - u_l|$, then $\mathbf{V}(u)$ would be precisely the total variation of u. However, to compensate the increase of the total variation we redefine the strength of a wave: the modified strength of shocks for which $u_l\, u_r < 0$ is *smaller* than the usual strength. Observe that a left state $u_l > 0$ can be connected by a single jump to a right state u_r iff $u_r \in \{\varphi(u_l)\} \cup [\alpha(u_l), +\infty)$. Hence, in the wave front tracking algorithm, we need to define $u_r \mapsto \sigma(u_l, u_r)$ only on this set.

We define σ by

$$\sigma(u_l, u_r) := \begin{cases} |u_r - u_l|, & \text{if } \; u_l u_r \ge 0 \; \text{ or } \; |u_l| \le \beta, \\ \theta(u_l) u_r + |u_l|, & \text{if } \; u_l u_r < 0 \; \text{ and } \; |u_r| \le \big|\alpha(u_l)\big| \; \text{ and } \; |u_l| > \beta, \\ \beta + \big|\alpha(u_l)\big|, & \text{if } \; u_r = \varphi(u_l) \; \text{ and } \; |u_l| > \beta, \end{cases}$$

where

$$\theta(u_l) := \frac{\beta - \big|\alpha(u_l)\big|}{\alpha(u_l)} \in [-1, 1].$$

The strength of a nonclassical shock having u_l as a left state is

$$\tilde{\sigma}(u_l) := \sigma\big(u_l, \varphi(u_l)\big) = \beta + |\alpha(u_l)|. \tag{76}$$

The graph of the function $u \mapsto \sigma(u_l, u)$, in the case $u_l > \beta$ and $\theta(u_l) > 0$, is shown in Figure 11. It is not hard to check that

$$|u_r - u_l| \le \sigma(u_l, u_r) \le C\, |u_r - u_l| \tag{77}$$

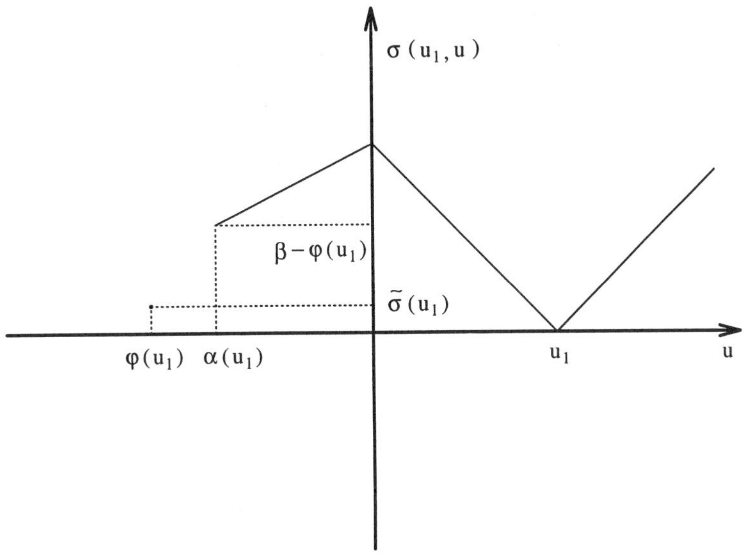

Fig. 11. Modified wave strength when $u_l > 0$.

for some $C > 1$ and precisely for every function $u = u(x)$

$$\mathbf{V}(u) \le \mathbf{TV}(u) \le \max\left\{3, \frac{4}{3\beta}\|u\|_\infty\right\}\mathbf{V}(u). \tag{78}$$

The key result of our analysis of the convergence of the front tracking scheme for nonclassical shocks is the following:

Proposition 14. *At the interaction of a wave connecting a state u_l to some state u_m and a wave connecting u_m to u_r, the sum of the modified wave strengths decreases, i.e.*

$$\sigma(u_l, u_r) \le \sigma(u_l, u_m) + \sigma(u_m, u_r). \tag{79}$$

Across an interaction, the modified wave strength of a rarefaction decreases.

The proof of this follows by analyzing about twenty different cases of possible interactions between classical and nonclassical shocks and rarefactions. Another more geometrical proof is based on the expression of σ and the monotonicity properties of the functions φ, α and $\tilde{\sigma}$.

From Proposition 14, it follows that the modified total variation is decreasing,

$$\mathbf{V}\big(u_\nu(t_2)\big) \le \mathbf{V}\big(u_\nu(t_1)\big) \qquad t_2 > t_1. \tag{80}$$

Combining this with the maximum principle in Proposition 13 and the estimate (78), we have

$$\begin{aligned}\mathbf{TV}\big(u_\nu(t)\big) &\le \max\Big\{3, \tfrac{4}{3\beta}\big\|u_\nu(t)\big\|_\infty\Big\}\,\mathbf{V}\big(u_\nu(t)\big)\\ &\le \max\Big\{3, \tfrac{4}{3\beta}\big\|u_\nu(0)\big\|_\infty\Big\}\,\mathbf{V}\big(u_\nu(0+)\big) \le C.\end{aligned} \tag{81}$$

By Helly's theorem there exists a subsequence of $\{u_\nu(x,t)\}$ that converges in $\mathbf{L}^1_{loc}(\mathbb{R}\times\mathbb{R}_+)$. It can be checked that the limiting solution is an entropy solution. We summarize the main result:

Theorem 15. *Consider the cubic conservation law* (20) *and a corresponding nonclassical Riemann solver defined from a kinetic function,* $\varphi:\mathbb{R}\to\mathbb{R}$*. Suppose that* φ *satisfies the assumptions* (32)-(69)-(70)-(74).

Given an initial data u_0 *with bounded total variation, the front tracking algorithm generates a sequence of approximate solutions* $u_\nu(x,t)$ *which satisfy the maximum principle and have uniformly bounded total variation. As* $\nu\to\infty$ *the scheme converges to a weak solution to the conservation law satisfying the entropy inequality for the quadratic entropy, the kinetic relation and the initial condition*

$$u(.,0) = u_0.$$

In the rest of this section we study the zero diffusion-dispersion limit for the Cauchy problem and prove the existence of nonclassical solutions for the scalar conservation law with cubic flux and the system of nonlinear elasticity. We rely on the method of compensated compactness of Murat and Tartar. The functional spaces relevant here are the Lebesgue spaces of measurable and integrable functions L^p rather than the space of functions of bounded variation, BV. For the lack of regularity of the solutions, it remains an open problem to formulate the kinetic relation for solely L^p functions.

We found that entropy inequalities are essential both in providing useful a priori estimates on the approximate solutions and in selecting nonclassical solutions. The convergence result below is due to Hayes and LeFloch [15] and extends an approach of Schonbek [38].

Theorem 16. *Let* $\{u^\varepsilon\}_{\varepsilon>0}$ *be a family of smooth solutions of the Cauchy problem*

$$\begin{aligned}&\partial_t u^\varepsilon + \partial_x u_\varepsilon^3 = \varepsilon\, u^\varepsilon_{xx} + \delta(\varepsilon)\, u^\varepsilon_{xxx},\\ &u^\varepsilon(x,0) = u_0^\varepsilon(x),\end{aligned} \tag{82}$$

where $\varepsilon>0$ *and* $\delta(\varepsilon)=\gamma\,\varepsilon^2$ *with* $\gamma>0$ *fixed. The initial data should satisfy*

$$\|u_0^\varepsilon\|_{L^2(\mathbb{R})} + \|u_0^\varepsilon\|_{L^4(\mathbb{R})} + \varepsilon\,\|\partial_x u_0^\varepsilon\|_{L^2(\mathbb{R})} \le C.$$

Then the solutions u^ε *remain uniformly bounded in* $L^\infty(\mathbb{R}_+, L^2(\mathbb{R}) \cap L^4(\mathbb{R}))$ *and converge in* $L^\infty_{loc}(L^p_{loc})$, $2 \le p < 4$, *to a solution* $\bar u \in L^\infty(\mathbb{R}_+, L^2(\mathbb{R}) \cap L^4(\mathbb{R}))$ *of (20).*

Denote by $\bar\mu = \lim_{\varepsilon\to 0} u_\varepsilon^4$, *which is a bounded Borel measure. Then we have*

$$\partial_t \bar u^2 + \partial_x \frac{3\,\bar\mu}{2} \le 0. \tag{83}$$

If u^ε *converges strongly in* L^4_{loc}, *then* $\bar\mu = \bar u^4$ *and the entropy inequality (8) holds.*

Let U *be a (smooth and sub-quadratic) convex entropy and* F *be the corresponding entropy flux. Then in general* $\bar u$ *does not satisfy the entropy inequality* (2).

The proof follows from the seminal work by Schonbek [38] where, in particular, the relevance of the space L^4 was pointed out and the notion of L^p Young measure, necessary to apply the compensated compactness method here, was introduced.

To derive the entropy estimates, we use first the entropy $U(u) = u^2$ and then $U(u) = u^4$. More precisely, we use the third time-invariant of the modified KdV equation, that is $u^4 + 2\,\delta(\varepsilon)\,|\partial_x u_\varepsilon|^2$.

Next we consider the augmented version of the elastodynamics system (15) together with the constitutive equation (16). We are interested in the hyperbolic case where $a > 0$. As the coefficient ε vanishes, the solutions of (15) converge to a nonclassical solution to the hyperbolic model (17). Observe that the presence of the dispersion term in the right-hand side of (15) prevents obtaining an L^∞ bound, so here we rely on L^p estimates, as we did to treat the scalar equation.

Define the internal energy W by $W'(w) = \sigma(w)$, so $W(w) = \big(w^4 + 2\,a\,w^2\big)/4$.

Theorem 17. *Let* $\big(v^\varepsilon, w^\varepsilon\big)$ *with* $\gamma \ge 0$ *fixed be a family of solutions to* (15) *assuming at* $t = 0$ *a Cauchy data* $\big(v_0^\varepsilon, w_0^\varepsilon\big)$ *satisfying*

$$v_0^\varepsilon \in L^1(\mathbb{R}) \cap L^2(\mathbb{R}), \qquad w_0^\varepsilon \in L^1(\mathbb{R}) \cap L^4(\mathbb{R}), \quad \varepsilon^{1/2}\,\partial_x w_0^\varepsilon \in L^2(\mathbb{R})$$

uniformly in ε. *Then the sequences* v^ε *and* w^ε *remain uniformly bounded in* $L^\infty\big(\mathbb{R}_+, L^2(\mathbb{R})\big)$ *and* $L^\infty\big(\mathbb{R}_+, L^4(\mathbb{R})\big)$ *respectively, and converge almost everywhere to solutions* $\bar v$ *and* $\bar w$ *of the hyperbolic system* (17).

The entropy pair (18) *is compatible with the diffusive-dispersive regularization* (15). *Limits of traveling wave solutions to* (15) *satisfy the entropy inequality* (19).

We do not expect the entropy inequalities

$$\partial_t U(v,w) + \partial_x F(v,w) \le 0, \tag{84}$$

with $U(v,w) \neq v^2/2 + W(w)$ (up to a linear function of v and w) to hold in general.

We end with the derivation of the main *a priori* estimates toward the proof of Theorem 17.

The bounds in L^2 and L^4 follow from a standard energy estimate. Multiplying the first equation in (15) by σ and the second one by v, we arrive at

$$\begin{aligned}\partial_t\big(W(w) + v^2/2\big) - \partial_x\big(v\,\sigma(w)\big) &= -\varepsilon\,|\partial_x v|^2 + \varepsilon\,\partial_x\big(v\,\partial_x v\big)\\ &\quad -\gamma\,\varepsilon^2\,\partial_x\big(v\,\partial_{xx}w\big) + \gamma\,\varepsilon^2\partial_x v\,\partial_{xx}w.\end{aligned}$$

Using the second equation in (15), we observe that

$$\partial_x v\,\partial_{xx}w = \partial_t w\,\partial_{xx}w = \partial_x\big(\partial_t w\,\partial_x w\big) - \partial_t\big(|\partial_x w|^2/2\big).$$

Therefore we have the entropy balance

$$\begin{aligned}&\partial_t\big(W(w) + v^2/2 + \gamma\,\varepsilon^2\,|\partial_x w|^2/2\big) - \partial_x\big(v\,\sigma(w)\big)\\ &= -\varepsilon\,|\partial_x v|^2 + \varepsilon\,\partial_{xx}\big(v^2/2\big) - \gamma\,\varepsilon^2\,\partial_x\big(v\,\partial_{xx}w\big) + \gamma\,\varepsilon^2\partial_x\big(\partial_x v\,\partial_x w\big).\end{aligned} \tag{85}$$

Integrating in space and time yields the uniform bounds

$$\begin{aligned}&\int_{\mathbb{R}}\big(W(w) + \frac{v^2}{2} + \frac{\gamma}{2}\,\varepsilon^2\,|\partial_x w|^2\big)(T)\,dx + \int_0^T\int_{\mathbb{R}}\varepsilon\,|\partial_x v|^2\,dxdt\\ &= \int_{\mathbb{R}}\big(W(w) + \frac{v^2}{2} + \frac{\gamma}{2}\,\varepsilon^2\,|\partial_x w|^2\big)(0)\,dx \leq C.\end{aligned} \tag{86}$$

Multiplying the first equation in (15) by $\partial_x w$ and integrating gives, on the one hand,

$$\begin{aligned}&\int_0^T\int_{\mathbb{R}}\big(\partial_x w\,\partial_t v - \partial_x w\,(3\,w^2 + a)\partial_x w\big)\,dxdt\\ &= \Big[\int_{\mathbb{R}}\partial_x w\,v\,dx\Big]_0^T - \int_0^T\int_{\mathbb{R}}\partial_{xx}v\,v\,dxdt - \int_0^T\int_{\mathbb{R}}(3\,w^2 + a)\,|\partial_x w|^2\,dxdt\\ &= \int_{\mathbb{R}}\partial_x w(T)\,v(T)\,dx - \int_{\mathbb{R}}\partial_x w(0)\,v(0)\,dx\\ &\quad + \int_0^T\int_{\mathbb{R}}|\partial_x v|^2\,dxdt - \int_0^T\int_{\mathbb{R}}\sigma_w(w)\,|\partial_x w|^2\,dxdt\end{aligned}$$

and, on the other hand,

$$\begin{aligned}&\int_0^T\int_{\mathbb{R}}\partial_x w\,\big(\varepsilon\,\partial_{xx}v - \gamma\,\varepsilon^2\,\partial_{xxx}w\big)\,dxdt\\ &= \int_0^T\int_{\mathbb{R}}\varepsilon\,\partial_x w\,\partial_{tx}w\,dxdt + \gamma\,\varepsilon^2\int_0^T\int_{\mathbb{R}}|\partial_{xx}w|^2\,dxdt\\ &= \Big[\varepsilon\int_{\mathbb{R}}|\partial_x w|^2/2\,dx\Big]_0^T + \gamma\,\varepsilon^2\int_0^T\int_{\mathbb{R}}|\partial_{xx}w|^2\,dxdt.\end{aligned}$$

Observe that

$$\left| \int_{\mathbb{R}} \partial_x w(T)\, v(T)\, dx \right| \le \varepsilon \int_{\mathbb{R}} |\partial_x w(T)|^2/2\, dx + (2\varepsilon)^{-1} \int_{\mathbb{R}} |v(T)|^2\, dx,$$

and similarly for the product $\partial_x w(0)\, v(0)$. Combining the above formulas, we find

$$\begin{aligned}
&\int_0^T \int_{\mathbb{R}} \varepsilon\, \sigma_w(w)\, |\partial_x w|^2\, dxdt + \gamma\, \varepsilon^2 \int_0^T \int_{\mathbb{R}} |\partial_{xx} w|^2\, dxdt \\
&\le \int_0^T \int_{\mathbb{R}} \varepsilon\, |\partial_x v|^2\, dxdt + \varepsilon^2 \int_{\mathbb{R}} |\partial_x w(0)|^2\, dx \\
&+ \int_{\mathbb{R}} |v(T)|^2/2\, dx, + \int_{\mathbb{R}} |v(0)|^2/2\, dx.
\end{aligned} \tag{87}$$

Combining (86) and (87), we have

$$\int_{\mathbb{R}} \left(v(T)^2 + w(T)^2 + w(T)^4\right) dx + \int_{\mathbb{R}} \gamma\, \varepsilon\, |\partial_x w(T)|^2\, dx \le C,$$

$$\int_0^T \int_{\mathbb{R}} \left(\varepsilon\, |\partial_x v|^2 + \varepsilon\, |\partial_x w|^2 + \gamma\, \varepsilon^2\, |\partial_{xx} w|^2\right) dxdt \le C.$$

7 Finite Difference Schemes

The wave front tracking algorithm and the random choice scheme do not smear out discontinuities in the approximate solutions. The shocks are represented by *sharp* jumps propagating with the correct Rankine-Hugoniot speed, the numerical error being only on the *location* of these jumps. These methods allow us to compute regularization sensitive shock waves as observed in [31,20,50,22].

On the other hand, a finite difference scheme contains actually diffusive-dispersive terms: the shocks are smeared out on a few computational cells and are possibly surrounded by oscillations, sometimes of large amplitude. (this is the case of the Lax-Wendroff scheme).

For the numerical analysis of such schemes, it is desirable to have at least one entropy inequality like (2), as it was in the case of the continuous approximations in Section 2 by balancing diffusion and dispersion. We demonstrate here, using in particular Tadmor's idea of entropy conservative scheme [44,45], that a single entropy inequality can be ensured for a class of finite difference schemes balancing diffusion and dispersion. For a discussion of the links between limits of difference schemes and limits of continuous approximations like (4), we refer to [17].

First of all we treat the general class of diffusive-dispersive approximations (14), for hyperbolic or hyperbolic-elliptic systems of N conservation laws. We search for a scheme that mimics the effects in the right hand side of (14), and that admits a discrete variant of the entropy inequality (2).

To obtain a scheme continuous in time, there are three terms to be discretized in (14). The discretization of $f(u)_x$ is based on a conservative $(2\,k+1)$-point numerical flux,

$$g^0 : \mathbb{R}^{2k+1} \to \mathbb{R}^N, \quad g^0(u, u, \cdots, u) := f(u) \qquad \text{for all } u. \tag{88}$$

For the diffusion and dispersion, we use high-order accurate, centered finite differences.

Denote by $u_j(t)$ an approximation of the solution at the point (x_j, t), where $x_j := j\,h$ describe a regular mesh of length $h \to 0$. Consider the continuous in time, conservative scheme

$$\frac{d}{dt} u_j(t) + \frac{1}{h}\big(g_{j+1/2}(t) - g_{j-1/2}(t)\big) = 0, \qquad t \geq 0, \tag{89}$$

where

$$g_{j+1/2} := g^0_{j+1/2} + g^1_{j+1/2} + g^2_{j+1/2}, \tag{90}$$

$$\begin{aligned} g^0_{j+1/2} &:= g^0(u_{j-k+1}, u_{j-k+2}, \cdots, u_{j+k}), \\ g^1_{j+1/2} &:= -\frac{\alpha}{2}\,\big(\hat{v}(u_{j+1}) - \hat{v}(u_j)\big), \\ g^2_{j+1/2} &:= -\frac{\beta}{6}\,\Big(\hat{v}(u_{j+2}) - \hat{v}(u_{j+1}) - \hat{v}(u_j) + \hat{v}(u_{j-1})\Big). \end{aligned} \tag{91}$$

The initial data is discretized in a standard fashion. The parameters $\alpha > 0$ and $\beta \in \mathbb{R}$ are *fixed,* but should be thought of as

$$\alpha\, h := \kappa_1\, \varepsilon, \qquad \beta\, h^2 := \kappa_2\, \gamma\, \varepsilon^2, \tag{92}$$

with precise constants κ_1 and κ_2 that can be calculated from the equivalent equation of the scheme. One anticipates that the $h \to 0$ *limit* of the scheme should be a good *approximation* to the $\varepsilon \to 0$ *limit* of (14). The scheme can also be studied here for its own sake; it naturally balances the effects of diffusion and dispersion. The accuracy of the discretization is discussed in [17].

A discrete entropy inequality has the form

$$\frac{d}{dt} U\big(u_j(t)\big) + \frac{1}{h}\big(G_{j+1/2}(t) - G_{j-1/2}(t)\big) \le 0, \qquad t \ge 0, \tag{93}$$

where $G_{j+1/2} := G(u_{j-m+1}, u_{j-m+2}, \cdots, u_{j+m})$ and the numerical entropy flux $G : \mathbb{R}^{2m+1} \to \mathbb{R}$ is consistent with the exact entropy flux,

$$G(u, u, \cdots, u) := F(u) \qquad \text{for all } u.$$

When (93) holds, the scheme (or the numerical flux) is said to be *entropy dissipative.* Following Tadmor [45], we say that a scheme is *entropy conservative* when (93) holds as an *equality* for all j.

Indeed, our proposed discretization of the diffusion and dispersion retains the entropy inequality of the continuous model.

Theorem 18. *Consider an entropy pair (U, F) for the system* (1). *Suppose that when $\alpha = \beta = 0$, the scheme* (89)-(91) *satisfies a local entropy inequality associated with the numerical entropy flux G^0. Then for all $\alpha \ge 0$ and β, the discrete entropy inequality* (93) *holds with*

$$\begin{aligned} G_{j+1/2} &:= G^0_{j+1/2} + G^1_{j+1/2} + G^2_{j+1/2}, \\ G^1_{j+1/2} &:= -\frac{\alpha}{2}\, \hat{v}(u_j)\, \Big(\hat{v}(u_{j+1}) - \hat{v}(u_j)\Big), \\ G^2_{j+1/2} &:= -\frac{\beta}{6}\, \Big(\hat{v}(u_{j+2})\, \hat{v}(u_j) + \hat{v}(u_{j+1})\, \hat{v}(u_{j-1}) - 2\, \hat{v}(u_{j+1})\, \hat{v}(u_j)\Big). \end{aligned} \tag{94}$$

When the scheme is L^∞ stable and convergent a.e. as $h \to 0$, the limiting function satisfies (1)-(2).

A uniform entropy estimate can also be obtained for the scheme. In Theorem 18, the entropy U need *not* be convex. Our result applies, for instance, when the flux term $f(u)_x$ is discretized via an *entropy conservative* numerical flux g^0.

The entropy inequality (2) satisfied at the limit is *independent* of the parameters α and β, and thus does not characterize a unique solution to the

Riemann problem. The limiting solutions generated by the scheme *depend* on α and β in general.

We now turn to the system of nonlinear elasticity. Previous activity on the numerical analysis of the model (17) was concerned with the case that σ is decreasing on some interval and phase transition phenomena appear. Cockburn and Gau [8] derived an energy bound (like (99) below) but no discrete entropy inequality for a high-order scheme approximating (15) when σ is piecewise linear. They also discussed the accuracy of their scheme. Slemrod and Flaherty [43], Affouf and Caflisch [3] and Jin [26] studied the discretization of (15) by the Lax-Friedrichs scheme.

Consider an initial data $(\bar{v}, \bar{w})$ for the hyperbolic or hyperbolic-elliptic problem (17)-(16). We discretize (15) in a similar spirit as done above for general systems but now using a specific *entropy conservative* flux for $\sigma(w)_x$ and v_x. For each α and β, we introduce the difference scheme

$$(v_j(0), w_j(0)) = \frac{1}{h} \int_{x_{j-1/2}}^{x_{j+1/2}} (\bar{v}(y), \bar{w}(y))\, dy, \tag{95}$$

and

$$\begin{aligned} &\tfrac{d}{dt} v_j(t) + \tfrac{1}{h}\big(g^v_{j+1/2}(t) - g^v_{j-1/2}(t)\big) = 0, \qquad t \geq 0, \\ &\tfrac{d}{dt} w_j(t) + \tfrac{1}{h}\big(g^w_{j+1/2}(t) - g^w_{j-1/2}(t)\big) = 0, \qquad t \geq 0, \end{aligned} \tag{96}$$

where

$$g^v_{j+1/2} := g^{v,0}_{j+1/2} + g^{v,1}_{j+1/2} + g^{v,2}_{j+1/2}, g^w_{j+1/2} := g^{w,0}_{j+1/2} + g^{w,1}_{j+1/2} + g^{w,2}_{j+1/2}, \tag{97}$$

$$\begin{aligned} g^{v,0}_{j+1/2} &:= -\big(w^3_{j+1} + w^3_j + a\, w_{j+1} + a\, w_j\big)/2, \qquad g^{w,0}_{j+1/2} := -(v_{j+1} + v_j)/2, \\ g^{v,1}_{j+1/2} &:= -\frac{\alpha}{2}\,(v_{j+1} - v_j), \qquad g^{w,1}_{j+1/2} := 0, \\ g^{v,2}_{j+1/2} &:= \frac{\beta}{6}\Big(w_{j+2} - w_{j+1} - w_j + w_{j-1}\Big), \qquad g^{w,2}_{j+1/2} := 0. \end{aligned} \tag{98}$$

Theorem 19. *For all $\alpha \geq 0$ and all β and a, the scheme* (96)-(98) *satisfies the discrete entropy inequality*

$$\frac{1}{2}\frac{d}{dt}\Big(v_j^2 + \frac{w_j^4}{4} + a\, w_j^2 + \beta\,(w_{j+1} - w_j)^2\Big) + \frac{1}{h}\big(G_{j+1/2}(t) - G_{j-1/2}(t)\big) \leq 0,$$

where $G_{j+1/2} := G^0_{j+1/2} + G^1_{j+1/2} + G^2_{j+1/2}$ and

$$\begin{aligned} G^0_{j+1/2} &:= -\frac{1}{2}\big(v_j\, \sigma(w_{j+1}) + v_{j+1}\, \sigma(w_j)\big), \\ G^1_{j+1/2} &:= -\frac{\alpha}{2}\, v_j\, \big(v_{j+1} - v_j\big), \\ G^2_{j+1/2} &:= \frac{\beta}{6}\big(v_j w_{j+2} - v_j w_{j+1} - 2 v_{j+1} w_j - v_{j+2} w_{j+1} \\ &\qquad + v_{j+2} w_j + v_{j+1} w_{j-1} + v_{j+1} w_{j+1}\big). \end{aligned}$$

When $a, \alpha, \beta > 0$, the scheme satisfies the uniform bounds

$$\sum_j \Big(v_j(t)^2 + w_j(t)^4 + a\, w_j(t)^2 + \beta\, (w_{j+1} - w_j)^2 \Big) \leq C \tag{99}$$

provided the initial data $\bar{v} \in L^2$ and $\bar{w} \in L^2 \cap L^4$.

The convergence result in Theorem 17 for the continuous model (15) remains valid for the discrete scheme (96)-(98), allowing us to prove that it converges strongly to a weak solution satisfying the entropy inequality (19).

8 Concluding Remarks

The convergence of a large class of diffusive-dispersive approximations in the regime where the dipersion is dominated by the diffusion was analysed by LeFloch and Natalini [33] for one-dimensional equations and by Correia and LeFloch [9] for multidimensional equations.

The wave front tracking algorithm was used in Section 6 to establish existence results for the Cauchy problem. It can also be implemented to effectively compute the nonclassical solutions. Note that the numerical analysis relative to the implementation of a kinetic relation, using a random choice scheme or a front tracking scheme, was discussed in the context of the phase transition dynamics, by LeFloch [31], Zhong, Hou and LeFloch [50] and Hou, Rosakis and LeFloch [21]. (See also the Proceedings paper [32].) A general algorithm of wave front tracking was presented and implemented in Hwang [22] for non-strictly hyperbolic systems.

The accuracy of finite difference schemes in computing nonclassical shocks is studied by Hayes and LeFloch [17]. In particlular the kinetic function associated with several approximation schemes is determined numerically. The kinetic relation is found to depend on the parameter γ in the continuous model, and on both the ratios of α/β and ϵ/h in the numerical schemes. The analysis is based on the equivalent equation obtained for each scheme from a formal Taylor expansion. A difference scheme whose equivalent equation best mimics the continuous model provides a better approximation. Still, for shocks with *large strength*, the numerical solution *diverges from* the one of the continuous model. In fact, the ε-limit of the solutions u^ε to the continuous model are distinct from the limit of the numerical solutions u^h, as the mesh is refined,

$$\lim_{\varepsilon\to 0} u^\varepsilon \neq \lim_{h\to 0} u^h. \tag{100}$$

This discrepancy is very subtle for shocks with *small strength*, and may remain undetected.

We want to point out that the continuous model may not be accurate anyway! The model may have been derived on the basis of various physical assumptions. Experimental tolerances stand on the coefficients and data. Thus we believe that the use of a difference scheme to determine nonclassical shocks is justified for practical purposes and that one may regard the discrete model as a (further) good approximation to the continuous one. Indeed, it may be very difficult in practice to determine the small-scale effects in the continuous model accurately, the experimental data being out of reach for various reasons.

Large-time calculations should be avoided however, since the error between the continuous and the discrete models may well accumulate: the corresponding solutions would then *differ substantially.* For strong shocks, the same effect may occur.

In our analysis of nonclassical shock, the entropy inequality

$$U(u)_t + F(u)_x \leq 0 \tag{101}$$

played a central role in understanding the properties of such shocks.

We want to observe that even tough (100) does not guarantee uniqueness for the Riemann problem, it does *severely restrict* the class of admissible solutions. First of all, *at most one* parameter is left undetermined for each wave family. More importantly, in light of our numerical experiments and the difficult assessment of the sensitivity of nonclassical shocks with respect to regularization parameters [17], we expect that for most practical applications the corresponding parameters occupy a *very limited range* of values. It would be interesting to characterize this range for specific physical applications, such as the dynamics of austenite-martensite phase transformations in solids (especially with *slow* phase boundaries), or the magnetohydrodynamics of the solar wind passing the earth's magnetosphere.

Finally we observe that these conclusions resemble those made by Hou and LeFloch [20] when studying nonconservative difference schemes. They should further extend to other types of regularization-sensitive shock waves.

9 Acknowledgments

I am very grateful to D. Kröner, M. Ohlberger and C. Rohde for their excellent organization of the International School hold in Freiburg University, and for having given me the opportunity to present this research.

The material in this paper originates from joint work with my collaborator B. Hayes, as well as with D. Amadori, P. Baiti, J. Correa and B. Piccoli. I especially thank R.V. Kohn who introduced me to the problems of the dynamics of phase transitions and to the work of Abeyaratne and Knowles. I benefited from helpful discussions with M. Shearer and am also grateful to R. Abeyaratne, C. Dafermos, P.D. Lax, B. Plohr and L. Truskinovsky for their interest.

The author was supported by the Centre National de la Recherche Scientifique (CNRS), the National Science Foundation (NSF) under grants DMS 95-02766 and a Faculty Early Career Development (CAREER) award, and by the Air Force Office (AFOSR) under grant F49620-94-1-0215.

References

1. R. Abeyaratne and J.K. Knowles, Kinetic relations and the propagation of phase boundaries in solids, Arch. Rat. Mech. Anal. 114 (1991), 119–154.
2. R. Abeyaratne and J.K. Knowles, Implications of viscosity and strain gradient effects for the kinetics of propagating phase boundaries in solids, SIAM J. Appl. Math. 51 (1991), 1205–1221.

3. M. Affouf and R. Caflisch, A numerical study of Riemann problem solutions and stability for a system of viscous conservation laws of mixed type, SIAM J. Appl. Math. 51 (1991), 605–634.
4. D. Amadori, P. Baiti, P.G. LeFloch and B. Piccoli, Nonclassical shocks and the Cauchy problem for nonconvex conservation laws, Jour. Diff. Equa., to appear.
5. A. Azevedo, D. Marchesin, B.J. Plohr, and K. Zumbrun, Non-uniqueness of solutions of Riemann problems caused by 2-cycles of shock waves, Proc. Fifth Internat. Conf. on Hyperbolic Problems: theory, numerics, applications, J. Glimm, M.J. Graham, J.W. Grove, and B.J. Plohr, ed., World Scientific Editions, 1996, pp. 43–51.
6. A. Azevedo, D. Marchesin, B.J. Plohr, and K. Zumbrun, Bifurcation of nonclassical viscous profiles from the constant state, to appear.
7. P. Baiti, P.G. LeFloch and B. Piccoli, Nonclassical shocks and the Cauchy problem, in preparation.
8. B. Cockburn and H. Gau, A model numerical scheme for the propagation of phase transitions in solids, SIAM J. Sci. Comput. 17 (1996), 1092-1121.
9. J. Correia and P.G. LeFloch, Diffusive-dispersive approximations of multidimensional conservation Laws, 'Nonlinear Partial Differential Equations", World Scientific Publishing, to appear.
10. C.M. Dafermos, Hyperbolic systems of conservation laws, Proceedings 'Systems of Nonlinear Partial Differential Equations", J.M. Ball editor, NATO Adv. Sci. Series C, 111, Dordrecht D. Reidel (1983), 25–70.
11. H.T. Fan and M. Slemrod, The Riemann problem for systems of conservation laws of mixed type, in 'Shock induces transitions and phase structures in general media", R. Fosdick, E. Dunn, and H. Slemrod ed., IMA Vol. Math. Appl. 52, Springer-Verlag (1993), pp. 61–91.
12. H. Freistühler, Dynamical stability and vanishing viscosity: A case study of a non-strictly hyperbolic system of conservation laws, Comm. Pure Appl. Math. 45 (1992), 561–582.
13. H. Freistühler, to appear.
14. H. Freistühler and T.P. Liu, Nonlinear stability of overcompressive shock waves in a rotationally invariant system of viscous conservation laws, Comm. Math. Phys. 153 (1993), 147–158.
15. B.T. Hayes and P.G. LeFloch, Nonclassical shocks and kinetic relations : Scalar conservation laws, Arch. Rat. Mech. Anal. 139 (1997), 1–56.
16. B.T. Hayes and P.G. LeFloch, Nonclassical shocks and kinetic relations : Strictly hyperbolic systems, SIAM J. Math. Anal., to appear.
17. B.T. Hayes and P.G. LeFloch, Nonclassical shocks and kinetic relations : Finite difference schemes, SIAM J. Numer. Anal., to appear.
18. B.T. Hayes, P.G. LeFloch and M. Shearer, in preparation.
19. B.T. Hayes and M. Shearer, Undercompressive shocks for scalar conservation laws with nonconvex fluxes, Proc. Royal Soc. Edinburgh A, to appear.
20. T.Y. Hou and P.G. LeFloch, Why nonconservative schemes converge to wrong solutions: error analysis, Math. of Comp. 62 (1994), 497–530.
21. T.Y. Hou, P. Rosakis, and LeFloch, A level set approach to the computation of twinning and phase transition dynamics, submitted to J. Comput. Phys.

22. H.C. Hwang, A front tracking method for regularization-sensitive shock wave, Ph.D. Thesis, State University of New York, Stony Brook, 1996.
23. E. Isaacson, D. Marchesin, C.F. Palmeira and B.J. Plohr, A global formalism for nonlinear waves in conservation laws, Comm. Math. Phys. 146 (1992), 505–552.
24. E. Isaacson, D. Marchesin and B. Plohr, Transitional waves for conservation laws, SIAM J. Math. Anal. 21 (1990), 837–866.
25. D. Jacobs, W.R. McKinney and M. Shearer, Traveling wave solutions of the modified Korteweg-deVries Burgers equation, J. Diff. Equa. 116 (1995), 448–467.
26. S. Jin, Numerical integrations of systems of conservation laws of mixed type, SIAM J. Appl. Math. 55 (1995), 1536-1551.
27. K.T. Joseph and P.G. LeFloch, Boubary layers in weak solutions to systems of conservation laws, Preprint Series 1402, Institute for Math. and its Appl., Minneapolis, May 1996. To appear in Arch. Rational Mech. Anal.
28. P.D. Lax, Hyperbolic systems of conservation laws, II, Comm. Pure Appl. Math. 10 (1957), 537–566.
29. P.D. Lax, Hyperbolic Systems of Conservation Laws and the Mathematical Theory of Shock Waves, Society for Industrial and Applied Mathematics, Philadelphia (1973).
30. P.D. Lax and C.D. Levermore, The small dispersion limit of the Korteweg-deVries equation, Comm. Pure Appl. Math. 36 (1983) I, 253–290, II, 571–593, III, 809–829.
31. P.G. LeFloch, Propagating phase boundaries: formulation of the problem and existence via the Glimm scheme, Arch. Rat. Mech. Anal. 123 (1993), 153–197.
32. P.G. LeFloch, Dynamics of solid-solid phase interfaces via a level set approach, Mathematica Contemporanea, to appear.
33. P.G. LeFloch and R. Natalini, Conservation laws with vanishing nonlinear diffusion and dispersion, Nonlinear Analysis, to appear.
34. T.P. Liu, The Riemann problem for general 2x2 conservation laws, Trans. Amer. Math. Soc. 199 (1974), 89–112.
35. T.P. Liu, Admissible solutions of hyperbolic conservation laws, Mem. Amer. Math. Soc. 30 (1981).
36. T.P. Liu and K. Zumbrun, On nonlinear stability of general undercompressive viscous shock waves, Comm. Math. Phys. 174 (1995), 319–345.
37. S. Schecter and M. Shearer, Undercompressive shocks for nonstrictly hyperbolic conservation laws, Dynamics Diff. Equa. 3 (1991), 199–271.
38. M.E. Schonbek, Convergence of solutions to nonlinear dispersive equations, Comm. Part. Diff. Equa. 7 (1982), 959–1000.
39. M. Shearer, The Riemann problem for a class of conservation laws of mixed type, Jour. Diff. Equa. 46 (1982), 426–443.
40. M. Shearer, D.G. Schaeffer, D. Marchesin, and P. Paes-Leme, Solution of the Riemann problem for a prototype 2×2 system of nonstrictly hyperbolic conservation laws, Arch. Rational Mech. Anal. 97 (1987), 299–320.

41. M. Shearer and Y. Yang, The Riemann problem for the p-system of conservation laws of mixed type with a cubic nonlinearity, Proc. Royal Soc. Edinburgh 125 A (1995), 675–699.
42. M. Slemrod, Admissibility criteria for propagating phase boundaries in a van der Waals fluid, Arch. Rational Mech. Anal. 81 (1983), 301–315.
43. M. Slemrod and J.E. Flaherty, Numerical integration of a Riemann problem for a van der Waals fluid, in 'Phase Transformations", E.C. Aifantis and J. Gittus, Eds., Elsevier Applied Science Publishers, 1986, pp. 203–212.
44. E. Tadmor, Numerical viscosity and the entropy condition for conservative difference schemes, Math. of Comp. 43 (1984), 217–235.
45. E. Tadmor, The numerical viscosity of entropy stable schemes for systems of conservation laws, Math. of Comp. 49 (1987), 91–103.
46. L. Truskinovsky, Dynamics of non-equilibrium phase boundaries in a heat conducting nonlinear elastic medium, J. Appl. Math. and Mech. (PMM) 51 (1987), 777–784.
47. L. Truskinovsky, Kinks versus shocks, in 'Shock induced transitions and phase structures in general media", R. Fosdick, E. Dunn, and H. Slemrod ed., IMA Vol. Math. Appl. 52, Springer-Verlag (1993).
48. C.C. Wu, New theory of MHD shock waves, in 'Viscous Profiles and Numerical Methods for Shock Waves", M. Shearer ed., SIAM Philadelphia (1991), pp. 231–235.
49. C.C. Wu and C.F. Kennel, Evolution of small-amplitude intermediate shocks in a dissipative and dispersive system, J. Plasma Physics 47 (1992), 85–109.
50. X. Zhong, T.Y. Hou, and P.G. LeFloch, Computational methods for propagating phase boundaries, J. Comput. Phys. 124 (1996), 192–216.

Viscosity and Relaxation Approximation for Hyperbolic Systems of Conservation Laws

Athanasios E. Tzavaras

Department of Mathematics, University of Wisconsin, Madison, WI 53706, and Institute of Applied and Computational Mathematics, FORTH, 711 10 Heraklion, Crete**

Abstract. These lecture notes deal with the approximation of conservation laws via viscosity or relaxation. The following topics are covered:

The general structure of viscosity and relaxation approximations is discussed, as suggested by the second law of thermodynamics, in its form of the Clausius-Duhem inequality. This is done by reviewing models of one dimensional thermoviscoelastic materials, for the case of viscous approximations, and thermomechanical theories with internal variables, for the case of relaxation.

The method of self-similar zero viscosity limits is an approach for constructing solutions to the Riemann problem, as zero-viscosity limits of an elliptic regularization of the Riemann operator. We present recent results on obtaining uniform BV estimates, in a context of strictly hyperbolic systems for Riemann data that are sufficiently close. The structure of the emerging solution, and the connection with shock admissibility criteria is discussed.

The problem of constructing entropy weak solutions for hyperbolic conservation laws via relaxation approximations is considered. We discuss compactness and convergence issues for relaxation approximations converging to the scalar conservation law, in a BV framework, and to the equations of isothermal elastodynamics, via compensated compactness.

Contents

** Research partially supported by the Office of Naval Research, the National Science Foundation, and the TMR programme HCL # ERBFMRXCT960033.

1 Introduction

These lecture notes deal with the approximation of hyperbolic systems of conservation laws via viscosity or relaxation. Despite recent successes with analyzing these questions for systems of one space dimension, their understanding remains incomplete and poses challenges to the theory of hyperbolic systems of conservation laws. A challenge amplified by the mere fact that theoretical understanding of such limiting processes reflects on the design and implementation of numerical algorithms for hyperbolic systems.

The interface between mechanical modeling and analytical theory has been a productive ground for the development of the theory of conservation laws. The problem of viscosity limits is intimately tied to the mechanical issue of passage from one continuum thermomechanical theory to another. In a similar fashion relaxation approximations, when viewed in the framework of continuum theories with internal variables, have analogous features. Therefore, we begin with the general structure of viscosity and relaxation approximations, as suggested by the second law of thermodynamics in its form of the Clausius-Duhem inequality. This presentation owes a lot to the point of view advocated by Dafermos [14]. Rather than stating the issues at an abstract level, we focus on the specific theories of thermoviscoelasticity, for the case of viscosity approximations, and thermomechanical theories with internal variables, for the case of relaxation.

We continue with a discussion of zero-viscosity limits for the scalar conservation law, in Section 3, and a presentation of self-similar viscosity limits in Section 4. The latter is an approach for constructing solutions of the Riemann problem, as zero-viscosity limits of an elliptic regularization of the Riemann operator. We present recent results on obtaining uniform BV estimates, in a context of strictly hyperbolic systems and for Riemann data that are sufficiently close [68]. The structure of the emerging solution, and the connection with shock admissibility criteria (in particular with the traveling wave criterion) is discussed.

In the last Section, we consider the problem of constructing entropy weak solutions for hyperbolic conservation laws via relaxation. Relaxation approximations exert a subtle dissipative effect on discontinuities as well as on oscillations, which is brought forth by analyzing their compactness and convergence properties. We present results of recent studies concerning relaxation limits to the scalar multi-d conservation law in a BV framework [28], and to the system of isothermal elastodynamics via compensated compactness [69].

2 The Structure induced by Continuum Thermomechanics

Continuum physical theories are described by field equations that are called balance laws. A body occupying a reference configuration $\mathcal{R} \subset \mathbb{R}^d$ is deforming through the action of a map $y(\cdot,t) : \mathcal{R} \to \mathcal{R}_t$, $t > 0$, which carries the typical point $x \in \mathcal{R}$ to the point $y = y(x,t)$ in the current configuration $\mathcal{R}_t \subset \mathbb{R}^d$. The map y, called motion, is required to be a bi-Lipschitz homeomorphism. The integral balance laws,

$$\partial_t \int_\Omega g(x,t)dx + \int_{\partial\Omega} \sum_{\alpha=1}^{d} f_\alpha(x,t) n_\alpha dS = \int_\Omega h(x,t)dx \quad \text{for } \Omega \subset \mathcal{R},\, t > 0, \tag{1}$$

describe the rate of change of the vector-quantity $\int_\Omega g dx$, in a control volume Ω, due to the effect of flux through the boundary $\partial\Omega$ and production (or absorption) in Ω. The number of equations reflect the number N of balance laws in the continuum theory, the vector densities $g = (g^i)$ and $h = (h^i)$ express the balanced and produced quantities, respectively, while the flux terms are expressed through flux densities, $f^i \cdot n = \sum_{\alpha=1}^{d} f^i_\alpha n_\alpha$, where $f = (f^i_\alpha)$ takes values in $\mathbb{R}^{N\times d}$ and n is the outer normal to $\partial\Omega$. The balance laws may be expressed in a Lagrangean description, in terms of density fields g, f, h defined for $x \in \mathcal{R}$ and t, or in an Eulerian description, by density fields $\bar{g}$, $\bar{f}$, $\bar{h}$ defined for $y \in \mathcal{R}_t$ and t. The fields are connected through the formulas

$$g(x,t) = \bar{g}(y(x,t),t)\,, \quad f(x,t) = \bar{f}(y(x,t),t)\,, \quad h(x,t) = \bar{h}(y(x,t),t)\,. \tag{2}$$

If g, f, and h are smooth, the balance laws can be described through the local form

$$\partial_t g + \sum_{\alpha=1}^{d} \partial_\alpha f_\alpha = h\,, \tag{3}$$

obtained from the integral form by using the Gauss Theorem and averaging. (The local form is still valid for fields of bounded variation - whose distributional derivatives are locally finite Borel measures - in which case (3) is interpreted as an equality of measures.)

The balance laws are supplemented with constitutive relations, characterizing the material response, and yield evolution equations that describe the process. For instance, when the state of the material is described by the state vector $U \in \mathbb{R}^N$ and the material response is determined by the constitutive relations

$$g = G(U)\,, \quad f_\alpha = F_\alpha(U)\,, \quad h = H(U)\,, \tag{4}$$

with $G,\ F_\alpha,\ H : \mathbb{R}^N \to \mathbb{R}^N,\ \alpha = 1,\dots,d$, then (3) give rise to the first order system of conservation laws

$$\partial_t G(U) + \sum_{\alpha=1}^{d} \partial_\alpha F_\alpha(U) = H(U)\,, \tag{5}$$

where $x \in \mathbb{R}^d$, $t > 0$ and $U(x,t)$ takes values in $\mathbb{R}^N$. The constitutive relations (4) are the typical, abstract example of homogeneous elastic response. The system (5) comprises the equations of compressible gas flow, the equations describing dynamic deformations of nonlinear elastic materials and certain models of the equations of magnetohydrodynamics.

While the above derivation is appealing in its conciseness, it fails to address several mechanical considerations. One such consideration is that constitutive theories are required to be consistent with the second law of thermodynamics, to comply with the principle of material frame indifference, and to reflect existing material symmetries. In the sequel, we expand on the restrictions imposed on constitutive relations by the principle of consistency with the second law of thermodynamics and the ensuing structure of viscosity and relaxation approximations. For simplicity, the presentation is done in the context of one-dimensional thermomechanical theories.

2.1 Thermomechanical theories in one space dimension

Thermomechanical theories seek to identify a pair of functions $(y(x,t), \theta(x,t))$ determining a thermomechanical process. The function $y(x,t)$ expresses the motion of the reference interval $[\alpha, \beta]$ while $\theta(x,t)$ stands for the temperature. The displacement $y(\cdot,t)$ is required, for each $t > 0$, to be a strictly increasing, bi-Lipschitz continuous map of the reference interval $[\alpha,\beta]$ onto the current configuration $[y(\alpha,t), y(\beta,t)]$ (see Fig. 1).

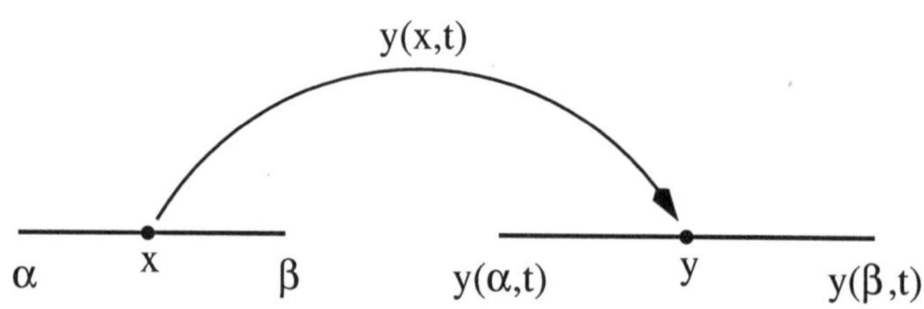

Fig. 1.

The list of quantities entering in a Lagrangean description of the thermomechanical process are: $\rho_0(x)$ the mass density in the reference configuration,

$\rho(y,t)$ the mass density in the current configuration, y the motion, $u = \frac{\partial y}{\partial x}$ the strain ($u > 0$), $v = \frac{\partial y}{\partial t}$ the velocity, τ the stress, f the body force per unit mass, θ the temperature ($\theta > 0$), e the specific internal energy, q the heat flux, r the radiating heat density and η the specific entropy. The equations

$$\rho(y,t)\frac{\partial y}{\partial x} = \rho_0(x) \tag{6}$$

$$\partial_t u - \partial_x v = 0 \tag{7}$$

$$\partial_t(\rho_0 v) - \partial_x \tau = \rho_0 f \tag{8}$$

$$\partial_t(\rho_0 \frac{1}{2}v^2 + \rho_0 e) = \partial_x(\tau v) + \partial_x q + \rho_0 f v + \rho_0 r \tag{9}$$

express the balance of mass, the kinematic compatibility relation, the balance of linear momentum, and the balance of energy (the first law of thermodynamics), respectively. They are supplemented with the Clausius-Duhem inequality, which reads, in integral form,

$$\frac{d}{dt}\int_a^b \rho_0 \eta \, dx \geq \frac{q}{\theta}(x,t)\Big|_{x=a}^{x=b} + \int_a^b \frac{\rho_0 r}{\theta} dx \quad \text{for } [a,b] \subset [\alpha,\beta] \text{ and } t > 0, \tag{10}$$

or, in local form,

$$\rho_0 \partial_t \eta \geq \partial_x \left(\frac{q}{\theta}\right) + \frac{\rho_0 r}{\theta}. \tag{11}$$

The Clausius-Duhem inequality expresses that the net production of entropy per unit time, in any control volume $[a,b]$, is positive, and manifests (a form of) the second law of thermodynamics.

The thermomechanical variables are connected through constitutive relations that characterize the material response. A constitutive theory is determined by assigning a class of independent (prime) variables and a class of dependent variables, derived from the prime variables via constitutive relations. In this separation, the set of thermodynamic variables is implicitly divided into "causes" and "effects". From the phenomenological standpoint of continuum thermomechanics, there is no a-priori reason why a cause in one constitutive relation should not be a cause in another. Therefore, in determining the general form of constitutive theories, one imposes Truesdell's *principle of equipresence*, which states that a quantity present as an independent variable in one constitutive relation should be present in all, except if its presence contradicts some law of physics or material symmetry [64]. Severe restrictions result from the second law of thermodynamics and the invariance under change of observers, called respectively *principle of consistency with the Clausius-Duhem inequality* and *principle of material frame indifference.*

The list of constitutive variables (prime and dependent) does not include the reference density ρ_0, the body force f, and the radiating heat transfer r, which are viewed as externally prescribed fields. Given a constitutive theory, the kinematic compatibility relation, and the balance laws of momentum and energy form a system of equations whose solution determines the thermomechanical process. In the Lagrangean description, the role of the balance of mass is to determine the current density ρ, once the process is identified. The role of the Clausius-Duhem inequality is more subtle: For *smooth processes*, the Clausius-Duhem inequality is viewed as restricting the form of constitutive relations. By contrast for *non-smooth processes*[1], it becomes an additional constraint that weak solutions must satisfy. These points will be clarified in the context of specific constitutive theories.

For smooth processes the balance of energy, balance of linear momentum and Clausius-Duhem inequality imply the energy dissipation inequality

$$\rho_0\big(\partial_t e - \theta\partial_t\eta\big) - \tau u_t - \frac{q\theta_x}{\theta} \le 0\,. \tag{12}$$

Upon introducing the Helmholtz free energy $\psi = e - \theta\eta$, the latter takes the form

$$\rho_0\partial_t\psi + \rho_0\eta\partial_t\theta - \tau u_t - \frac{q\theta_x}{\theta} \le 0\,. \tag{13}$$

2.2 The constitutive theory of thermoviscoelasticity

In the constitutive theory of thermoviscoelasticity, the prime variables are the strain $u = y_x$, the strain rate $w = u_t = y_{xt}$, the temperature θ, and the temperature gradient $g = \theta_x$, while the dependent variables ψ, η, τ and q are determined through constitutive relations of the general form

$$\begin{aligned} \psi &= \psi(u,w,\theta,g)\,, \quad & \eta &= \eta(u,w,\theta,g)\,, \\ \tau &= \tau(u,w,\theta,g)\,, \quad & q &= q(u,w,\theta,g)\,, \end{aligned} \tag{14}$$

following the principle of equipresence.

It is postulated that every smooth process is realizable and must be consistent with the Clausius-Duhem inequality. The postulate that smooth processes are realizable is compatible with both the balance laws and the constitutive relations, in the following sense. Given a smooth process $(y(x,t),\theta(x,t))$ and the referential density ρ_0, one computes u, w, θ, g and, in turn, ψ, η, τ, q and the internal energy $e = \psi + \theta\eta$. Then, the balance of mass determines the current density ρ, (7) is trivial, while the balance of momentum and balance of energy equations are satisfied by externally regulating the body force f and radiating heat r.

The energy dissipation inequality (13) implies that constitutive relations be constrained so that

$$(\rho_0\psi_u - \tau)\dot u + (\rho_0\psi_w)\dot w + \rho_0\big(\eta + \psi_\theta\big)\dot\theta + (\rho_0\psi_g)\dot g - \big(\frac{qg}{\theta}\big) \le 0 \tag{15}$$

[1] The term non-smooth processes is used in a loose sense to signify processes containing shocks. It is a question of analysis to precise the smoothness class in each specific context.

is satisfied for any smooth process. One constructs test processes, defined in the vicinity of (x_0, t_0), such that the local values of u, w, θ and g at (x_0, t_0) are assigned arbitrarily and, in addition, the local values of $\dot{w}$, $\dot{\theta}$ and $\dot{g}$ are assigned independently of the former. It follows that to comply with (15) the constitutive relations must be of the reduced form

$$\begin{aligned} \psi &= \psi(u,\theta) \\ \eta &= -\tfrac{\partial\psi}{\partial\theta}(u,\theta) \\ \tau &= \rho_0 \tfrac{\partial\psi}{\partial u} + Z(u,w,\theta,g) \\ q &= Q(u,w,\theta,g) \end{aligned} \tag{16}$$

where Q and Z are subject to the constraint

$$Zw + \frac{Qg}{\theta} \geq 0\,, \qquad \text{for any } u,\ w,\ \theta \text{ and } g. \tag{17}$$

Further analysis shows that $Z(u,0,\theta,0) = 0$ and $Q(u,0,\theta,0) = 0$. Hence,

$$\sigma(u,\theta) := \rho_0 \frac{\partial\psi}{\partial u} \tag{18}$$

is interpreted as the elastic part of the stress, Z as the viscous part of the stress, and the elastic part of the stress is derived from a potential. In practice, frequent use is made of the constitutive relations

$$Q = k(u,\theta)g\,, \quad Z = \mu(u,\theta)w\,, \tag{19}$$

where the viscous and heat conducting effects are decoupled. In that case, (17) dictates that the heat conductivity $k(u,\theta)$ and viscosity $\mu(u,\theta)$ coefficients are positive.

The constitutive relations of an ideal, viscous, heat conducting gas

$$\tau = -\rho_0 R\frac{\theta}{u} + \mu\frac{v_x}{u}\,, \quad e = c\theta\,, \quad q = k\frac{\theta_x}{u}\,, \qquad u,\theta > 0, \tag{20}$$

where R, c, μ and k are positive constants, are an example within the constitutive theory of thermoviscoelasticity. The free energy ψ and entropy η are given by

$$\psi(u,\theta) = -R\theta\ln u - c(\theta\ln\theta - \theta)\,, \quad \eta = -\frac{\partial\psi}{\partial\theta} = R\ln u + c\ln\theta\,, \tag{21}$$

and η is a concave function.

Constitutive theories should also comply with the principle of material frame indifference. In one-space dimension, the resulting restrictions are independence of the constitutive relations from the displacement y and the velocity v, and they have already been factored in (14). In several space dimensions, the restrictions are far more severe because of rotating frames, *c.f.* [64].

The equations of one-dimensional thermoviscoelasticity take the form

$$
\begin{aligned}
\partial_t u - \partial_x v &= 0 \\
\rho_0 \partial_t v - \partial_x \sigma(u,\theta) &= (\mu v_x)_x + \rho_0 f \\
\partial_t\Big(\frac{1}{2}\rho_0 v^2 + \rho_0 e(u,\theta)\Big) - \partial_x(\sigma(u,\theta)v) &= (\mu v_x v)_x + (k\theta_x)_x + \rho_0 f v + \rho_0 r
\end{aligned}
\tag{22}
$$

where we took Z, Q as in (19). The constitutive class is determined by the free energy function $\psi(u,\theta)$, in conjunction with the viscosity and conductivity coefficients $\mu = \mu(u,\theta) \geq 0$ and $k = k(u,\theta) \geq 0$. The remaining constitutive functions are determined by the thermodynamic relations

$$
\sigma = \rho_0 \frac{\partial \psi}{\partial u}, \quad \eta = -\frac{\partial \psi}{\partial \theta}, \quad e = \psi + \theta\eta \tag{23}
$$

Various derivative thermodynamic relations, like $\frac{\partial e}{\partial \theta} = \theta \frac{\partial \eta}{\partial \theta} = -\theta \frac{\partial^2 \psi}{\partial \theta^2}$, $\frac{\partial \sigma}{\partial \theta} = -\rho_0 \frac{\partial \eta}{\partial u}$, connect the thermodynamic functions.

By requirement, any smooth process is consistent with the Clausius-Duhem inequality. It is instructive to derive (11) directly from (22), (16) and (19). A calculation shows

$$
\rho_0 \partial_t \eta(u,\theta) - \Big(\frac{k\theta_x}{\theta}\Big)_x = \frac{\mu v_x^2}{\theta} + \frac{k\theta_x^2}{\theta^2} + \frac{\rho_0 r}{\theta}\,. \tag{24}
$$

The identity captures the dissipative structure of thermoviscoelastic materials, and is instrumental in global existence of smooth solutions for the system of thermoviscoelasticity [13].

2.3 A hierarchy of thermomechanical theories

The theory of thermoviscoelasticity is equipped with sufficiently strong dissipative structure to guarantee the persistence of smooth processes. On the other extreme is the theory of thermoelastic non-conducting materials ($Z = 0$ and $Q = 0$), described by the system of equations

$$
\begin{aligned}
\partial_t u - \partial_x v &= 0 \\
\rho_0 \partial_t v - \partial_x \sigma(u,\theta) &= \rho_0 f \\
\partial_t\Big(\frac{1}{2}\rho_0 v^2 + \rho_0 e(u,\theta)\Big) - \partial_x(\sigma(u,\theta)v) &= \rho_0 f v + \rho_0 r
\end{aligned}
\tag{25}
$$

with constitutive relations (23). If $\frac{\partial \eta}{\partial \theta} > 0$ and $\frac{\partial \sigma}{\partial u} > 0$, then (25) is hyperbolic with characteristic speeds $\lambda_\pm = \pm\big(\frac{\sigma_u}{\rho_0} + \frac{\sigma_\theta^2}{\rho_0^2 \eta_\theta}\big)^{1/2}$, $\lambda_0 = 0$. Under conditions of compression smooth processes can break down and develop shock waves. The theory of thermoelastic nonconductors of heat is regarded as a limiting theory of thermoviscoelasticity as the viscosity and heat conductivity tend to zero. Accordingly, non-smooth thermomechanical processes inherit the constraint

$$\rho_0 \partial_t \eta(u, \theta) \geq \frac{\rho_0 r}{\theta}, \tag{26}$$

and the Clausius-Duhem inequality becomes a restriction on admissible non-smooth processes.

The constitutive theory of thermoelasticity is an intermediate theory, appropriate for materials where the stress and the heat flux are independent of the strain rate. Thermoelastic materials are characterized by the constitutive relations

$$\begin{aligned}
&\psi = \psi(u, \theta) \\
&\eta = -\frac{\partial \psi}{\partial \theta}(u, \theta) \quad \tau = \rho_0 \frac{\partial \psi}{\partial u}(u, \theta) \\
&q = Q(u, \theta, g), \quad \text{subject to } \; Qg \geq 0,
\end{aligned} \tag{27}$$

that are consistent for smooth processes with the Clausius-Duhem inequality (11). Processes of thermoelastic materials, with a Fourier law $Q = kg$, are described by the system (22) with $\mu = 0$. Again non-smooth processes inherit (11) as an admissibility restriction.

Isothermal motions of thermoelastic materials are processes $(y(x,t),\theta(x,t))$, where the temperature is kept constant, $\theta = \theta_0$, and accordingly $Q = Q(u,\theta_0,0)$ $= 0$. They are described by the equations

$$\begin{aligned}
\partial_t u - \partial_x v &= 0 \\
\rho_0 \partial_t v - \partial_x \sigma(u, \theta_0) &= \rho_0 f,
\end{aligned} \tag{28}$$

that are pertinent to a purely mechanical process. When $\sigma_u > 0$ the system (28) is strictly hyperbolic, with characteristic speeds $\lambda_\pm = \pm\left(\frac{\sigma_u}{\rho_0}\right)^{1/2}$, and non-smooth processes can appear due to formation of shock waves. It is instructive to regard this situation as a limiting case of the theory of thermoviscoelasticity.

From a mechanical viewpoint, isothermal processes are attained by externally controlling the radiation heat transfer r so that $\theta = \theta_0$ and $Q = 0$. The balance of energy $(22)_3$ and (24) imply

$$\begin{aligned}
&\rho_0 \partial_t \left(\tfrac{1}{2} v^2 + [e(u, \theta_0) - \theta_0 \eta(u, \theta_0)]\right) \\
&\qquad - \partial_x \left(\sigma(u, \theta_0)\, v\right) + \mu v_x^2 \; = \; (\mu v_x v)_x + \rho_0 f v.
\end{aligned} \tag{29}$$

In the zero-viscosity limit, non-smooth mechanical processes inherit from thermodynamics the admissibility constraint

$$\rho_0 \partial_t \left(\frac{1}{2} v^2 + \psi(u, \theta_0)\right) - \partial_x \left(\sigma(u, \theta_0)\, v\right) \leq \rho_0 f v. \tag{30}$$

In gas dynamics, it is customary to express the pressure in terms of the specific volume u and the entropy η. This can be attained by assuming that $\frac{\partial\eta}{\partial\theta} > 0$, inverting the equation $\eta = -\frac{\partial\psi}{\partial\theta}$, and writing the constitutive theory (23) in the form

$$e = e(u,\eta)\,, \quad \theta = \frac{\partial e}{\partial \eta}(u,\eta)\,, \quad \sigma = -p = \rho_0 \frac{\partial e}{\partial u}\,, \tag{31}$$

where p is the pressure function. Non-smooth processes of thermoelastic nonconductors of heat have to comply with (26), which of course is still valid under the expression (31) of the constitutive relations. Isentropic motions of thermoelastic nonconductors ($\eta = \eta_0$ constant) are described by the system of equations

$$\begin{aligned} \partial_t u - \partial_x v &= 0 \\ \rho_0 \partial_t v + \partial_x p(u,\eta_0) &= \rho_0 f\,, \end{aligned} \tag{32}$$

which is a strictly hyperbolic system when $p_u < 0$. Non-smooth isentropic processes inherit from the expression (26) of the second law of thermodynamics the admissibility constraint

$$\rho_0 \partial_t \big(\frac{1}{2}v^2 + e(u,\eta_0)\big) + \partial_x \big(p(u,\eta_0)\, v\big) \le \rho_0 f v\,. \tag{33}$$

2.4 Materials with internal variables

Viscosity and heat conduction are one of the possible ways of prescribing dissipative mechanisms. Complementary descriptions of dissipation are supplied by the theory of simple materials with fading memory and the theory of materials with internal state variables. The class of *simple materials* consists of those materials for which the free energy, entropy, stress and heat flux at any point x and time t can be described in terms of the present value of the temperature gradient g at (x,t) and the history of strain u and temperature θ at the point x at all times prior to t. Under conditions of fading memory, simple materials are equipped with a subtle dissipative mechanism, brought forth by analyzing their thermodynamics [6].

The class of materials with internal state variables is a subclass of materials with fading memory, which is appealing in its simplicity and encompasses some interesting models (like the ideal gas with vibrational relaxation). In a theory with internal variables, the thermomechanical process is described by a vector function $(y(x,t), \theta(x,t), \alpha(x,t))$, where y is the motion, θ the temperature, and the internal vector-variable α evolves according to a differential law

$$\partial_t \alpha = F(u,\theta,\alpha)\,. \tag{34}$$

The remaining thermodynamic quantities are determined by constitutive relations of the form

$$\begin{aligned} \psi = \Psi(u,\theta,g,\alpha)\,, \quad & \eta = H(u,\theta,g,\alpha)\,, \\ \tau = S(u,\theta,g,\alpha)\,, \quad & q = Q(u,\theta,g,\alpha)\,. \end{aligned} \tag{35}$$

In rough terms, such models have fading memory when the differential equation (34) is exponentially dissipative.

We pursue the implications of the Clausius-Duhem inequality on the form of the constitutive functions. A remark is in order: while (35) satisfies the principle of equipresence, the differential constraint (34) does not. In fact, (34) is not viewed here as a constitutive relation but rather as defining the class of admissible processes. This simplifies somewhat the reduction process, while it is compatible with specific examples that motivate this theory. We refer to [8] for the case that F also depends on g.

Consistency with the Clausius-Duhem inequality is tested against all admissible processes, that is all smooth processes that are compatible with the differential constraint (34). A count of equations and unknowns indicates that all admissible processes can be realized, by externally regulating f and r so as to fulfill the balance of momentum and energy. Then (13), (34) and (35) imply

$$(\rho_0\Psi_u - S)\dot{u} + \rho_0(\Psi_\theta + H)\dot{\theta} + \rho_0\Psi_g\,\dot{g} + \rho_0\Psi_\alpha \cdot F(u,\theta,\alpha) - \frac{Qg}{\theta} \le 0 \tag{36}$$

for all admissible processes. Since the local values of u, θ, α, g, θ_t, u_t and g_t can be assigned independently, the constitutive relations have the reduced form

$$\begin{aligned} \psi &= \Psi(u,\theta,\alpha) \\ \tau &= S = \rho_0 \tfrac{\partial\Psi}{\partial u} \\ \eta &= H = -\tfrac{\partial\Psi}{\partial\theta} \\ q &= Q(u,\theta,g,\alpha) \end{aligned} \tag{37}$$

subject to the constraint

$$-\frac{\partial\Psi}{\partial\alpha} \cdot F(u,\theta,\alpha) + \frac{1}{\theta} Q(u,\theta,g,\alpha)g \ge 0 \quad \text{for all } u,\theta,g,\alpha. \tag{38}$$

It follows from (38) that $Q(u,\theta,0,\alpha) = 0$ and

$$-\frac{\partial\Psi}{\partial\alpha} \cdot F(u,\theta,\alpha) \ge 0 \quad \text{for all } u,\theta,\alpha. \tag{39}$$

If Q is given by a Fourier law for heat conduction, $Q = k(u,\theta,\alpha)g$, then (38) is equivalent to (39) and $k \ge 0$.

The thermomechanical process $(y(x,t), \theta(x,t), \alpha(x,t))$ is described by (7-9) supplemented with (34) and the constitutive relations (37-39). For smooth processes with Fourier heat conduction, a direct computation yields

$$\rho_0 \partial_t H(u,\theta,\alpha) - \left(\frac{k\theta_x}{\theta}\right)_x = -\rho_0 \frac{1}{\theta} \Psi_\alpha \cdot F(u,\theta,\alpha) + \frac{k\theta_x^2}{\theta^2} + \frac{\rho_0 r}{\theta}. \tag{40}$$

Equation (40) captures the dissipative structure of a heat conducting thermoelastic material with internal variables.

2.5 A thermomechanical model with stress relaxation

Thermomechanical theories with internal variables provide a natural framework to consider the structure of relaxation approximations to conservation laws, in the continuum physics context. To develop the connections, consider a theory with one scalar internal variable α evolving according to the differential law

$$\partial_t \alpha = -\lambda(\alpha - h(u,\theta)). \tag{41}$$

This law is of exponential dissipative type with relaxation time $\frac{1}{\lambda}$ and equilibrium states $\alpha_{eq} = h(u,\theta)$. The internal variable theory is completed with constitutive relations $\psi = \Psi(u,\theta,\alpha)$ for the free energy, $\tau = S(u,\theta,\alpha)$ for the stress, $\eta = H(u,\theta,\alpha)$ for the entropy and a Fourier law, $q = Q = k(u,\theta,\alpha)g$, for the heat flux. The constitutive functions are required to satisfy (37-39), with $F = -\lambda(\alpha - h(u,\theta))$, so that the internal variable theory is consistent with the second law of thermodynamics and is equipped with the dissipation estimate (40). Then the function $-H(u,\theta,\alpha)$ provides, in the terminology of [4], a (possibly not convex) "entropy" function for the emerging relaxation process. We are interested to explore the relation of the thermomechanical model corresponding to $\lambda > 0$ with the model emerging in the small-relaxation time limit $\lambda \to \infty$.

In practice, one is often faced with the question: under what conditions is a given set of constitutive functions Ψ, S and H achieved from a theory consistent with the second law of thermodynamics. For example, suppose we are given a stress distribution $S(u,\theta,\alpha)$. Then the question becomes to investigate if there a free energy function $\Psi(u,\theta,\alpha)$ such that

$$\begin{aligned} \frac{\partial \Psi}{\partial u} &= \frac{1}{\rho_0} S(u,\theta,\alpha) \\ \text{subject to} \quad \frac{\partial \Psi}{\partial \alpha}(\alpha - h(u,\theta)) &\geq 0 \quad \text{for all } u,\theta,\alpha. \end{aligned} \tag{42}$$

Note that (42) implies in particular that Ψ satisfies

$$\begin{cases} \frac{\partial \Psi}{\partial \alpha} \geq 0 & \text{for } \alpha > h(u,\theta) \\ \frac{\partial \Psi}{\partial \alpha} = 0 & \text{for } \alpha = \alpha_{eq} = h(u,\theta) \\ \frac{\partial \Psi}{\partial \alpha} \leq 0 & \text{for } \alpha < h(u,\theta), \end{cases} \tag{43}$$

and that, since solutions of $(42)_1$ are given by

$$\rho_0 \Psi(u,\theta,\alpha) = G(\theta,\alpha) + \int_0^u S(\xi,\theta,\alpha)\, d\xi\,, \tag{44}$$

the inequality $(42)_2$ is satisfied if and only if there is a function $G(\theta,\alpha)$ such that

$$\Big(G_\alpha(\theta,\alpha) + \int_0^u S_\alpha(\xi,\theta,\alpha)\, d\xi\Big)\Big(\alpha - h(u,\theta)\Big) \geq 0 \quad \text{for all } u,\theta,\alpha. \tag{45}$$

We emphasize that solving (45) is equivalent to deciding whether the given model with internal variables is consistent with the second law of thermodynamics, and that, for (45) to admit solutions, conditions must be imposed on the functions S and h. For instance, (43) implies

$$G_\alpha(\theta, h(u,\theta)) = -\int_0^u S_\alpha(\xi,\theta,h(u,\theta))\, d\xi \tag{46}$$

Given a solution G, the associated free energy function is given by (44).

We next consider a special case, where the given stress distribution is

$$S(u,\theta,\alpha) = f(u,\theta) + \alpha\,. \tag{47}$$

This case is completely solvable. Indeed, (45) reads: is there a function $G(\theta,\alpha)$ such that $j(\theta,\alpha) := -G_\alpha(\theta,\alpha)$ satisfies

$$\Big(u - j(\theta,\alpha)\Big)\Big(\alpha - h(u,\theta)\Big) \geq 0 \quad \text{for all } u,\theta,\alpha. \tag{48}$$

It is easy to see that this happens if and only if $h(u,\theta)$ is strictly decreasing in u, $j(\theta,\alpha)$ is strictly decreasing in α, and $j = h^{-1}$ is the inverse function of h for θ fixed,

$$j(\theta, h(u,\theta)) = u\,, \quad h(j(\theta,\alpha),\theta) = \alpha\,.$$

For simplicity, we assume the slightly stronger condition $h_u(u,\theta) < 0$ and note that the associated G is given by the formula

$$G(\theta,\alpha) = -\int_0^\alpha j(\theta,\zeta) d\zeta - \int_1^\theta s(z)\, dz\,, \tag{49}$$

where s is an arbitrary function of θ. In turn, the constitutive functions of the internal variable theory are $S(u,\theta,\alpha) = f(u,\theta) + \alpha$, as requested, and

$$\begin{aligned} \rho_0\psi = \rho_0\Psi(u,\theta,\alpha) &= -\int_0^\alpha j(\theta,\zeta) d\zeta - \int_1^\theta s(z)\, dz + \alpha u + \int_0^u f(\xi,\theta) d\xi\,, \\ \rho_0\eta = \rho_0 H(u,\theta,\alpha) &= \int_0^\alpha j_\theta(\theta,\zeta) d\zeta + s(\theta) - \int_0^u f_\theta(\xi,\theta) d\xi\,. \end{aligned} \tag{50}$$

As an application, consider a model for a viscoelastic material where the total stress τ is decomposed into a viscoelastic part, evolving according to stress relaxation, and a viscous part with Newtonian viscosity,

$$\begin{aligned} \tau &= \sigma + \mu v_x\,, \quad \mu \geq 0 \\ \partial_t(\sigma - f(u,\theta)) &= -\lambda(\sigma - g(u,\theta))\,. \end{aligned} \tag{51}$$

The viscoelastic part of the stress may be put into the integral form,

$$\sigma(\cdot,t) = f(u,\theta)(\cdot,t) + \int_{-\infty}^{t} \lambda e^{-\lambda(t-s)} \big(g(u,\theta) - f(u,\theta)\big)(\cdot,s)\,ds\,, \tag{52}$$

of a Maxwell type viscoelastic fluid with memory. The function $f(u,\theta)$ describes the instantaneous elastic stress-strain response, while $g(u,\theta)$ describes the equilibrium stress-strain response.

The inviscid version of (51) is formulated in the context of internal variables by setting

$$\begin{aligned} \sigma &= f(u,\theta) + \alpha \\ \partial_t \alpha &= -\lambda(\alpha - h(u,\theta)) \qquad \text{with } h(u,\theta) := g(u,\theta) - f(u,\theta). \end{aligned} \tag{53}$$

The model is achieved from a theory consistent with the second law of thermodynamics if and only if the functions f and g satisfy $(g-f)(u,\theta)$ is strictly decreasing in u. Henceforth, we focus on functions satisfying

$$g_u(u,\theta) < f_u(u,\theta) \tag{54}$$

while the free energy ψ and entropy η are determined by (50) for $\alpha = \sigma - f(u,\theta)$.

The thermomechanical process $\big(y(x,t),\theta(x,t),\sigma(x,t)\big)$, associated to the material model (51), is described by the system of equations

$$\begin{aligned} \partial_t u - \partial_x v &= 0 \\ \rho_0 \partial_t v - \partial_x \sigma &= (\mu v_x)_x + \rho_0 f \\ \partial_t\Big(\tfrac{1}{2}\rho_0 v^2 + \rho_0 e\Big) - \partial_x(\sigma v) &= (\mu v_x v)_x + (k\theta_x)_x + \rho_0 f v + \rho_0 r \\ \partial_t\big(\sigma - f(u,\theta)\big) &= -\lambda(\sigma - g(u,\theta)) \end{aligned} \tag{55}$$

where the internal energy is determined by

$$\begin{aligned} \rho_0 e &= \rho_0\big(\psi + \theta\eta\big) = \rho_0(\Psi + \theta H)(u,\theta,\sigma - f(u,\theta)) \\ &= \int_0^{\sigma - f(u,\theta)} \big(\theta j_\theta - j\big)(\theta,\zeta)d\zeta + \big(\theta s(\theta) - \int_1^\theta s(z)\,dz\big) \\ &\quad + (\sigma - f(u,\theta))u + \int_0^u (f - \theta f_\theta)(\xi,\theta)d\xi\,. \end{aligned} \tag{56}$$

A direct computation using (55), in conjunction with (37), (47) and (50), shows that the thermomechanical process is equipped with the dissipation estimate

$$\rho_0 \partial_t \Big(H(u,\theta,\sigma - f(u,\theta)) \Big) - \Big(\frac{k\theta_x}{\theta} \Big)_x = \qquad (57)$$

$$\lambda \frac{1}{\theta}(u - h^{-1}(\theta,\alpha))(\alpha - h(u,\theta)) \Big|_{\alpha=\sigma - f(u,\theta)} + \frac{k\theta_x^2}{\theta^2} + \frac{\mu v_x^2}{\theta} + \frac{\rho_0 r}{\theta} ,$$

which, in view of (54) and (48), implies that smooth processes satisfy the Clausius-Duhem inequality, for all values of $\lambda > 0$ and μ, $k \geq 0$, and yields an estimate for the amount of dissipation.

The model (51), for materials with stress relaxation, gives rise to a hierarchy of thermomechanical theories as the parameters describing the viscosity μ and heat-conductivity k tend to zero, and to a second hierachy of theories as the relaxation parameter λ tends to infinity. In the limit $\lambda \to \infty$, one formally obtains the theory of thermoviscoelasticity presented in Section 2.2. As both $\lambda \to \infty$ and μ and/or k tend to zero one can obtain the various thermomechanical theories mentioned in Section 2.3. Any non-smooth limit processes inherit the limit form of the dissipation estimate (57).

We close by considering the case of isothermal motions, that is processes along which $\theta = \theta_0$ is kept constant and $Q = 0$. The process is described now by the equations

$$\begin{aligned} \partial_t u - \partial_x v &= 0 \\ \rho_0 \partial_t v - \partial_x \sigma &= (\mu v_x)_x + \rho_0 f \\ \partial_t \big(\sigma - f(u,\theta_0)\big) &= -\lambda(\sigma - g(u,\theta_0)) \end{aligned} \qquad (58)$$

and inherits from thermodynamics the dissipative structure

$$\partial_t \Big(\frac{1}{2}\rho_0 v^2 + \rho_0 \Psi(u,\theta_0,\sigma - f(u,\theta_0)) \Big) - \partial_x(\sigma v) + \mu v_x^2 \qquad (59)$$

$$+ \lambda(u - h^{-1}(\theta_0,\alpha))(\alpha - h(u,\theta_0)) \Big|_{\alpha=\sigma - f(u,\theta_0)} = (\mu v_x v)_x + \rho_0 f v \, .$$

The limiting theory $\mu \to 0$ is described by

$$\begin{aligned} \partial_t u - \partial_x v &= 0 \\ \rho_0 \partial_t v - \partial_x \sigma &= \rho_0 f \\ \partial_t \big(\sigma - f(u,\theta_0)\big) &= -\lambda(\sigma - g(u,\theta_0)) \, . \end{aligned} \qquad (60)$$

It is known that the stress relaxation equation exerts a subtle dissipative effect on smooth processes, and as a result the system admits smooth solutions for initial data close to equilibrium. By contrast, for data away from equilibrium shock waves can develop in finite time, [14]. The inviscid theory inherits the dissipative structure

$$\partial_t\Big(\frac{1}{2}\rho_0 v^2 + \rho_0\Psi(u,\theta_0,\sigma - f(u,\theta_0))\Big) - \partial_x(\sigma v) \qquad (61)$$
$$+ \lambda(u - h^{-1}(\theta_0,\alpha))(\alpha - h(u,\theta_0))\Big|_{\alpha=\sigma-f(u,\theta_0)} \le \rho_0 f v\,,$$

with equality for smooth isothermal processes.

In the limit $\lambda \to \infty$, the internal variable theory (60) yields the equations of one-dimensional isothermal elasticity,

$$\begin{aligned}\partial_t u - \partial_x v &= 0\\ \rho_0\partial_t v - \partial_x g(u,\theta_0) &= \rho_0 f\,,\end{aligned} \qquad (62)$$

a strictly hyperbolic system when $g_u > 0$. If $f_u > g_u$ the internal variable theory is consistent with the Clausius-Duhem inequality. (Remarkably, this is precisely the subcharacteristic condition for the associated relaxation process, *i.e.* consistency with the second law of thermodynamics implies, in this context, the subcharacteristic condition.) The function

$$\rho_0\Psi(u,\theta_0,\alpha) = -\int_0^\alpha h^{-1}(\theta_0,\zeta)d\zeta + \alpha u + \int_0^u f(\xi,\theta_0)d\xi \qquad (63)$$

provides an "entropy" function for the associated relaxation process, which is convex in (u,α) if $-\partial_\alpha h^{-1}\partial_u f \ge 1$ for all u and α.

Bibliographic remarks. We refer to books on Continuum Mechanics, *e.g.* Truesdell and Noll [64], on the topics of consistency of constitutive relations with the second law of thermodynamics, the principle of material frame indifference, and the effect of material symmetries. The requirements imposed by consistency with the Clausius-Duhem inequality are developed in Coleman-Noll [5] and Coleman-Mizel [7] for the theory of thermoviscoelasticity, in Coleman [6] for simple materials with fading memory, and in Coleman-Gurtin [8] for materials with internal state variables. The thermodynamical structure of mechanical theories with internal variables has been extensively investigated in the mechanics literature, *c.f.* Coleman-Gurtin [8], Gurtin-Williams-Suliciu [23], Faciu and Mihailescu-Suliciu [20], Suliciu [61] and references therein. Since consistency with the second law of thermodynamics leads to "entropy" functions for the relaxation process, the issue is important in both the design of numerical relaxation schemes, Coquel-Perthame [9], as well as for theoretical investigations of relaxation, Tzavaras [69]. There is an extensive literature on the classification of the strength of dissipation, for various mechanical theories, and the related issue of global existence of smooth processes. We refer to Dafermos [14] for a survey of results prior to 1985.

3 Zero-Viscosity Limits for the Scalar Conservation Law

The problem of zero-viscosity limits consists of constructing weak solutions of the hyperbolic system

$$\partial_t U + \partial_x F(U) = 0\,, \qquad x \in \mathbb{R} \;\; t > 0\,, \tag{64}$$

as $\varepsilon \to 0$ limits of the viscous system

$$\partial_t U + \partial_x F(U) = \varepsilon\, \partial_x (B(U) U_x)\,, \tag{65}$$

where $U(x,t)$ takes values in $\mathbb{R}^N$ and $B(U)$ is a positive semidefinite diffusion matrix expressing the viscous structure.

Most of the analysis regarding this question is based on the notion of entropy-entropy flux pairs. A scalar-valued function $\eta(U)$ is called an entropy with corresponding entropy flux $q(U)$ if every smooth solution of the conservation law (64) satisfies the additional conservation law

$$\partial_t \eta(U) + \partial_x q(U) = 0\,. \tag{66}$$

Pairs $(\eta(U), q(U))$ are generated by solving the system of linear differential equations

$$\nabla q(U) = \nabla \eta(U) \cdot \nabla F(U)\,. \tag{67}$$

Trivial solutions are provided by $(c \cdot U\,,\, c \cdot F(U))$, with c any constant vector in $\mathbb{R}^N$. Since (67) is underdetermined for $N = 1$, determined for $N = 2$ and overdetermined for $N \geq 3$, for systems of two equations there exist many entropies, but for larger systems the existence of nontrivial entropies is the exception rather than the rule. Nevertheless, specific systems arising in applications are often endowed with some entropy-entropy flux pairs.

In the sequel we present the convergence of viscosity limits

$$u_t + f(u)_x = \varepsilon u_{xx} \tag{68}$$

to the scalar conservation law

$$u_t + f(u)_x = 0\,. \tag{69}$$

Let $\lambda(u) = f'(u)$. Consider a family of approximate solutions u^ε emanating from initial data u_0^ε that are stable in $L^2 \cap L^\infty$. By the maximum principle the family u^ε is stable in L^∞

$$|u^\varepsilon| \leq C\,, \tag{70}$$

and, by the representation theorem for Young measures [63], there exists a subsequence (denoted again by u^ε) and a measurable family of probability measures $\nu = \nu_{(x,t)}$ such that

$$f(u^\varepsilon) \rightharpoonup < \nu, f(k) >\,, \qquad \text{for any continuous } f. \tag{71}$$

Solutions of the viscosity problem satisfy the uniform bound

$$\int_{\mathbb{R}} \frac{1}{2} u^2 dx + \varepsilon \int_0^t \int_{\mathbb{R}} u_x^2 \le \int_{\mathbb{R}} \frac{1}{2} {u_0^\varepsilon}^2 dx \le O(1) \tag{72}$$

Let (η, q) be any C^2 entropy pair, $q'(u) = \lambda(u)\eta'(u)$. Along solutions of the viscosity problem

$$\partial_t \eta(u^\varepsilon) + \partial_x q(u^\varepsilon) = \varepsilon \partial_x (\eta_u(u^\varepsilon) u_x^\varepsilon) - \varepsilon \eta_{uu}(u^\varepsilon) {u_x^\varepsilon}^2 = I_1 + I_2 \,. \tag{73}$$

The term I_1 converges to zero in H^{-1}, the term I_2 is uniformly bounded in L^1, and the sum $I_1 + I_2$ is uniformly bounded in $W^{-1,\infty}$. It follows from [48] that $I_1 + I_2$ lies in a compact of H^{-1}_{loc}. One concludes with the following theorem due to Tartar [63]. The proof presented here is taken out of [52].

Theorem 1. *Suppose that*

$$\partial_t \eta(u^\varepsilon) + \partial_x q(u^\varepsilon) \quad \textit{lies in a compact of } H^{-1}_{loc} \tag{74}$$

for any (η, q) with $\eta_u \in C_c^1(\mathbb{R})$. Then either the support of the Young measure ν is a point or else it is contained in an interval where f is linear,

$$supp\ \nu_{(x,t)} \subset \{\xi \in I : \lambda(\xi) = const.\} \tag{75}$$

Proof. Consider the following classes of entropy-entropy flux pairs (motivated by the kinetic formulation of the scalar conservation law):

$$\eta_1(u) = \int_{-\infty}^{u} \phi(\xi) d\xi = \int_{\mathbb{R}} \mathbb{1}_{u>\xi} \phi(\xi) d\xi$$

$$q_1(u) = \int_{-\infty}^{u} \lambda(\xi)\phi(\xi) d\xi = \int_{\mathbb{R}} \mathbb{1}_{u>\xi} \lambda(\xi)\phi(\xi) d\xi$$

$$\eta_2(u) = \int_{u}^{\infty} \psi(\theta) d\theta = \int_{\mathbb{R}} \mathbb{1}_{u<\theta} \psi(\theta) d\theta$$

$$q_2(u) = \int_{u}^{\infty} \lambda(\theta)\psi(\theta) d\theta = \int_{\mathbb{R}} \mathbb{1}_{u<\theta} \lambda(\theta)\psi(\theta) d\theta$$

where ϕ, $\psi \in C_c^1(\mathbb{R})$. Both pairs are constant near infinity, the first pair represents entropies that vanish at $-\infty$ and the second entropies vanishing at ∞.

Applying the usual compensated compactness bracket,

$$< \nu, \eta_1 q_2 - \eta_2 q_1 > = < \nu, \eta_1 >< \nu, q_2 > - < \nu, \eta_2 >< \nu, q_1 > , \tag{76}$$

to the pairs, we obtain, following an application of Fubini's Theorem,

$$\int_{\mathbb{R}} \int_{\mathbb{R}} \Big[[\lambda(\theta) - \lambda(\xi)] (\overline{\mathbb{1}_{u>\xi} \mathbb{1}_{u<\theta}} - \overline{\mathbb{1}_{u>\xi}}\ \overline{\mathbb{1}_{u<\theta}}) \Big] \phi(\xi)\psi(\theta) d\xi d\theta = 0 \,, \tag{77}$$

where the notation

$$\overline{\mathbb{1}_{u>\xi}} = \int \mathbb{1}_{u>\xi} d\nu \,.$$

From (77) we deduce

$$[\lambda(\theta) - \lambda(\xi)]\,(\overline{\mathbb{1}_{u>\xi}\mathbb{1}_{u<\theta}} - \overline{\mathbb{1}_{u>\xi}}\;\overline{\mathbb{1}_{u<\theta}}) = 0 \quad \text{a.e } \xi, \theta \tag{78}$$

and in turn

$$[\lambda(\theta) - \lambda(\xi)]\,\overline{\mathbb{1}_{u>\xi}}\;\overline{\mathbb{1}_{u<\theta}} = 0 \quad \text{for a.e } \xi, \theta \text{ with } \xi > \theta \,. \tag{79}$$

Let $F(\xi)$ be the distribution function of ν, defined by $F(\xi) = \nu\big((-\infty, \xi]\big)$. Then F is right continuous, increasing and

$$F(\xi) = \int \mathbb{1}_{u<\xi} d\nu \quad \text{a.e } \xi.$$

Using (79), we conclude (upon taking limits)

$$[\lambda(\theta) - \lambda(\xi)]\,(1 - F(\xi))\,F(\theta) = 0 \quad \text{for all } \xi, \theta \text{ with } \xi > \theta. \tag{80}$$

If the supp ν is not a point, let ξ, θ be two points in supp ν with $\xi > \theta$. Then $0 < F(\theta) \leq F(\xi) < 1$ and (80) implies $\lambda(\xi) = \lambda(\theta)$ and concludes the proof. □

Bibliographic Remarks. The notion of entropy-entropy flux pairs of Lax [35] and the theory of compensated compactness of Murat [47] and Tartar [63] play an important role in the analysis of viscosity limits - in one space dimension - for the scalar conservation law, Tartar [63], for several systems of two equations, *e.g.* DiPerna [17,18], Serre [53], Chen [3], Lin [39], Shearer [56], Serre-Shearer [54], Lions-Perthame-Tadmor [41] and Lions-Perthame-Souganidis [42], and for systems containing rich families of entropies, *e.g.* Heibig [24].

4 The Riemann Problem and Self-Similar Viscosity Limits

We consider the strictly hyperbolic system of conservation laws

$$\partial_t U + \partial_x F(U) = 0\,, \qquad x \in \mathbb{R} \;\; t > 0\,, \tag{81}$$

where $U(x,t)$ takes values in $\mathbb{R}^N$ and the Jacobian matrix $\nabla F(U)$ has real and distinct eigenvalues $\lambda_1(U) < \lambda_2(U) < \ldots < \lambda_N(U)$. The right and left eigenvectors $r_i(U)$ and $l_i(U)$ are linearly independent and are normalized,

$$\nabla F\, r_i = \lambda_i r_i\,, \quad l_i \cdot \nabla F = \lambda_i l_i\,, \quad l_i \cdot r_j = \delta_{ij} = \begin{cases} 0 & i \neq j \\ 1 & i = j \end{cases}\,; \tag{82}$$

$\{r_i\}$ and $\{l_i\}$ form a pair of local bases in the state space $\mathbb{R}^N$.

The Riemann problem consists of solving (81) with initial data a single jump discontinuity

$$U(x,0) = \begin{cases} U_- & x < 0, \\ U_+ & x > 0. \end{cases} \tag{83}$$

Due to the invariance of (81), (83) under dilations of coordinates $(x,t) \mapsto (\alpha x\,,\alpha t)$, $\alpha > 0$, solutions of the Riemann problem are sought in the form of functions $U(\frac{x}{t})$ of the single variable $\xi = \frac{x}{t}$, where $U = U(\xi)$ is a weak solution of the boundary value problem

$$(\mathcal{P}) \qquad \begin{aligned} -\,\xi U' + F(U)' &= 0\,, \\ U(\pm\infty) &= U_\pm\,. \end{aligned}$$

In solving $(\mathcal{P})$ one encounters lack of uniqueness that is accounted for by imposing admissibility restrictions on solutions. We refer to Dafermos [15] for a detailed discussion of the issue of admissibility together with historical references.

For weak waves in strictly hyperbolic systems it suffices to impose admissibility restrictions only at shocks. The classical solution of the Riemann problem proceeds in two steps: First, special solutions of rarefaction waves, shock waves or contact discontinuities are constructed and are in turn used for constructing the elementary wave curves. There is one elementary wave curve associated with each characteristic field with the parametrization of the curve serving as a measure of the strength of the associated wave. Second, it is shown that the compound curves emanating from a fixed left state U_- give rise to an invertible map that covers a full neighborhood of right end states U_+. The construction provides a unique solution for the Riemann problem, in the class of weak waves, for genuinely nonlinear systems (Lax [34]) as well as for a large class of non-genuinely nonlinear systems (Liu [43,44]).

4.1 The problem of self-similar viscosity limits

The method of self-similar viscosity limits, introduced in Dafermos [11], provides a complementary approach for solving the Riemann problem in the spirit of viscosity approximations. An elliptic regularization of the Riemann operator in $(\mathcal{P})$ is introduced

$$(\mathcal{P}_\varepsilon)_B \qquad \begin{aligned} -\,\xi U' + F(U)' &= \varepsilon\Big(B(U)U'\Big)' \\ U(\pm\infty) &= U_\pm\,, \end{aligned}$$

where $\varepsilon > 0$ and $B(U)$ is a positive matrix accounting for the viscous structure. The admissible solutions of $(\mathcal{P})$ are selected as $\varepsilon \searrow 0$ limit-points of solutions to the problem $(\mathcal{P}_\varepsilon)_B$. In contrast to shock admissibility criteria, self-similar viscosity limits penalize the whole wave-fan simultaneously, and the resulting admissibility criterion is called *viscous wave-fan admissibility criterion.*

In this section we review the method, in the framework of weak waves for strictly hyperbolic $N \times N$ systems with $B(U) = Id$,

$$(\mathcal{P}_\varepsilon) \qquad \begin{aligned} -\xi U' + F(U)' &= \varepsilon U'' \\ U(\pm\infty) &= U_\pm . \end{aligned}$$

We start with a summary of the result [68].

Theorem 2. *Let (81) be strictly hyperbolic, $B(U) = Id$ and suppose the jump of the Riemann data $|U_+ - U_-|$ is small.*
(i) There exists a family $\{U_\varepsilon\}$ of smooth solutions to $(\mathcal{P}_\varepsilon)$ such that U_ε satisfy the uniform bounds

$$(V) \qquad |U_\varepsilon| + TV U_\varepsilon \leq C ,$$

and $|U_\varepsilon'(\xi)| \leq \frac{C}{\varepsilon} e^{-\frac{\alpha}{\varepsilon}\xi^2}$, $|\xi| \geq \Lambda$, for some α and Λ independent of ε.
(ii) Let U_{ε_n} be a subsequence of $\{U_\varepsilon\}$ such that $U_{\varepsilon_n} \to U(\xi)$ pointwise for $\xi \in \mathbb{R}$. Then U is a BV function that satisfies

$$-\xi U' + F(U)' = 0 \tag{84}$$

in the sense of measures and the Rankine Hugoniot conditions,

$$-\xi \left[U(\xi+) - U(\xi-)\right] + \left[F(U(\xi+)) - F(U(\xi-)\right] = 0 , \tag{85}$$

at any point of discontinuity $\xi \in \mathcal{S}_U$.
(iii) The function U consists of N wave fans separated by constant states. Each wave fan consists of an alternating sequence of shocks and rarefactions so that each shock adjacent to a rarefaction on one side is a contact on that side. At a shock $\xi \in \mathcal{S}_U$ belonging to the k-th wave fan, a weak form of the Lax shock conditions is satisfied

$$\lambda_k(U(\xi+)) \leq \xi \leq \lambda_k(U(\xi-)) . \tag{86}$$

Finally, if $\xi \in \mathcal{S}_U$ then the sequence U_{ε_n} has at ξ the internal structure of a shock profile.

The theorem provides an alternative route to the solution of the Riemann problem, requiring strict hyperbolicity but no geometric conditions on the wave curves. In that sense it provides a general theory within the class of strictly hyperbolic systems and weak waves. The proof is analytical, replacing the construction of the wave curves with the construction of a class of solutions to $(\mathcal{P}_\varepsilon)$, that we call approximate wave curves. The main issue towards proving (V) is to construct a framework for measuring the total variation of approximate solutions that persists in the $\varepsilon \to 0$ limit. This construction may provide insight to the understanding of the corresponding issue in the (harder) problem of viscosity limits.

The study of self-similar viscosity limits may be decomposed into three steps:

(i) Construction of smooth solutions to the problem $(\mathcal{P}_\varepsilon)_B$, $\varepsilon > 0$.
(ii) Performing the passage to the limit $\varepsilon \searrow 0$, from $(\mathcal{P}_\varepsilon)_B$ to $(\mathcal{P})$.
(iii) Study of the structure of the emerging solution.

Step (i) is technical but routine, and general results can be established under weak assumptions: If (81) is equipped with an L^p estimate then $(\mathcal{P}_\varepsilon)$ has smooth solutions for each $\varepsilon > 0$, [68]. This applies in particular to the class of *symmetric hyperbolic* systems.

The usual framework for Step (ii) is uniform stability in $BV\big([a,b];\mathbb{R}^N\big)$,

$$(V) \qquad |U_\varepsilon| + TVU_\varepsilon \le C\,.$$

If (V) holds on an interval $[a,b]$ then Helly's Theorem implies that there exists a subsequence $\{U_{\varepsilon_n}\}$, with $\varepsilon_n \to 0$, and a function $U \in BV$ such that $U_{\varepsilon_n} \to U$ for $\xi \in [a,b]$.

Suppose now that the family $\{U_\varepsilon\}$ of solutions to $(\mathcal{P}_\varepsilon)_B$ satisfies the uniform BV-bound (V) and, for some C, $\alpha > 0$ and Λ independent of ε, the uniform estimates

$$(D) \qquad |U_\varepsilon'(\xi)| \le \frac{C}{\varepsilon} e^{-\frac{\alpha}{\varepsilon}\xi^2}\,, \qquad \text{for } |\xi| \ge \Lambda\,,$$

$$(M) \qquad \varepsilon \int_{\mathbb{R}} |U'|^2\, d\xi \le C\,.$$

We show how to construct solutions of the Riemann problem. (Note that if (81) is strictly hyperbolic and the family $\{U_\varepsilon\}$ is bounded in L^∞, then (D) can be proved in both the case $B(U) = Id$ as well as in some cases with singular diffusion matrices [67,68,32]. Also, that (M) follows, if there is an entropy-entropy flux pair (η, q) such that $\nabla^2\eta \cdot B \ge c\,Id$ for some $c > 0$.) Consider a subsequence $\{U_{\varepsilon_n}\}$ such that

$$U_{\varepsilon_n}(\xi) \to U(\xi) \quad \text{for } \xi \in \mathbb{R}\,. \tag{87}$$

From $(\mathcal{P}_\varepsilon)_B$ we obtain, for a test function ψ,

$$\int_{\mathbb{R}} U_\varepsilon \cdot (\xi\psi)' - F(U_\varepsilon)\cdot\psi'\, d\xi = -\varepsilon \int_{\mathbb{R}} B(U_\varepsilon)U_\varepsilon' \cdot \psi'\, d\xi\,. \tag{88}$$

Using (V), (M) and (87) we pass to the limit $\varepsilon_n \to 0$ and obtain

$$\int_{\mathbb{R}} U \cdot (\xi\psi)' - F(U)\cdot\psi'\, d\xi = 0\,. \tag{89}$$

As $U \in BV$, its domain can be decomposed into two disjoint sets : $\mathcal{C}_U$ the set of points of continuity of U and $\mathcal{S}_U$ the set of points of discontinuity, respectively. The set $\mathcal{S}_U$ is at most countable, and the right and left limits $U(\xi+)$, $U(\xi-)$ exist at each ξ. The equation (84) is satisfied in the sense of measures. In particular, at $\xi \in \mathcal{S}_U$, the Rankine-Hugoniot conditions (85) are satisfied. Finally, (D) implies that $U = U_-$ on the interval $(-\infty, -\Lambda)$ and $U = U_+$ on $(\Lambda, +\infty)$. The function $U(\frac{x}{t})$ is a weak solution of the Riemann problem (81), (83), and the set $\mathcal{S}_U$ is the set of shocks for this wave-fan solution.

In Sections 4.3 and 4.4, we outline the derivation of uniform variation estimates for the problem $(\mathcal{P}_\varepsilon)$ with $B(U) = Id$, first for the single conservation law and then for a strictly hyperbolic system. In Section 4.2, we show that stable BV-families of solutions to $(\mathcal{P}_\varepsilon)$ have, near shocks, the internal structure of shock profiles.

4.2 The connection with shock profiles

First, we investigate the relation between self-similar viscosity limits and shock profiles. Let $\{U_\varepsilon\}$ be a family of solutions to $(\mathcal{P}_\varepsilon)$ satisfying (V), (D) and (87).

Fix a point of discontinuity ξ of U and note that $U(\xi\pm)$ satisfy the Rankine-Hugoniot conditions (85). Consider a sequence of points $\xi_\varepsilon \to \xi$ as $\varepsilon \to 0$. Define the functions

$$V_\varepsilon(\zeta) = U_\varepsilon(\xi_\varepsilon + \varepsilon\, \zeta), \qquad -\infty < \zeta < \infty\,. \tag{90}$$

This transformation introduces a stretching of the independent variable centered around ξ_ε; the point ξ_ε is a shift of the shock speed ξ. The functions V_ε are uniformly bounded in BV,

$$TV_\zeta\ V_\varepsilon(\cdot) = TV_\zeta\ U_\varepsilon(\xi_\varepsilon + \varepsilon\ \cdot) = TV_\xi\ U_\varepsilon(\cdot) \le C\,. \tag{91}$$

Using Helly's theorem and a diagonal argument we establish the existence of a subsequence and a function V such that

$$U_\varepsilon(\xi_\varepsilon + \varepsilon\, \zeta) \to V(\zeta) \quad \text{pointwise for } -\infty < \zeta < \infty\,. \tag{92}$$

Proposition 3. *Let $\xi \in \mathcal{S}_U$ and suppose that $\{\xi_\varepsilon\}$ is a sequence of points with $\xi_\varepsilon \to \xi$. Then $V(\zeta)$, defined in* (92)*, is continuously differentiable and satisfies on $(-\infty, \infty)$ the traveling wave equations*

$$-\xi\,[\,V - U(\xi-)\,] + [\,F(V) - F(U(\xi-))\,] = \frac{dV}{d\zeta} \tag{93}$$

with initial condition $V(0) = \lim_{\varepsilon\to 0} U_\varepsilon(\xi_\varepsilon)$. The limits $\lim\limits_{\zeta\to\pm\infty} V(\zeta) =: V_\pm$ exist, are finite, and V_+, V_- solve the algebraic equations

$$-\xi\,[\,V - U(\xi-)\,] + [\,F(V) - F(U(\xi-))\,] = 0\,. \tag{94}$$

Proof. We integrate $(\mathcal{P}_\varepsilon)$ between the points $\xi_\varepsilon + \varepsilon\zeta$ and θ and then integrate the resulting equation in θ between ξ and $\xi + \delta$, for some $\delta \neq 0$, to arrive at

$$\begin{aligned}
&[-(\xi_\varepsilon + \varepsilon\,\zeta)U_\varepsilon(\xi_\varepsilon + \varepsilon\,\zeta) + F(U_\varepsilon(\xi_\varepsilon + \varepsilon\,\zeta))\,] \\
&\qquad -\frac{1}{\delta}\int_\xi^{\xi+\delta} [\,-\theta U_\varepsilon(\theta) + F(U_\varepsilon(\theta))\,]\,d\theta + \frac{1}{\delta}\int_\xi^{\xi+\delta}\int_\theta^{\xi_\varepsilon+\varepsilon\zeta} U_\varepsilon(\tau)\,d\tau d\theta \\
&\quad = \frac{d}{d\zeta}\big(U_\varepsilon(\xi_\varepsilon + \varepsilon\zeta)\big) - \varepsilon\frac{1}{\delta}\int_\xi^{\xi+\delta} U_\varepsilon'(\theta)\,d\theta\;.
\end{aligned}$$

After an integration in ζ we get

$$\begin{aligned}
&\int_0^\zeta [\,-(\xi_\varepsilon + \varepsilon s)U_\varepsilon(\xi_\varepsilon + \varepsilon s) + F(U_\varepsilon(\xi_\varepsilon + \varepsilon s))\,]\,ds \\
&\qquad -\zeta\frac{1}{\delta}\int_\xi^{\xi+\delta} [-\theta U_\varepsilon(\theta) + F(U_\varepsilon(\theta))]\,d\theta + \frac{1}{\delta}\int_0^\zeta\int_\xi^{\xi+\delta}\int_\theta^{\xi_\varepsilon+\varepsilon s} U_\varepsilon(\tau)\,d\tau d\theta ds \\
&\quad = U_\varepsilon(\xi_\varepsilon + \varepsilon\zeta) - U_\varepsilon(\xi_\varepsilon) - \frac{\varepsilon\zeta}{\delta}\int_\xi^{\xi+\delta} U_\varepsilon'(\theta)\,d\theta\;.
\end{aligned}$$

Letting $\varepsilon \to 0$ and using (87), (92) and (V), we deduce

$$\begin{aligned}
\int_0^\zeta [\,-\xi V(s) + F(V(s))\,]\,ds - \zeta\frac{1}{\delta}\int_\xi^{\xi+\delta} [\,-\theta U(\theta) + F(U(\theta))\,]\,d\theta \\
+ \zeta\frac{1}{\delta}\int_\xi^{\xi+\delta}\int_\theta^\xi U(\tau)\,d\tau d\theta = V(\zeta) - V(0)\;.
\end{aligned}$$

Letting consecutively $\delta \to 0+$ and $\delta \to 0-$, we obtain

$$\int_0^\zeta \Big[-\xi\big(V(s) - U(\xi\pm)\big) + F(V(s)) - F(U(\xi\pm))\Big]\,ds = V(\zeta) - V(0). \quad (95)$$

It follows that $V(\zeta)$ is a continuously differentiable function that satisfies the traveling wave equations (93). Since V is of bounded variation on $\mathbb{R}$, the limits $\lim_{\zeta\to\pm\infty} V(\zeta) =: V_\pm$ exist and are finite. Also, for any integer n

$$\int_n^{n+1} \Big[-\xi\big(V(s) - U(\xi-)\big) + F(V(s)) - F(U(\xi-))\Big]\,ds = V(n+1) - V(n)\;.$$

Taking the i-th component and using the mean value theorem, we see hat there are t_n^i with $n \le t_n^i \le n+1$ such that for $i = 1, \ldots, N$

$$-\xi\big(V^i(t_n^i) - U^i(\xi-)\big) + F^i(V(t_n^i)) - F^i(U(\xi-)) = V^i(n+1) - V^i(n).$$

Letting $n \to \infty$ shows that V_+ is an equilibrium of (93). Similarly, V_- satisfies (94). □

The function V as well as the limiting values $V_\pm$ depend on the choice of the sequence $\{\xi_\varepsilon\}$. For several choices of $\{\xi_\varepsilon\}$ it may happen that the traveling wave disintegrates to a constant solution. Two questions arise: (i) Is it always possible to choose $\{\xi_\varepsilon\}$ so that the resulting V does not disintegrate to a constant solution of (93). (ii) What is the relation of $U(\xi-)$, $U(\xi+)$ and nontrivial heteroclinic orbits. These questions are taken up in [68], Sec 9. It turns out:

Proposition 4. *Let $\xi \in \mathcal{S}_U$ be fixed and suppose the set of solutions of* (94) *is not connected. There exists a sequence of shock shifts $\{\xi_\varepsilon\}$ such that the resulting V in* (92) *is a nontrivial heteroclinic (or homoclinic) orbit.*

The hypothesis in Proposition 4 is violated only for shocks associated with a linearly degenerate characteristic field : $\nabla\lambda_k(U)\cdot r_k(U) = 0$ for all U. Addressing (ii) is quite complicated at the full level of generality. We give one result indicating what can happen if there is a finite number of equilibria in B_C, the ball of radius C where the functions U_ε take values.

Proposition 5. *Let $\xi \in \mathcal{S}_U$ and suppose that* (94) *has a finite number of solutions in B_C. There exists a subsequence $\varepsilon_n \to 0$ and choices $\{\xi_{1\varepsilon_n}\}$, $\{\xi_{2\varepsilon_n}\}$ of the shock shifts such that $\xi_{1\varepsilon_n} \le \xi_{2\varepsilon_n}$, $\xi_{1\varepsilon_n} \to \xi$, $\xi_{2\varepsilon_n} \to \xi$,*

$$\left.\begin{aligned} &U_{\varepsilon_n}(\xi_{1\varepsilon_n} + \varepsilon_n\zeta) \to V_1(\zeta)\,, \\ &U_{\varepsilon_n}(\xi_{2\varepsilon_n} + \varepsilon_n\zeta) \to V_2(\zeta)\,, \end{aligned}\right\} \quad \textit{pointwise for } -\infty < \zeta < \infty, \tag{96}$$

and V_1, V_2 are nontrivial solutions of (93) *that satisfy $V_1(-\infty) = U(\xi-)$, $V_2(+\infty) = U(\xi+)$.*

Associated to characteristic fields that are not linearly degenerate, there exists one heteroclinic orbit of (93) that emanates from $U(\xi-)$ and one that concludes at $U(\xi+)$. If more than two states in B_C satisfy the Rankine-Hugoniot conditions at a given $\xi \in \mathcal{S}_U$, or if multiple heteroclinic connections between two equilibria are possible, then the precise relation between self-similar limits and shock profiles requires a detailed analysis of the shock profiles. The structure of traveling wave solutions is well understood for weak shocks, even for general diffusion matrices (Majda and Pego [46]). By contrast, relatively little is known for strong shocks. In general, it is possible that there are intermediate states V_j, $j = 1, ..., J$, finitely many or even countable, satisfying (94) and a chain of shock profiles, at the same shock speed ξ, that connect successively $U(\xi-)$ to V_1, each of the points V_j to the next, and V_J to $U(\xi+)$. The latter situation occurs for the equations of isothermal elasticity in the presence of multiple inflection points in the stress-strain relation, for specific positions of the Riemann data relative to the stress-strain curve [67].

4.3 The scalar conservation law

In this section we consider the problem of self-similar viscosity limits for the scalar conservation law, and discuss the proof of the uniform bounds (V) and the structure of the emerging solution. Let $\{u_\varepsilon\}_{\varepsilon>0}$ be a family of scalar-valued functions satisfying

$$\begin{aligned} \varepsilon\, u_\varepsilon'' &= -\xi\, u_\varepsilon' + f(u_\varepsilon)' \\ u_\varepsilon(\pm\infty) &= u_\pm\,. \end{aligned} \tag{97}$$

It is easy to see that solutions of (97) satisfy the representation formula

$$u_\varepsilon'(\xi) = (u_+ - u_-)\ \frac{\exp\left\{-\frac{1}{\varepsilon}\int_\rho^\xi s - \lambda\big(u_\varepsilon(s)\big)\,ds\right\}}{\int_{-\infty}^{\infty}\exp\left\{-\frac{1}{\varepsilon}\int_\rho^\zeta s - \lambda\big(u_\varepsilon(s)\big)\,ds\right\}d\zeta} = \tau\varphi_\varepsilon(\xi)\,, \tag{98}$$

where $\lambda(u) = f'(u)$ denotes the characteristic speed and ρ is any real number. Above, we used the notations $\tau = (u_+ - u_-)$, as a measure of the strength of the wave, and

$$\varphi_\varepsilon(\xi) = \varphi_\varepsilon[u_\varepsilon](\xi) = \frac{e^{-\frac{1}{\varepsilon}g_\varepsilon(\xi)}}{\int_{-\infty}^{\infty} e^{-\frac{1}{\varepsilon}g_\varepsilon(\zeta)}d\zeta} = \frac{1}{I_\varepsilon}e^{-\frac{1}{\varepsilon}g_\varepsilon(\xi)} \tag{99}$$

$$\text{where} \qquad g_\varepsilon(\xi) = g[u_\varepsilon](\xi) = \int_\rho^\xi s - \lambda\big(u_\varepsilon(s)\big)\,ds$$

$$\text{and} \qquad I_\varepsilon = \int_{-\infty}^{\infty} e^{-\frac{1}{\varepsilon}g_\varepsilon(\zeta)}d\zeta$$

Note that φ_ε and g_ε depend implicitly on the solution u_ε, through the dependence on the characteristic speed $\lambda(u_\varepsilon)$.

It follows from (98) that u_ε' has a sign, and thus

$$\min\{u_-, u_+\} \le u_\varepsilon \le \max\{u_-, u_+\}\,, \qquad TV u_\varepsilon = |u_+ - u_-|\,.$$

Another way to see (V) is to observe that φ_ε are positive functions and uniformly bounded in L^1, hence $\{u_\varepsilon\}$ is of uniformly bounded variation. Given the bound (V), we can pass to the $\varepsilon \to 0$ limit and obtain a solution of the problem $(\mathcal{P})$ for the scalar case.

In the sequel, we study the quantities φ_ε in (99), under various frameworks of uniform bounds. For a family of solutions $\{u_\varepsilon\}$ bounded in L^∞, we have

$$(A) \qquad\qquad \lambda_- \le \lambda(u_\varepsilon) \le \lambda_+\,.$$

Lemma 6. *Under Hypothesis (A), as $\varepsilon \to 0$:*
(i) If $d = \lambda_+ - \lambda_- > 0$, then $\frac{1}{O(1)}\frac{\varepsilon}{d} \le I_\varepsilon \le d + \sqrt{2\pi\varepsilon}$, and

$$0 < \varphi_\varepsilon(\xi) \le \begin{cases} O(1)\,\frac{d}{\varepsilon}\,e^{-\frac{1}{2\varepsilon}(\xi-\lambda_-)^2} & \xi < \lambda_-, \\ O(1)\,\frac{d}{\varepsilon} & \xi \in \mathbb{R}, \\ O(1)\,\frac{d}{\varepsilon}\,e^{-\frac{1}{2\varepsilon}(\xi-\lambda_+)^2} & \xi > \lambda_+. \end{cases} \tag{100}$$

(ii) If $d = \lambda_+ - \lambda_- = 0$*, then* $I_\varepsilon = \sqrt{2\pi\varepsilon}$ *and* $\varphi_\varepsilon = \frac{1}{\sqrt{2\pi\varepsilon}} e^{-\frac{1}{2\varepsilon}(\xi-\lambda_-)^2}$.

Proof. The estimates for φ_ε reflect the property that, under Hypothesis (A), g_ε has the form of a potential-well function (*cf.* Figure 2). We select ρ as the point where g_ε achieves its global minimum. Then ρ satisfies $\lambda_- \leq \rho = \lambda(u_\varepsilon(\rho)) \leq \lambda_+$, and $g_\varepsilon(\xi) \geq g_\varepsilon(\rho) = 0$.

Assume first that $d > 0$. Then

$$g_\varepsilon(\zeta) = \int_\rho^\zeta s - \lambda(u_\varepsilon(s))ds \leq \begin{cases} \frac{1}{2}(\zeta-\rho)^2 + d(\zeta-\rho) & \text{for } \zeta > \rho \\ \frac{1}{2}(\zeta-\rho)^2 - d(\zeta-\rho) & \text{for } \zeta < \rho \end{cases}$$

In turn,

$$\begin{aligned} I_\varepsilon &= \int_{-\infty}^{\rho} e^{-\frac{1}{\varepsilon}g_\varepsilon} d\zeta + \int_\rho^{+\infty} e^{-\frac{1}{\varepsilon}g_\varepsilon} d\zeta \\ &\geq \sqrt{\varepsilon}\, e^{\frac{d^2}{2\varepsilon}} \int_{-\infty}^{0} e^{-\frac{1}{2}\left(\eta - \frac{d}{\sqrt{\varepsilon}}\right)^2} d\eta + \sqrt{\varepsilon}\, e^{\frac{d^2}{2\varepsilon}} \int_0^\infty e^{-\frac{1}{2}\left(\eta + \frac{d}{\sqrt{\varepsilon}}\right)^2} d\eta \\ &\geq \frac{1}{O(1)} \frac{\varepsilon}{d}, \qquad \text{for } \varepsilon \text{ small.} \end{aligned}$$

On the other hand, estimating g_ε from below yields

$$g_\varepsilon(\zeta) \geq \begin{cases} \frac{1}{2}(\zeta-\lambda_-)^2 & \text{for } \zeta < \lambda_- \\ 0 & \text{for } \lambda_- < \zeta < \lambda_+ \\ \frac{1}{2}(\zeta-\lambda_+)^2 & \text{for } \zeta > \lambda_+ \end{cases}$$

whence

$$I_\varepsilon \leq \int_{-\infty}^{\lambda_-} e^{-\frac{1}{2\varepsilon}(\zeta-\lambda_-)^2} d\zeta + d + \int_{\lambda_+}^{\infty} e^{-\frac{1}{2\varepsilon}(\zeta-\lambda_+)^2} d\zeta = d + \sqrt{2\pi\varepsilon}\,.$$

The proof of (100) now follows from (99). Finally, if $d = 0$ then $\lambda(u_\varepsilon)$ remains constant, say λ_-, and part (ii) follows from a direct calculation. □

The family $\{u_\varepsilon\}$ is uniformly bounded in BV, while $\{\varphi_\varepsilon\}$ is uniformly bounded in L^1. There is a subsequence u_{ε_n}, φ_{ε_n} and a finite positive Borel measure ϕ such that

$$\begin{aligned} u_{\varepsilon_n} &\to u\,, && \text{pointwise in } \mathbb{R}, \\ \varphi_{\varepsilon_n} &\rightharpoonup \phi\,, && \text{weak-}\star \text{ in measures.} \end{aligned} \tag{101}$$

By (100) no mass escapes at infinity and the total mass of the measure ϕ is one.

The distribution function of ϕ is the right continuous function $\frac{1}{\tau}(u(\xi+)-u_-)$. Along the same sequence

$$g_{\varepsilon_n}(\xi) = \int_{\rho_{\varepsilon_n}}^{\xi} s - \lambda(u_{\varepsilon_n}(s))\, ds \to \int_{\rho}^{\xi} s - \lambda(u(s))\, ds =: g(\xi) \tag{102}$$

uniformly on compact subsets of $(-\infty, \infty)$. We show that points in the support of ϕ are global minima for the function g.

Proposition 7. *If $\xi \in$ supp ϕ then $g(\zeta) \geq g(\xi)$ for all $\zeta \in (-\infty, \infty)$.*

Proof. Fix $\xi \in \mathbb{R}$, $\alpha > 0$ and consider the set

$$\mathcal{A} = \{\zeta \in \mathbb{R} \ : \ g(\zeta) - g(\xi) < -\alpha < 0\}\,. \tag{103}$$

Step 1 : If the Lebesgue measure $m(\mathcal{A}) > 0$, then $\xi \notin$ supp ϕ.

Since $g(\zeta) \to \infty$ as $|\zeta| \to \infty$, $\mathcal{A}$ is contained in some compact interval $[a, b]$. By (102) and the continuity of g there are δ and ε_0 such that

$$g_{\varepsilon_n}(\zeta) - g_{\varepsilon_n}(\theta) < -\frac{\alpha}{2}$$

for $\varepsilon < \varepsilon_0$, $\zeta \in \mathcal{A}$ and $\theta \in J = (\xi - \delta, \xi + \delta)$. Hence,

$$0 < \varphi_{\varepsilon_n}(\theta) \leq \frac{1}{\int_{\mathcal{A}} \exp\{-\frac{1}{\varepsilon_n}(g_{\varepsilon_n}(\zeta) - g_{\varepsilon_n}(\theta))\}\, d\zeta} \leq \frac{e^{-\frac{\alpha}{2\varepsilon_n}}}{m(\mathcal{A})}\,. \tag{104}$$

Let $\chi \in C_c(J)$. Then (104) and (101) give

$$\int_{(\xi-\delta,\xi+\delta)} \varphi_{\varepsilon_n}(\theta)\, \chi(\theta)\, d\theta \to 0\,, \qquad \text{as } \varepsilon_n \to 0\,,$$

and thus $\xi \notin$ supp ϕ.

Step 2 : If $\xi \in$ supp ϕ, then $m(\mathcal{A}) = 0$ and thus $\mathcal{A}$ is empty for any $\alpha > 0$. Hence, $g(\zeta) \geq g(\xi)$ for $\zeta \in \mathbb{R}$. □

The minimization property for g provides information on the structure of the BV-function u. In particular, a weak form of the Lax shock conditions is induced at points of discontinuity.

Corollary 8. *Let ξ, $\xi' \in$ supp $\phi \subset [\lambda_-, \lambda_+]$ with $\xi < \xi'$.*
(a) If $\xi \in C_u$, then $\xi = \lambda(u(\xi))$.
(b) If $\xi \in S_u$, then u satisfies at ξ the jump conditions (85) and the inequalities

$$\lambda(u(\xi+)) \leq \xi \leq \lambda(u(\xi-))\,. \tag{105}$$

(c) If ξ, $\xi' \in$ supp ϕ then $\lambda(u(\xi+)) = \xi$, $\lambda(u(\xi'-)) = \xi'$. Moreover, at any $\theta \in (\xi, \xi')$,

$$\begin{aligned} \theta = \lambda(u(\theta)) \qquad & \textit{if} \quad \theta \in C_u\,, \\ \lambda(u(\theta+)) = \theta = \lambda(u(\theta-)) \qquad & \textit{if} \quad \theta \in S_u\,. \end{aligned} \tag{106}$$

Proof. The function g is continuous, satisfies $g(\xi) \to \infty$ as $|\xi| \to \infty$, and the limits

$$\lim_{\zeta \to \xi \pm} \frac{g(\zeta) - g(\xi)}{\zeta - \xi} = \lim_{\zeta \to \xi \pm} \frac{1}{\zeta - \xi} \int_{\xi}^{\zeta} s - \lambda(u(s))\, ds = \xi - \lambda(u(\xi \pm))$$

exist. Proposition 7 implies that if $\xi \in \operatorname{supp} \phi$ then $\xi - \lambda(u(\xi+)) \geq 0$ and $\xi - \lambda(u(\xi-)) \leq 0$. In turn, this implies (a) if $\xi \in \mathcal{C}_u$ and (b) if $\xi \in \mathcal{S}_u$.

It remains to show (c). Let ξ, $\xi' \in \operatorname{supp} \phi$ with $\xi < \xi'$. Then ξ, ξ' are both global minima for g with $g(\xi) = g(\xi')$. We claim

$$g(\theta) = g(\xi) \quad \text{for any} \quad \theta \in (\xi, \xi') \,. \tag{107}$$

If (107) is violated, there exist a, b with $\xi \leq a < b \leq \xi'$ such that

$$g(a) = g(b) = g(\xi)\,, \quad g(\theta) > g(\xi) \quad \text{for } a < \theta < b \,.$$

At the points a, b we have

$$\begin{aligned} \lambda(u(a+)) &\leq a \leq \lambda(u(a-)) \\ \lambda(u(b+)) &\leq b \leq \lambda(u(b-)) \,. \end{aligned} \tag{108}$$

On the other hand, at any $\theta \in (a, b)$ the set $\mathcal{A} = \{\zeta \in \mathrm{I\!R} : g(\zeta) - g(\theta) < -\alpha\}$ is nonempty for some $\alpha > 0$. Proposition 7 implies that $\theta \notin \operatorname{supp} \phi$ and the function $u(\xi)$ remains constant on the interval (a, b). Hence $\lambda(u(a+)) = \lambda(u(b-))$ and the inequalities (108) yield $b \leq a$. This contradicts $a < b$ and (107) follows. □

In summary, the region where u is nonconstant consists of one closed interval I_λ (which could degenerate to one single point). The solution u takes the values u_- and u_+ on the complement of I_λ and looks like a wave-fan consisting of rarefactions, shocks and contacts at points of I_λ.

4.4 BV stability for self-similar viscosity limits

Next, we outline the derivation of the uniform BV bounds, for weak waves in $N \times N$ strictly hyperbolic systems:

Theorem 9. *Let* (81) *be strictly hyperbolic and U_- be fixed. If $|U_+ - U_-|$ is sufficiently small, the problem $(\mathcal{P}_\varepsilon)$ admits a smooth solution U_ε for each $\varepsilon > 0$. Moreover, the family of solutions $\{U_\varepsilon\}_{\varepsilon > 0}$ is of uniformly bounded (and small) oscillation and total variation.*

Sketch of Proof. First, $(\mathcal{P}_\varepsilon)$ is recast into an alternative formulation. Let U_ε be a solution to $(\mathcal{P}_\varepsilon)$ connecting U_- to U_+ and consider the decomposition of U'_ε in the basis $\{r_k(U_\varepsilon)\}$,

$$U'_\varepsilon(\xi) = \sum_{k=1}^{N} a_{k\varepsilon}(\xi)\, r_k\big(U_\varepsilon(\xi)\big) \,. \tag{109}$$

The amplitudes $a_{k\varepsilon}$ can be recovered from the formula

$$a_{k\varepsilon}(\xi) = l_k(U_\varepsilon(\xi)) \cdot U_\varepsilon'(\xi)\,, \tag{110}$$

and a simple calculation, taking the inner product of the system in $(\mathcal{P}_\varepsilon)$ with $l_k(U_\varepsilon)$, shows that $a_{k\varepsilon}$ satisfy the equations

$$\begin{aligned} &\varepsilon\, a_{k\varepsilon}' + \big[\xi - \lambda_k(U_\varepsilon(\xi))\big]\, a_{k\varepsilon} \\ &\quad = \varepsilon \sum_{m,n=1}^{N} \big[\nabla l_k(U_\varepsilon(\xi))\, r_m(U_\varepsilon(\xi)) \cdot r_n(U_\varepsilon(\xi))\big]\, a_{m\varepsilon} a_{n\varepsilon}\,. \end{aligned} \tag{111}$$

Integrating (109) over $(-\infty, \infty)$, we have

$$U_+ - U_- = \sum_{k=1}^{N} \int_{-\infty}^{\infty} a_{k\varepsilon}(\zeta)\, r_k\big(U_\varepsilon(\zeta)\big)\, d\zeta\,. \tag{112}$$

Equations (111)-(112) provide an equivalent formulation of the problem $(\mathcal{P}_\varepsilon)$. Henceforth we suppress the ε-dependence of functions and introduce the notation

$$\begin{aligned} \lambda_k &= \lambda_k(U_\varepsilon(\xi)) \\ \beta_{k,mn} &= \beta_{k,mn}(U_\varepsilon(\xi)) = \nabla l_k(U_\varepsilon(\xi))\, r_m(U_\varepsilon(\xi)) \cdot r_n(U_\varepsilon(\xi))\,. \end{aligned} \tag{113}$$

The functions a_k satisfy the coupled system of ordinary differential equations with variable coefficients

$$\varepsilon\, a_k' + (\xi - \lambda_k)\, a_k = \varepsilon \sum_{m,n=1}^{N} \beta_{k,mn}\, a_m\, a_n\,. \tag{114}$$

We consider the following question: Assume we are given a family $\{U_\varepsilon\}_{\varepsilon>0}$ of solutions that are of uniformly bounded, small oscillation

$$(C_o) \qquad \sup_{-\infty<\xi<+\infty} |U_\varepsilon(\xi) - U_-| \le \mu.$$

Examine under what conditions the given family is of uniformly bounded variation

$$(S) \qquad TV_{(-\infty,+\infty)}(U_\varepsilon) \le C.$$

Note that (C_o) imposes the restriction $|U_+ - U_-|$ small on the Riemann data, and dictates that U_ε satisfy the uniform L^∞-bound, $\sup_{-\infty<\xi<+\infty} |U_\varepsilon(\xi)| \le M$, with the constants M and μ independent of ε and μ also small. Along the family $\{U_\varepsilon\}$, each wave speed is bounded

$$\lambda_{k-} \le \lambda_k\big(U_\varepsilon(\xi)\big) \le \lambda_{k+} \tag{115}$$

by constants λ_{k-}, λ_{k+} independent of ε. If the oscillation of U_ε is sufficiently small, then the wave speeds are ***totally separated***, that is

$$\begin{aligned}\lambda_{1-} \le \lambda_1\big(U_\varepsilon(\xi)\big) \le \lambda_{1+} < \lambda_{2-} \le \lambda_2\big(U_\varepsilon(\xi)\big) \le \lambda_{2+} < \dots \\ < \lambda_{(N-1)-} \le \lambda_{N-1}\big(U_\varepsilon(\xi)\big) \le \lambda_{(N-1)+} < \lambda_{N-} \le \lambda_N\big(U_\varepsilon(\xi)\big) \le \lambda_{N+}.\end{aligned} \tag{116}$$

Finally, the coefficients $\beta_{k,mn}$ are uniformly bounded, $|\beta_{k,mn}| \le B$, by a constant B depending on μ but not on ε.

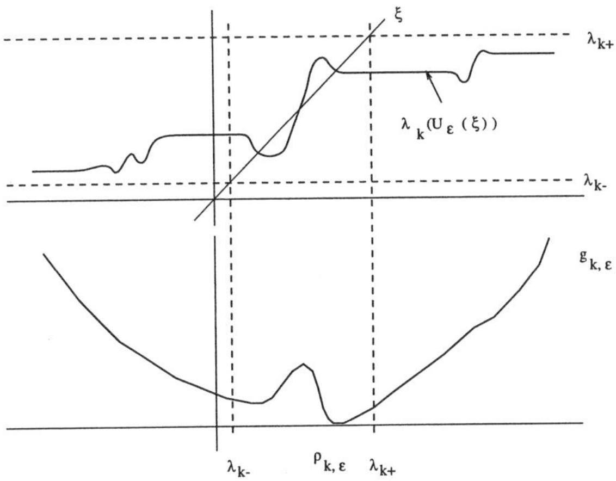

Fig. 2.

The L^1 norm of the function $\sum_{k=1}^N |a_k|$ provides a natural measure of the variation of U_ε. Hence, in order to prove (S) it suffices to estimate in L^1 the solutions $a_{k\varepsilon}$ of the system (114), under the hypotheses that the wave speeds λ_k are totally separated and the coefficients $\beta_{k,mn}$ are bounded. The quadratic terms in (114) represent the effect induced on the k-family by interactions of waves of all the families, and $\beta_{k,mn}$ measure the weights of such contributions. There are three problems to be resolved : First, to find a natural framework for measuring the L^1 norm of $\sum_{k=1}^N |a_k|$. Second, to understand the effect of the quadratic terms. Third, differential systems like (114) are best amenable to analysis under pointwise conditions. On the other hand the existing information connecting a_k with the data is of integral type. Therefore, a scheme is needed that connects pointwise to integral information.

Let g_k be an antiderivative of $g_k' = \xi - \lambda_k(U_\varepsilon(\xi))$. By (115), $g_k' > 0$ for $\xi > \lambda_{k+}$, $g_k' < 0$ for $\xi < \lambda_{k-}$, and thus g_k looks like a potential-well function (see Fig. 2). Let $\rho_{k\varepsilon}$ be a point where g_k attains its global minimum. If we set

$$g_k(\xi) = g_k[U_\varepsilon](\xi) := \int_{\rho_{k\varepsilon}}^{\xi} s - \lambda_k(U_\varepsilon(s))\,ds \tag{117}$$

then $\lambda_{k-} \le \rho_{k\varepsilon} = \lambda_k(U_\varepsilon(\rho_{k\varepsilon})) \le \lambda_{k+}$, $g_k(\xi) \ge g_k(\rho_{k\varepsilon}) = 0$, and $g_k(\xi) = O(|\xi|^2)$ as $|\xi| \to \infty$.

Consider the linearization of the system (114), consisting of the decoupled system of equations

$$\varepsilon \varphi_k' + (\xi - \lambda_k)\varphi_k = 0\,. \tag{118}$$

The solutions of (118) are constant multiples of

$$\begin{aligned}\varphi_k &= \frac{\exp\{-\frac{1}{\varepsilon}\int_{\rho_{k\varepsilon}}^{\xi} s - \lambda_k(U_\varepsilon(s))\,ds\}}{\int_{-\infty}^{\infty}\exp\{-\frac{1}{\varepsilon}\int_{\rho_{k\varepsilon}}^{\zeta} s - \lambda_k(U_\varepsilon(s))\,ds\}\,d\zeta} \\ &= \frac{e^{-\frac{1}{\varepsilon}g_k}}{\int_{-\infty}^{\infty} e^{-\frac{1}{\varepsilon}g_k}\,d\zeta} = \frac{1}{I_k}e^{-\frac{1}{\varepsilon}g_k}\,.\end{aligned} \tag{119}$$

The functions $\{\varphi_{k\varepsilon}\}$ are strictly positive and uniformly bounded in L^1, independently of ε. Due to (115) and Lemma 6, $\varphi_{k\varepsilon}$ satisfy, in the case $d_k = \lambda_{k+} - \lambda_{k-} > 0$, the estimates

$$0 < \varphi_{k\varepsilon}(\xi) \le \begin{cases} O(1)\,\frac{d_k}{\varepsilon}\,e^{-\frac{1}{2\varepsilon}(\xi-\lambda_{k-})^2} & \xi < \lambda_{k-}, \\ O(1)\,\frac{d_k}{\varepsilon} & \xi \in \mathbb{R}, \\ O(1)\,\frac{d_k}{\varepsilon}\,e^{-\frac{1}{2\varepsilon}(\xi-\lambda_{k+})^2} & \xi > \lambda_{k+}, \end{cases} \tag{120}$$

and can be evaluated as in Lemma 6 (ii), in case $d_k = 0$.

The functions φ_k serve as a yardstick to estimate the amplitudes a_k, as follows. First, one constructs solutions of (114) that admit the representation

$$a_k = \tau_k\,\varphi_k + \theta_k(\cdot\,;\tau), \tag{121}$$

where $\tau = (\tau_1, \tau_2, \dots, \tau_N)$ is a vector-parameter in $\mathbb{R}^N$ and $\theta_k(\xi;\tau)$ is of second order in τ in the sense that, for some constant C independent of ε, it satisfies the estimate

$$|\theta_k(\cdot\,;\tau)| \le C\,|\tau|^2 \sum_{j=1}^{N} \varphi_j. \tag{122}$$

The decomposition (121)-(122) can be thought as an asymptotic expansion of the amplitudes a_k in a parameter τ representing the strength of elementary waves. Since "most" of the ε-dependence is carried by the φ_k's, the expansion is uniform in ε in the L^1-norm.

The key step in validating the expansion (121)-(122) concerns the pointwise behavior of the integrals

$$F_{k,mn} = e^{-\frac{1}{\varepsilon}g_k}\int_{c_k}^{\xi} e^{\frac{1}{\varepsilon}g_k}\,\varphi_m\,\varphi_n\,d\zeta\,,$$

which express the contributions on the k-th family effected by interactions between elementary waves of the m-th and n-th families. As $\varepsilon \to 0$, the terms $F_{k,mn}$ behave as follows [68, Lemmas 4.3, 4.4]: $F_{k,mk}$, $F_{k,km}$ and $F_{k,kk}$ have non-zero limiting contributions supported on the k-th wave speed. $F_{k,mn} \to 0$ as $\varepsilon \to 0$ when $m \neq n$, $m \neq k$ and $n \neq k$, which suggests that diffusion induced interactions of two distinct families have no contribution as $\varepsilon \to 0$ on a third family. (Recall that we are dealing with Riemann data solutions.) By contrast, the terms $F_{k,mm}$, $m \neq k$, accounting for the effect of two interacting waves of the m-th family on the k-th family, have a non-zero contribution in the $\varepsilon \to 0$ limit which is supported on the m-th wave speed.

The second problem is to connect the parameters τ with the data U_-, U_+ in order to fulfill (112). To this end, for U_- fixed, one considers a map S_ε that takes τ in a neighborhood of $0 \in \mathbb{R}^N$ to the vector

$$S_\varepsilon(\tau) = U_- + \sum_{k=1}^{N} \int_{-\infty}^{+\infty} \left[\tau_k \, \varphi_{k\varepsilon}(\zeta) + \theta_{k\varepsilon}(\zeta;\, \tau)\right] r_k(U_\varepsilon(\zeta))\, d\zeta\,. \tag{123}$$

It is shown that S is locally invertible in a neighborhood of $\tau = 0$ and that the inverse map S_ε^{-1} is uniformly bounded, independently of ε, for ε small.

Finally, the formulation (111-112) suggests a construction scheme for proving existence of solutions U_ε of $(\mathcal{P}_\varepsilon)$ in weighted spaces, so that the constructed solutions satisfy the asymptotic expansion (121)-(122). We refer to [68] for details and state the final result:

Theorem 10. *Assume* (81) *is strictly hyperbolic and let U_- be fixed. There exists a (sufficiently small) r such that, for $\varepsilon > 0$ and for any U_+ satisfying $|U_+ - U_-| \leq r$, the problem $(\mathcal{P}_\varepsilon)$ admits a solution U_ε with the following properties:*
(i) The family $\{U_\varepsilon\}_{\varepsilon>0}$ satisfies (C_o) with some μ independent of ε.
(ii) The solutions U_ε satisfy the representation formula

$$U_\varepsilon' = \sum_{k=1}^{N} \left[\tau_{k,\varepsilon}\varphi_{k\varepsilon} + \theta_{k\varepsilon}(\cdot;\tau_\varepsilon)\right] r_k(U_\varepsilon), \tag{124}$$

where $\varphi_{k\varepsilon}$ is given by (119), *$a_{k\varepsilon}(\cdot;\tau)$ of the form* (121) *satisfy* (122), *and τ_ε solves $S(\tau_\varepsilon) = U_+$.*
(iii) The family $\{U_\varepsilon'\}_{\varepsilon>0}$ is uniformly bounded in $L^1(\mathbb{R})$ and $\{U_\varepsilon\}_{\varepsilon>0}$ is of uniformly bounded (and small) total variation.

We conclude by indicating the proof of the variation bounds from the representation formula (124). Let $\{U_\varepsilon\}_{\varepsilon>0}$ be a family of solutions to $(\mathcal{P}_\varepsilon)$, of uniformly bounded oscillation (C_o) and satisfying (124). By the construction process,

$$U_\varepsilon'(\xi) = \sum_k a_{k\varepsilon}(\xi;\tau_\varepsilon) r_k(U_\varepsilon(\xi))$$

$$
\begin{aligned}
& a_{k\varepsilon}(\cdot;\tau_\varepsilon) \ \text{ satisfies the asymptotic expansion (121)-(122)} \\
& S_\varepsilon(\tau_\varepsilon) = U_+ \quad \text{and there exists } C \text{ such that } |\tau_\varepsilon| \le C|U_+ - U_-|
\end{aligned}
$$

Using (121)-(122),

$$
|a_{k\varepsilon}(\xi;\tau_\varepsilon)| \le |\tau_{k,\varepsilon}|\varphi_{k\varepsilon} + C|\tau_\varepsilon|^2 \sum_j \varphi_{j\varepsilon} \le C\,|U_+ - U_-|\Big(\varphi_{k\varepsilon} + \sum_j \varphi_{j\varepsilon}\Big)
$$

and thus

$$
|U_\varepsilon'(\xi)| \le K \sum_{j=1}^N \varphi_{j\varepsilon}. \tag{125}
$$

where the constant K is of order $O(|U_+ - U_-|)$ and independent of ε. As $\{\varphi_{j\varepsilon}\}$ are uniformly bounded in $L^1(\mathbb{R})$, we deduce $\{U_\varepsilon'\}$ is uniformly bounded in $L^1(\mathbb{R})$.

4.5 The relation with the problem of viscosity limits

It is interesting to see how the problem of self-similar viscosity limits relates to viscosity approximations for Riemann data solutions. For the system of viscous conservation laws

$$
\partial_t U + \partial_x F(U) = \varepsilon\, \partial_x^2 U \tag{126}
$$

subject to Riemann data, the invariance under dilations $(x,t) \mapsto (\alpha x, \alpha t)$, $\alpha > 0$, no longer holds. Due to uniqueness results for parabolic systems, the solution U^ε of (126)-(83) can be expressed as

$$
U^\varepsilon(x,t) = V\Big(\frac{x}{t}, -\frac{\varepsilon}{t}\Big) \tag{127}
$$

where $V(\xi,s)$ is independent of ε and satisfies

$$
V_s - V_{\xi\xi} = \frac{1}{s}\big(-\xi V_\xi + F(V)_\xi\big) \tag{128}
$$

for $-\infty < \xi < \infty$, $-\infty < s < 0$. We see that the zero-viscosity limits problem for Riemann data is a two parameter problem and that studying the limit of U^ε as $\varepsilon \downarrow 0$ amounts to studying the limit of $V(\xi,s)$ as $s \uparrow 0-$. The problem $(\mathcal{P}_\varepsilon)$ arises when replacing the parabolic operator in (128) by an elliptic operator and solving on the collapsed domain $\xi \in \mathbb{R}$.

Bibliographic remarks. Elliptic regularizations of the Riemann problem operator appear in [26,70,11]. Tupciev [70] uses $(\mathcal{P}_\varepsilon)$ as a starting point to motivate that admissible shocks should have an associated viscous shock profile. Dafermos [11] proposed this regularization as a devise to select the admissible solutions of the Riemann problem. The procedure is carried out in [11,12,16,59] for strictly hyperbolic 2×2 systems, and in [68] for weak waves of $N \times N$ systems. These studies concern the case $B(U) = Id$. As the equations of continuum thermomechanics involve singular diffusion matrices, there are investigations of the systems of isothermal elasticity [67] and isentropic gas dynamics [32] with singular diffusion matrices. A comparison of self-similar viscosity limits with viscosity approximations is carried out in [58] for Burgers's equation. Self-similar viscosity limits serve as a tool for investigating wave admissibility in situations involving loss of strict hyperbolicity, or when "exotic" phenomena are at play. There are a number of such investigations concerning: large shocks or even delta shock waves [30,31,62,19], mixed hyperbolic-elliptic systems [57,21,22], Riemann type solutions for fully nonlinear systems [55], and fluid dynamic limits for the Broadwell model [60,66]. We point out that self-similar limits provide a notion of solution and an existence theory for the Riemann problem in the class of non-conservative, strictly-hyperbolic systems [37,38].

5 Relaxation Approximations of Hyperbolic Conservation Laws

The presence of relaxation mechanisms is widespread in both the continuum mechanics as well as the kinetic theory contexts. Relaxation provides a subtle "dissipative" mechanism against the destabilizing effect of nonlinear response, as well as a damping effect on oscillations (at least when assisted by nonlinear response). The objective of this section is to bring up these properties, by examining the zero-relaxation limit in two examples, concerning respectively a single conservation law in several space dimensions and a system of two conservation laws in one space dimension.

5.1 The structure of relaxation approximations

We begin with an outline of the general structure of relaxation approximations. For $\varepsilon > 0$, a system of semilinear hyperbolic equations,

$$\partial_t U + \sum_{i=1}^{d} A_i \partial_{x_i} U = \frac{1}{\varepsilon} R(U) , \tag{129}$$

governs the dynamics of a function $U = U(x,t)$, $x \in \mathbb{R}^d$, $t > 0$. The state variable U takes values in $\mathbb{R}^N$ and will be called the mesoscopic variable. The matrices A_i are assumed constant $N \times N$ matrices such that (129) is hyperbolic. (All examples considered in this section are semilinear systems. Relaxation of quasilinear systems is also of interest for applications, but we will not pursue it here).

It is assumed that (129) is equipped with m conservation laws, *i.e.* there are linearly independent vectors $q_j \in \mathbb{R}^N$, $j = 1, \dots, m$, such that the variables $u_j = q_j \cdot U$ satisfy the conservation laws

$$\partial_t \big(q_j \cdot U \big) + \sum_{i=1}^{d} \partial_{x_i} \big(q_j \cdot A_i U \big) = 0 \,. \tag{130}$$

The variables u_i are called macroscopic variables, and $u = (u_1, \dots, u_m)$ stands for the vector of all the macroscopic variables.

For the system of ordinary differential equations

$$U_t = \frac{1}{\varepsilon} R(U) \tag{131}$$

it is assumed : (i) It is equipped with m conservation laws for the variables $u_j = q_j \cdot U$, that is $q_j \cdot R(U) = 0$, $j = 1, \dots, m$. (ii) There is an m-dimensional manifold of equilibria $\mathcal{M}$ parametrized by the m macroscopic variables u_j, the set of equilibria $\mathcal{M}$ is described in the form $U = \mathcal{E}(u)$. (iii) The flow of the system of ordinary differential equations (131) is attracted to $\mathcal{M}$. This is a minimum set of hypotheses. Additional hypotheses are imposed in the examples.

Finally, we assume that the system (129) is equipped with an entropy function $\Psi(U)$, *i.e.* there is a multiplier Ψ_U such that $\Psi_U \cdot R(U) \leq 0$ and

$$\partial_t \Psi(U) + \sum_{i=1}^{d} \partial_{x_i} \Phi_i(U) = \frac{1}{\varepsilon} \Psi_U \cdot R(U) \leq 0 \,. \tag{132}$$

This structure is common in several systems of relaxation type, whose origin is in the kinetic theory of gases [4], and is induced by the second law of thermodynamics for models with internal variables originating in the continuum physics context.

Under these hypotheses, it is conceivable that $u_\varepsilon = q \cdot U_\varepsilon \to u$ as $\varepsilon \to 0$, where u is a solution of the system of m conservation laws

$$u_t + \sum_{i=1}^{d} \partial_{x_i} \big(q \cdot A_i \mathcal{E}(u) \big) = 0 \tag{133}$$

In the sequel, we discuss two examples concerning convergence of relaxation systems to a scalar conservation law in several space dimensions ($m = 1$ with $d = n$) and to the system of isothermal elastodynamics in one space dimension ($m = 2$ with $d = 1$).

5.2 The scalar multi-dimensional conservation law via relaxation

It is a classical result that the Cauchy problem for the scalar conservation law,

$$\begin{cases} \partial_t u + \sum_{i=1}^{n} \partial_{x_i} F_i(u) = 0, \quad x \in \mathbb{R}^n, t > 0, \\ u(x,0) = u_0(x), \end{cases} \tag{134}$$

with $u_0 \in L^1(\mathbb{R}^n) \cap L^\infty(\mathbb{R}^n)$ admits a unique global weak solution $u(x,t)$ satisfying the Kruzhkov entropy conditions [33],

$$\partial_t |u-k| + \sum_{i=1}^{n} \partial_{x_i} \left[(F_i(u) - F_i(k)) \text{sign}\,(u-k) \right] \leq 0, \text{ in } \mathcal{D}', \text{ for all } k \in \mathbb{R}. \tag{135}$$

Entropy weak solutions of (134) are constructed as viscosity limits for parabolic regularizations, [71], [33], or as the small mean-free-path limit for kinetic equations, [51]. Here, we review the construction of entropy solutions for (134) via relaxation approximations, [28].

This problem is decomposed in two steps. First, we consider the semilinear hyperbolic system of relaxation type,

$$\begin{aligned} \partial_t w + U_0 \cdot \nabla w &= \frac{1}{\varepsilon} \sum_{j=1}^{n} (z_j - h_j(w)) \\ \partial_t z_i + U_i \cdot \nabla z_i &= -\tfrac{1}{\varepsilon}(z_i - h_i(w)), \quad i = 1, 2, ..., n. \end{aligned} \tag{136}$$

The system governs the dynamics of the state vector (w, Z), $Z = (z_1, ..., z_n)$, $U_0, U_1, \ldots, U_n$ are given convective velocities, and $h_i(w)$ are strictly increasing smooth functions with $h_i(0) = 0$. It may be interpreted as a discrete velocity system with an unconventional collision operator, or as a model in chemical kinetics. Solutions of (136) satisfy the conservation law

$$\partial_t (w + \sum_i z_i) + \text{div}\big(U_0 w + \sum_i U_i z_i\big) = 0\,. \tag{137}$$

As $\varepsilon \to 0$, the local equilibria $z_i = h_i(w)$ are enforced and the limiting dynamics is described by the conservation law

$$\partial_t (w + \sum_i h_i(w)) + \text{div}\big(U_0 w + \sum_i U_i h_i(w)\big) = 0\,. \tag{138}$$

This convergence is justified provided that $h_i(w)$ are strictly increasing.

The question arises under what circumstances a given conservation law (134) can be realized as a relaxation limit. In view of the convergence of (136) to (138), the question becomes whether (134) can be mapped into the form

(138). Suppose first that the velocities $U_1, \dots U_n$ are in the coordinate directions, $U_i = V_i \hat{e}_i$, $i = 1, \dots, n$, and that U_0 is expressed as $U_0 = \sum_i \omega_i V_i \hat{e}_i$. Then mapping (134) into (138) leads to the algebraic problem: Given a curve $(u, F_1(u), \dots, F_n(u))$, is it possible to find w and strictly increasing functions $h_i(w)$ such that

$$w + \sum_{i=1}^{n} h_i(w) = u\,, \qquad \omega_i w + h_i(w) = \frac{1}{V_i} F_i(u)\,. \tag{139}$$

Solving (139) explicitly, we see that this happens if and only if the following multidimensional analog of the subcharacteristic condition,

$$1 + \sum_i \frac{1}{V_i}\frac{dF_i}{du} > 0\,, \qquad \frac{1}{V_i}\frac{dF_i}{du} < \frac{\omega_i}{1 + \sum_i \omega_i}\Big(1 + \sum_i \frac{1}{V_i}\frac{dF_i}{du}\Big)\,, \tag{140}$$

is satisfied. Clearly, (140) holds if $\omega_i > 0$ and the speeds V_i are selected sufficiently large. The general problem, when $U_1, \dots, U_n$ are linearly independent, can be transformed into the above special case, by performing a linear transformation of coordinates.

Theorem 11. *Let ω_i, V_i be such that (140) is satisfied, let $U_i = V_i \hat{e}_i$, $i = 1, \dots, n$ and $U_0 = \sum_i \omega_i V_i \hat{e}_i$. Suppose that the initial data $w_{0\varepsilon}, z_{i0\varepsilon}$ lie in a bounded set of $BV \cap L^\infty(\mathbb{R}^n)$ and are tight in $L^1(\mathbb{R}^n)$. If $\|u_{0\varepsilon} - u_0\|_{L^1} = o(1)$, then there exists a function*

$$u \in L^\infty([0,T]; BV \cap L^\infty(\mathbb{R}^n)) \cap Lip([0,T]; L^1(\mathbb{R}^n))$$

such that

$$u_\varepsilon = w_\varepsilon + \sum_{i=1}^{n} z_{i\varepsilon} \to u \quad in \quad L^1(\mathbb{R}^n \times \mathbb{R}^+)\,, \tag{141}$$

and u is a weak solution of (134) satisfying the Kruzhkov entropy conditions (135).

The main step is to show that, for any $U_0, \dots, U_n$ and for h_i strictly increasing, solutions of (136) converge to an entropy solution of (138). We present an outline of this proof and refer to [28] for the details as well as for further properties of rates of convergence, and convergence to measure-valued solutions under weaker hypotheses on the data.

First, if (w, z_i) and $(\bar{w}, \bar{z}_i)$ are two solutions of (136), then they satisfy

$$\begin{aligned}
&\partial_t \Big(|w - \bar{w}| + \sum_i |z_i - \bar{z}_i| \Big) + \operatorname{div}\Big(U_0 |w - \bar{w}| + \sum_i U_i |z_i - \bar{z}_i| \Big) \\
&\quad = \frac{1}{\varepsilon} \sum_i \Big(\text{sign}\,(w - \bar{w}) - \text{sign}\,(z_i - \bar{z}_i) \Big) \Big((z_i - \bar{z}_i) - (h_i(w) - h_i(\bar{w})) \Big) \\
&\quad \le 0
\end{aligned} \tag{142}$$

In particular, if $(\bar{w}, \bar{z}_i)$ is the equilibrium solution $(\kappa, h_i(\kappa))$ then (142) yields the inequalities

$$\partial_t\Big(|w-\kappa| + \sum_i |z_i - h_i(\kappa)|\Big) \tag{143}$$
$$+ \operatorname{div}\Big(U_0|w-\kappa| + \sum_i U_i|z_i - h_i(\kappa)|\Big) \le 0 \quad \text{for } \kappa \in \mathbb{R},$$

which is a version of the Kruzhkov entropy inequalities for the relaxation system, and turn out to provide the Kruzhkov entropy conditions for the conservation law (138) in the limit $\varepsilon \to 0$.

Using (142) and the conservation law (137) as key ingredients, we have the following theorem.

Theorem 12. *Let h_i be strictly increasing. If $w_0, z_{i0} \in L^1 \cap L^\infty(\mathbb{R}^n)$ then there exists a unique globally defined weak solution (w, Z) of (136), which satisfies:*
(i) If (w, Z) and $(\hat{w}, \hat{Z})$ are two solutions then

$$\int |w(x,t) - \hat{w}(x,t)| + \sum_{i=1}^n |z_i(x,t) - \hat{z}_i(x,t)|\, dx \le \int |w_0 - \hat{w}_0| + \sum_{i=1}^n |z_{i0} - \hat{z}_{i0}|\, dx\,.$$

(ii) For any $a < b$ the sets $\mathcal{R}_{a,b} := [a,b] \times \prod_{i=1}^n \big[h_i(a), h_i(b)\big]$ are positively invariant.
(iii) If w_0, $z_{i0} \in BV(\mathbb{R}^n)$ then $w(\cdot,t)$, $z_i(\cdot,t) \in BV(\mathbb{R}^n)$.

For h_i strictly increasing, the relaxation system (136) is equipped with a globally defined entropy function

$$\partial_t\Big(\frac{1}{2}w^2 + \sum_{i=1}^n \Psi_i(z_i)\Big) \tag{144}$$
$$+ \operatorname{div}\Big(U_0 \frac{1}{2} w^2 + \sum_{i=1}^n U_i \Psi_i(z_i)\Big) + \frac{1}{\varepsilon}\sum_{i=1}^n \phi_i(w, z_i) = 0\,,$$

where

$$\Psi_i(z_i) = \int_0^{z_i} h_i^{-1}(\xi)\, d\xi\,,$$

is positive and strictly convex, while

$$\phi_i(w, z_i) = (w - h_i^{-1}(z_i))(h_i(w) - z_i)$$

satisfies $\phi_i \ge 0$ and $\phi_i = 0$ if and only if $(w, Z) \in \mathcal{M}$. The "dissipation" estimate (144) provides control of the distance of solutions from the equilibrium curve $\mathcal{M}$. Since $\phi_i \ge c(h_i(w) - z_i)^2$, it leads to

$$\int_0^\infty \int_{\mathbb{R}^n} (h_i(w) - z_i)^2\, dx dt \le C\varepsilon\,. \tag{145}$$

Let now (w, z_i) be a solution of (136) and $(\kappa, h_i(\kappa))$, $\kappa \in \mathbb{R}$, be an equilibrium. Then, (143) gives

$$\partial_t\Big(|w-\kappa| + \sum_i |h_i(w) - h_i(\kappa)|\Big) + \operatorname{div}\Big(U_0|w-\kappa| + \sum_i U_i|h_i(w) - h_i(\kappa)|\Big)$$
$$\leq \partial_t \sum_i g_i + \operatorname{div} \sum_i U_i g_i \tag{146}$$

where g_i can be estimated in terms of the distance of each solution from the Maxwellian values,

$$|g_i| = \Big||h_i(w) - h_i(\kappa)| - |z_i - h_i(\kappa)|\Big| \leq |h_i(w) - z_i| \,. \tag{147}$$

If we set

$$u = w + \sum_i h_i(w) \,, \quad k = \kappa + \sum_i h_i(\kappa) \tag{148}$$

we see that $u > k$ if and only if $w > \kappa$. Letting

$$F(u) = U_0 w + \sum_i U_i h_i(w) \,, \quad F(k) = U_0 \kappa + \sum_i U_i h_i(\kappa) \,, \tag{149}$$

be the fluxes of (138), it follows that the left hand side of (146) is written

$$\partial_t |u-k| + \operatorname{div}\Big((F(u) - F(k))\text{sign }(u-k)\Big) \leq \partial_t \sum_i g_i + \operatorname{div} \sum_i U_i g_i \,. \tag{150}$$

The formula indicates that u in (148) is an approximate solution of (138).

To complete the proof, consider a family of solutions $(w^\varepsilon, z_i^\varepsilon)$ of the relaxation system. The L^1 contraction property together with the conservation law (137) enables us to deduce precompactness of $w^\varepsilon + \sum_i z_i^\varepsilon$ in $L^1(\mathbb{R}^n \times [0,T])$. There exists a subsequence w^{ε_n} and $z_i^{\varepsilon_n}$ and a function u such that

$$u^{\varepsilon_n} = w^{\varepsilon_n} + \sum_i z_i^{\varepsilon_n} \to u \,, \quad \text{a.e. } (x,t).$$

Since $g_i^{\varepsilon_n} \to 0$ and the functions h_i are strictly increasing, it follows that w^{ε_n} also converges a.e. to some function w, with $u = w + \sum_i h_i(w)$. Passing to the limit $\varepsilon \to 0$ in (150) we deduce that the limiting u is an entropy solution of (138).

5.3 A relaxation limit to the equations of isothermal elastodynamics

In this section, we address the problem of constructing weak solutions of the equations of isothermal elasticity with $g_u > 0$,

$$\begin{aligned} \partial_t u - \partial_x v &= 0 \\ \partial_t v - \partial_x g(u) &= 0 \,, \end{aligned} \tag{151}$$

as $\varepsilon \to 0$ limits of the relaxation system

$$\begin{aligned} \partial_t u - \partial_x v &= 0 \\ \partial_t v - \partial_x \sigma &= 0 \\ \partial_t(\sigma - Eu) &= -\frac{1}{\varepsilon}(\sigma - g(u)) . \end{aligned} \tag{152}$$

The model (152) is suggested as an approximating model for the equations of isothermal elastodynamics in [20].

We work under the standing hypotheses $g(0) = 0$ and $0 < g_u < E$, in which case (152) admits globally defined smooth solutions, if the initial data are smooth. The hypothesis $g_u < E$ can be motivated in two ways. On the one hand, it guarantees that the internal variable theory described by (152) is consistent with the Clausius-Duhem inequality (see Sec. 2.5). On the other hand, it can be motivated by the analog of the Chapman-Enskog expansion for the relaxation process.

In the Chapman-Enskog expansion one seeks to identify the effective response of the relaxation process as it approaches the surface of local equilibria. It is postulated that the relaxing variable σ^ε can be described in an asymptotic expansion that involves *only* the local macroscopic values u^ε, v^ε and their derivatives, *i.e.*

$$\sigma^\varepsilon = g(u^\varepsilon) + \varepsilon S(u^\varepsilon, v^\varepsilon, u^\varepsilon_x, v^\varepsilon_x, \dots) + O(\varepsilon^2) \tag{153}$$

To calculate the form of S, we use (152),

$$\begin{aligned} \partial_t u^\varepsilon - \partial_x v^\varepsilon &= 0 \\ \partial_t v^\varepsilon - \partial_x g(u^\varepsilon) &= \varepsilon S_x + O(\varepsilon^2) \\ \partial_t(g(u^\varepsilon) - Eu^\varepsilon) + O(\varepsilon) &= -S + O(\varepsilon) , \end{aligned} \tag{154}$$

whence we obtain

$$S = [E - g_u(u^\varepsilon)] v^\varepsilon_x + O(\varepsilon) , \tag{155}$$

and we conclude that the effective equations describing the process are

$$\begin{aligned} &\partial_t u^\varepsilon - \partial_x v^\varepsilon = 0 \\ &\partial_t v^\varepsilon - \partial_x g(u^\varepsilon) = \varepsilon \partial_x \big([E - g_u(u^\varepsilon)] v^\varepsilon_x \big) + O(\varepsilon^2) . \end{aligned} \tag{156}$$

This is a stable parabolic system provided the condition $g_u < E$ is satisfied.

According to Section 2.5, when $g_u < E$ the system (152) describes a theory with internal variables that is consistent with the second law of thermodynamics. Smooth solutions (u, v, σ) satisfy the energy dissipation identity

$$\partial_t \Big(\frac{1}{2} v^2 + \Psi(u, \sigma - Eu) \Big) - \partial_x(\sigma v) + \frac{1}{\varepsilon}(u - h^{-1}(\alpha))(\alpha - h(u)) \Big|_{\alpha = \sigma - Eu} = 0 \tag{157}$$

where

$$\Psi(u,\alpha) = -\int_0^\alpha h^{-1}(\zeta)d\zeta + \alpha u + \int_0^u E\xi d\xi \tag{158}$$

$h(u) = g(u) - Eu$ and h^{-1} is the inverse function of h. The function Ψ provides an "entropy" function for the associated relaxation process, which is convex in (u,α) if $-\partial_\alpha h^{-1}\partial_u f \geq 1$ for all u and α.

Henceforth, we assume that the initial data (u_0, v_0, σ_0) are smooth (of compact support or decaying fast at infinity) and the function $g(u) \in C^3$ satisfies

$$(h) \qquad 0 < \gamma \leq g_u(u) \leq \Gamma < E\,.$$

for some positive constants γ and Γ. It is easy to check that (152) admits global smooth solutions, and we proceed to study the $\varepsilon \to 0$ relaxation process. Equation (157) provides stability in L^2 for the relaxation process.

Lemma 13. *Under hypothesis* (h),

$$\begin{aligned}\int_{\mathbb{R}} (u^2+v^2+\sigma^2)dx + \frac{1}{\varepsilon C}\int_0^t\int_{\mathbb{R}} \big(\sigma - g(u)\big)^2 dxdt \\ \leq C\int_{\mathbb{R}} (u_0^2+v_0^2+\sigma_0^2)dx\end{aligned} \tag{159}$$

for some C *independent of* ε *and* t.

Proof. From (158) we have

$$\begin{aligned}\Psi(u,\sigma - Eu) &= -\int_0^{\sigma-Eu} h^{-1}(\zeta)d\zeta + \frac{\sigma^2}{2E} - \frac{1}{2E}(\sigma-Eu)^2 \\ &= \int_0^{\sigma-Eu} \kappa(\zeta)d\zeta + \frac{\sigma^2}{2E}\end{aligned} \tag{160}$$

where $\kappa(\alpha) = -\frac{1}{E}\alpha - h^{-1}(\alpha)$. Hypothesis (h) implies

$$\frac{\gamma}{E(E-\gamma)} \leq \frac{d\kappa}{d\alpha} = \frac{g_u}{E(E-g_u)} \leq \frac{\Gamma}{E(E-\Gamma)}$$

and thus there is a constant C, depending only on γ, Γ and E, so that

$$\frac{1}{C}\big((\sigma-Eu)^2+\sigma^2\big) \leq \Psi(u,\sigma-Eu) \leq C\big((\sigma-Eu)^2+\sigma^2\big) \tag{161}$$

Furthermore, since $-\frac{d}{d\alpha}h^{-1}(\alpha) = \frac{1}{E-g_u} \geq \frac{1}{E}$, we have

$$(u-h^{-1}(\alpha))(\alpha - h(u)) \geq \frac{1}{E}(\alpha-h(u))^2 \tag{162}$$

The result now follows from (157), upon using (161) and (162). □

We proceed with some estimations that capture the dissipative structure of the relaxation process. In preparation, note that solutions of (152) satisfy

$$\begin{aligned} &\partial_t u - \partial_x v = 0 \\ &\partial_t v - \partial_x g(u) = \partial_x(\sigma - g(u)) = \varepsilon(E v_{xx} - v_{tt}) \end{aligned} \tag{163}$$

The problem under consideration is thus an approximation of (151) via the wave equation.

Lemma 14. *Suppose that the initial data satisfy*

$$(a) \qquad \begin{aligned} &\int_{\mathbb{R}} v_0^2 + u_0^2 + \sigma_0^2 dx \leq O(1)\,, \\ \varepsilon^2 &\int_{\mathbb{R}} u_{0x}^2 + v_{0x}^2 + \sigma_{0x}^2 dx \leq O(1)\,. \end{aligned}$$

Under hypothesis (h), solutions (u, v, σ) of (152) satisfy the ε independent estimates

$$\varepsilon \int_0^t \int_{\mathbb{R}} u_x^2 + v_x^2 + \sigma_x^2 dx dt \leq O(1)\,. \tag{164}$$

Proof. We multiply $(163)_1$ by $g(u)$ and $(163)_2$ by v. Adding and rearranging the terms we obtain the energy identity

$$\partial_t\big(\frac{1}{2}v^2 + W(u) + \varepsilon v v_t\big) - \partial_x\big(v g(u)\big) + \varepsilon(E v_x^2 - v_t^2) = \varepsilon \partial_x(E v v_x)\,, \tag{165}$$

where the stored energy function $W(u)$ is given by

$$W(u) = \int_0^u g(\xi) d\xi\,. \tag{166}$$

The problem is that the term $E v_x^2 - v_t^2$ is not positive definite. To compensate for that, we first multiply $(163)_2$ by v_t to obtain

$$v_t^2 - g_u u_x v_t = \varepsilon\Big[(E v_t v_x)_x - \partial_t\big(\frac{1}{2}E v_x^2 + \frac{1}{2}v_t^2\big)\Big]$$

and, in turn

$$\varepsilon^2 \partial_t(E v_x^2 + v_t^2) + \varepsilon(2v_t^2 - 2g_u u_x v_t) = 2\varepsilon^2 \partial_x(E v_t v_x)\,. \tag{167}$$

Using once again $(163)_2$ and the identity $a_x b_t - a_t b_x = \partial_t(a_x b) - \partial_x(a_t b)$, we have

$$\begin{aligned} g_u u_x^2 &= u_x \partial_t(v + \varepsilon v_t) - \varepsilon E u_x v_{xx} \\ &= \Big[u_t \partial_x(v + \varepsilon v_t) + \partial_t\big(u_x(v + \varepsilon v_t)\big) - \partial_x\big(u_t(v + \varepsilon v_t)\big)\Big] - \varepsilon \partial_t\big(\frac{1}{2}E u_x^2\big)\,, \end{aligned}$$

which in turn yields

$$\varepsilon^2\partial_t\big(\frac{1}{2}E^2u_x^2 - \frac{1}{2}Ev_x^2\big) - \varepsilon\partial_t\big(Eu_x(v+\varepsilon v_t)\big) \\ + \varepsilon(Eg_uu_x^2 - Ev_x^2) = -\varepsilon\partial_x\big(Eu_t(v+\varepsilon v_t)\big)\,. \tag{168}$$

Adding (165), (167) and (168), we arrive at

$$\partial_t\Big(\tfrac{1}{2}(v+\varepsilon v_t - \varepsilon E u_x)^2 + \tfrac{1}{2}\varepsilon^2(v_t^2 + Ev_x^2) + W(u)\Big) - \partial_x(vg(u)) \\ + \varepsilon\big[v_t^2 - 2g_uu_xv_t + Eg_uu_x^2\big] = \varepsilon^2(Ev_tv_x)_x. \tag{169}$$

Because of (h) the third term in (169) is positive definite

$$\varepsilon\big[v_t^2 - 2g_uu_xv_t + Eg_uu_x^2\big] \geq \varepsilon g_u(E-g_u)u_x^2 \geq 0\,. \tag{170}$$

Therefore, we conclude

$$\int_{\mathbb{R}} \frac{1}{2}(v+\varepsilon v_t - \varepsilon Eu_x)^2 + \frac{1}{2}\varepsilon^2(v_t^2 + Ev_x^2) + W(u)dx \\ + \varepsilon\int_0^t\int_{\mathbb{R}} g_u(E-g_u)u_x^2 dxdt \\ \leq \int_{\mathbb{R}} \frac{1}{2}(v_0 + \varepsilon\sigma_{0x} - \varepsilon Eu_{0x})^2 + \frac{1}{2}\varepsilon^2(\sigma_{0x}^2 + Ev_{0x}^2) + W(u_0)dx \leq O(1) \tag{171}$$

and, due to (h) and (a),

$$\varepsilon\int_0^t\int_{\mathbb{R}} g_u(E-g_u)u_x^2 dxdt \leq O(1).$$

In turn, (167) and (165) imply

$$\varepsilon\int_0^t\int_{\mathbb{R}} \sigma_x^2 dxdt \leq O(1)$$
$$\varepsilon\int_0^t\int_{\mathbb{R}} v_x^2 dxdt \leq O(1)$$

and (164) follows. □

We come next to the convergence Theorem.

Theorem 15. *Let $g(u)$ be a smooth function satisfying (h) such that g_{uu} vanishes at exactly one point. If $(u^\varepsilon, v^\varepsilon, \sigma^\varepsilon)$ is a family of smooth solutions of (152) that are uniformly stable in L^∞,*

$$|u^\varepsilon| + |v^\varepsilon| + |\sigma^\varepsilon| \leq C\,, \tag{H}$$

and emanate from initial data satisfying the uniform bounds (a), then, along a subsequence if necessary,

$$u^\varepsilon \to u\,, \quad v^\varepsilon \to v\,, \quad a.e.\ (x,t)\,. \tag{172}$$

If in addition $u_0^\varepsilon \to u_0$, $v_0^\varepsilon \to v_0$ *a.e* x, *and*

$$(b) \qquad \varepsilon^2 \int_{\mathbb{R}} u_{0x}^2 + v_{0x}^2 + \sigma_{0x}^2 dx = o(1)\,, \quad as\ \varepsilon \to 0,$$

then (u,v) *is a weak solution of (151) and*

$$\partial_t\big(\frac{1}{2}v^2 + W(u)\big) - \partial_x\big(g(u)v\big) \le 0\,, \quad in\ \mathcal{D}'. \tag{173}$$

The hypothesis of uniform L^∞ stability is artificial. The convergence in the natural framework of L^2 stability is pursued in [69]. It is worth noting that while Hypothesis (a) is sufficient to establish (172), Hypothesis (b) is necessary to justify the energy dissipation (173) relative to the initial data (u_0, v_0).

Proof. Let $\eta(u,v)$, $q(u,v)$ be an entropy pair for the equations of isothermal elasticity. Using (163) we obtain

$$\begin{aligned} \partial_t\eta(u^\varepsilon,v^\varepsilon) + \partial_x q(u^\varepsilon,v^\varepsilon) &= \eta_v\partial_x(\sigma - g(u)) \\ &= \partial_x\big(\eta_v(\sigma-g(u))\big) - (\eta_{vu}\varepsilon^{\frac{1}{2}}u_x + \eta_{vv}\varepsilon^{\frac{1}{2}}v_x)\frac{\sigma-g(u)}{\varepsilon^{\frac{1}{2}}} \\ &= I_1 + I_2. \end{aligned} \tag{174}$$

In view of (159), (164) and (H), the term I_1 lies in a compact of H^{-1}, the term I_2 is uniformly bounded in L^1, and the sum I_1+I_2 is uniformly bounded in $W^{-1,\infty}$. One concludes from a lemma of Murat [48] that $I_1 + I_2$ lies in a compact of H^{-1}_{loc}. Then from a theorem of DiPerna [17] we obtain, along a subsequence, $u^\varepsilon \to u$ and $v^\varepsilon \to v$ a.e. (x,t).

It remains to prove (173). Let φ be a positive test function with compact support in $[0,T)\times\mathbb{R}$. From (169) we have

$$\begin{aligned} &-\int_0^T\int_{\mathbb{R}} \varphi_t\big[\frac{1}{2}(v^\varepsilon + \varepsilon v_t^\varepsilon - \varepsilon E u_x^\varepsilon)^2 + \frac{1}{2}\varepsilon^2({v_t^\varepsilon}^2 + E{v_x^\varepsilon}^2) + W(u^\varepsilon)\big]\, dxdt \\ &\quad + \int_0^T\int_{\mathbb{R}} \varphi_x(v^\varepsilon g(u^\varepsilon))\, dxdt + \varepsilon\int_0^T\int_{\mathbb{R}} \varphi\big[{v_t^\varepsilon}^2 - 2g_u u_x^\varepsilon v_t^\varepsilon + E g_u {u_x^\varepsilon}^2\big]\, dxdt \\ &\quad - \int_{\mathbb{R}} \varphi(x,0)\big[\frac{1}{2}(v_0^\varepsilon + \varepsilon\sigma_{0x}^\varepsilon - \varepsilon E u_{0x}^\varepsilon)^2 + \frac{1}{2}\varepsilon^2({\sigma_{0x}^\varepsilon}^2 + E{v_{0x}^\varepsilon}^2) + W(u_0^\varepsilon)\big]\, dx \\ &= -\varepsilon^2\int_0^T\int_{\mathbb{R}} \varphi_x(E v_t^\varepsilon v_x^\varepsilon)\, dxdt\,. \end{aligned} \tag{175}$$

We use $u_{0\varepsilon} \to u_0$, $v_{0\varepsilon} \to v_0$ a.e, Hypotheses (a) and (b) for the data, together with (164) and (170), to conclude

$$- \int_0^T \int_{\mathbb{R}} \varphi_t \big[\frac{1}{2}v^2 + W(u)\big] - \varphi_x v g(u) \, dxdt \\ \leq \int_{\mathbb{R}} \varphi(x,0) \big[\frac{1}{2}v_0^2 + W(u_0)\big] \, dx \,. \tag{176}$$

The convergence of (152) to (151) follows from a similar argument, passing to the limit in (163) and using (159). □

Bibliographic remarks. Weak solutions of the scalar multidimensional conservation law can be constructed as zero-viscosity limits of parabolic regularizations, Volpert [71], Kruzhkov [33], via fluid-dynamic limits for BGK models, Perthame-Tadmor [51], or via relaxation approximations, Katsoulakis-Tzavaras [27,28], Natalini [50]. There are two equivalent notions of solution, the Kruzhkov entropy solution [33], and the kinetic formulation of Lions-Perthame-Tadmor [40]; the solution operator defines a contraction semigroup in L^1, [10].

The importance of the Chapman-Enskog expansion and the subcharacteristic condition was recognized in early studies of relaxation phenomena, see Liu [45]. The Hilbert expansion is very useful for studying relaxation to smooth solutions and initial layers, see Caflisch-Papanicolaou [2], Yong [72]. A general framework for investigating relaxation to processes containing shocks is proposed in Chen-Levermore-Liu [4], and the mechanism is exploited in Jin-Xin [25] to construct a class of nonoscillatory numerical schemes. There are a number of studies establishing convergence to scalar conservation laws in one-space dimension, [4], [49], [65], and relaxation can be used as an intermediate step to establish convergence of stochastic interacting particle systems to scalar equations [29]. Concerning relaxation limits to systems, we refer to Coquel-Perthame [9], Brenier-Corrias-Natalini [1], and Tzavaras [69] (from where the material of Sections 2.4, 2.5 and 5.3 is taken).

References

1. Brenier Y., Corrias L. Natalini R., Relaxation limits for a class of balance laws with kinetic formulation, (1997) (preprint).
2. Caflisch R.E and Papanicolaou G.C., The fluid dynamic limit of a nonlinear model of the Boltzmann equation, *Comm. Pure Appl. Math.* **32** (1979), 589-616.
3. Chen G.-Q., Propagation and cancellation of oscillations for hyperbolic systems of conservation laws, *Comm. Pure Appl. Math.* **44** (1991), 121-139.
4. Chen G.-Q,, Levermore C.D. and Liu T.-P., Hyperbolic conservation laws with stiff relaxation terms and entropy, *Comm. Pure Appl. Math.* **47** (1994), 789-830.
5. Coleman B.D. and Noll W., The thermodynamics of elastic materials with heat conduction and viscosity, *Arch. Rational Mech. Anal.* **13** (1963), 167-178.

6. Coleman B.D., Thermodynamics of materials with memory, *Arch. Rational Mech. Anal.* **17** (1964), 1-46.
7. Coleman B.D. and Mizel V.J., Existence of caloric equations of state in thermodynamics, *J. Chem. Phys.* **40** (1964), 1116-1125.
8. Coleman B.D. and Gurtin M.E., Thermodynamics with internal state variables, *J. Chem. Physics* **47** (1967), 597-613.
9. Coquel F. and Perthame B., Relaxation of energy and approximate Riemann solvers for general pressure laws in fluid dynamics, *SIAM Num. Anal.* (to appear).
10. Crandall M., The semigroup approach for first order quasilinear equations in several space variables, *Israel J. Math.*, **12** (1972), 108-132.
11. Dafermos C.M., Solution of the Riemann problem for a class of hyperbolic conservation laws by the viscosity method, *Arch. Rational Mech. Analysis* **52** (1973), 1-9.
12. Dafermos C.M., Structure of solutions of the Riemann problem for hyperbolic systems of conservations laws, *Arch. Rational Mech. Analysis* **53** (1974), 203-217.
13. Dafermos, C.M., Global smooth solutions to the initial-boundary value problem for the equations of one-dimensional nonlinear thermoviscoelasticity. *SIAM J. Math. Anal.* **13** (1982), 397-408.
14. Dafermos C.M., Contemporary issues in the dynamic behavior of continuous media, Lecture Notes, Brown University, 1985.
15. Dafermos C.M., Admissible wave fans in nonlinear hyperbolic systems, *Arch. Rational Mech. Analysis* **106** (1989), 243-260.
16. Dafermos C.M. and DiPerna R.J., The Riemann problem for certain classes of hyperbolic systems of conservation laws, *J. Diff. Equations* **20** (1976), 90-114.
17. DiPerna R., Convergence of approximate solutions to conservation laws, *Arch. Rational Mech. Analysis* **60** (1983), 75-100.
18. DiPerna R., Measure-valued solutions to conservation laws, *Arch. Rational Mech. Analysis* **88** (1985), 223-270.
19. Ercole G., Delta-shock waves as self-similar viscosity limits, (1997) (preprint).
20. Faciu C. and Mihailescu-Suliciu M., The energy in one-dimensional rate-type semilinear viscoelasticity, *Int. J. Solids Structures* **23** (1987), 1505-1520.
21. Fan H.-T., A limiting "viscosity" approach to the Riemann problem for materials exhibiting change of phase (II), *Arch. Rational Mech. Analysis* **116** (1992), 317-338.
22. Fan H.-T., One-phase Riemann problem and wave interactions in systems of conservation laws of mixed type *SIAM J. Math. Anal.* **24** (1993), 840-865.
23. Gurtin M.E., Williams W.O. and Suliciu I., On rate type constitutive equations and the energy of viscoelastic and viscoplastic materials, *Int. J. Solids Structures* **16** (1980), 607-617.
24. Heibig A., Existence and uniqueness of solutions for some hyperbolic systems of conservation laws. *Arch. Rational Mech. Anal.* **126** (1994), 79-101.
25. Jin S. and Xin Z., The relaxing schemes for systems of conservation laws in arbitrary space dimensions, *Comm. Pure Appl. Math.* **48** (1995), 235-277.

26. Kalasnikov A.S., Construction of generalized solutions of quasi-linear equations of first order without convexity conditions as limits of solutions of parabolic equations with a small parameter, *Dokl. Akad. Nauk SSSR* **127** (1959), 27-30 (in Russian).
27. Katsoulakis M.A. and Tzavaras A.E., Contractive relaxation systems and interacting particles for scalar conservation laws, *C. R. Acad. Sci. Paris, Sér. I Math.* **323** (1996), 865-870.
28. Katsoulakis M.A. and Tzavaras A.E., Contractive relaxation systems and the scalar multidimensional conservation law, *Comm. Partial Differential Equations* **22** (1997), 195-233.
29. Katsoulakis M.A. and Tzavaras A.E., Multiscale analysis of interacting particles: Relaxation schemes and scalar conservation laws, (1998) (submitted)
30. Keyfitz B. and Kranzer H., A viscosity approximation to a system of conservation laws with no classical Riemann solution, in " Proceedings of International Conference on Hyperbolic Problems", Bordeaux, 1988.
31. Keyfitz B. and Kranzer H., Spaces of weighted measures for conservation laws with singular shock solutions, *J. Differential Equations* **118** (1995), 420–451.
32. Kim Y.-J., A self-similar viscosity approach for the Riemann problem in isentropic gas dynamics and the structure of its solutions, (1998) (preprint).
33. Kruzhkov S.N., First order quasilinear equations with several independent variables, *Math. USSR Sbornik* **10** (1970), 217-243.
34. Lax P.D., Hyperbolic systems of conservation laws II, *Comm. Pure Appl. Math.* **10** (1957), 537-566.
35. Lax P.D., Shock waves and entropy, in: "Contributions to Nonlinear Functional Analysis." E.H. Zarantonello, ed. New York: Academic Press, 1971, pp. 603-634.
36. LeFloch P.G. and Tzavaras A.E., Existence theory for the Riemann problem for non-conservative hyperbolic systems, *C.R. Acad. Sci., Paris, Série I* **323** (1996), 347-352.
37. LeFloch P.G. and Tzavaras A.E., Representation of weak limits and definition of nonconservative products, (1997) (preprint).
38. LeFloch P.G. and Tzavaras A.E., (in preparation).
39. Lin P., Young measures and an application of compensated compactness to one-dimensional nonlinear elastodynamics. *Trans. Amer. Math. Soc.* **329** (1992), 377-413.
40. Lions P.L., Perthame B. and Tadmor E., A kinetic formulation of scalar multidimensional conservation laws, *J. AMS* **7** (1994), 169-191.
41. Lions P.L., Perthame B. and Tadmor E., Kinetic formulation of the isentropic gas dynamics and p-systems, *Comm. Math. Physics* **163** (1994), 415-431.
42. Lions P.L., Perthame B. and Souganidis P.E., Existence and stability of entropy solutions for the hyperbolic systems of isentropic gas dynamics in Eulerian and Lagrangian coordinates, *Comm. Pure Appl. Math.* **49** (1996), 599-638.
43. Liu T.-P., The Riemann problem for general 2×2 conservation laws, *Trans. Amer. Math. Society* **199** (1974), 89-112.
44. Liu T.-P., The Riemann problem for general systems of conservation laws, *J. Diff. Equations* **18** (1975), 218-234.

45. Liu T.-P., Hyperbolic conservation laws with relaxation, *Comm. Math. Phys.* **108** (1987), 153-175.
46. Majda A. and Pego R.L., Stable viscosity matrices for systems of conservation laws, *J. Diff. Equations* **56** (1985), 229-262.
47. Murat F., Compacité par compensation. *Ann. Scuola Norm. Sup. Pisa Cl. Sci.* **5** (1978), 489-507.
48. Murat F., L'injection du cone positif de H^{-1} dans $W^{-1,q}$ est compacte pour tout $q < 2$. *J. Math. Pures Appl.* **60** (1981), 309-322.
49. Natalini R., Convergence to equilibrium for the relaxation approximations of conservation laws, *Comm. Pure Appl. Math.* **49** (1996), 795-823.
50. Natalini R., A discrete kinetic approximation of entropy solutions to multidimensional scalar conservation laws, J. Diff. Equations (to appear).
51. Perthame B. and Tadmor E., A kinetic equation with kinetic entropy functions for scalar conservation laws, *Comm. Math. Phys.* **136** (1991), 501-517.
52. Perthame B. and Tzavaras A.E., (in preparation).
53. Serre D., La compacitè par compensation pour les systèmes hyperboliques non linéaires de deux équations a une dimension d' espace. *J. Maths. Pures et Appl.* **65** (1986), 423-468.
54. Serre D. and Shearer J., Convergence with physical viscosity for nonlinear elasticity, (1993) (preprint).
55. Shearer, M. and Schaeffer, D.G., Fully nonlinear hyperbolic systems of partial differential equations related to plasticity. *Comm. Partial Differential Equations* **20** (1995), 1133-1153.
56. Shearer J.W., Global existence and compactness in L^p for the quasi-linear wave equation. *Comm. Partial Differential Equations* **19** (1994), 1829-1877.
57. Slemrod M., A limiting "viscosity" approach to the Riemann problem for materials exhibiting change of phase, *Arch. Rational Mech. Analysis* **105** (1989), 327-365.
58. Slemrod M., A comparison of two viscous regularizations of the Riemann problem for Burgers's equation, *SIAM J. Math. Analysis* **26** (1995), 1415-1424.
59. Slemrod M. and Tzavaras A.E., A limiting viscosity approach for the Riemann problem in isentropic gas dynamics, *Indiana Univ. Math. J.* **38** (1989), 1047-1074.
60. Slemrod M. and Tzavaras A.E., Self-similar fluid-dynamic limits for the Broadwell system, *Arch. Rational Mech. Anal.* **122** (1993), 353-392.
61. Suliciu I., On the thermodynamics of rate-type fluids and phase transitions. I-Rate-type fluids and II-Phase transitions, (1997) (preprint).
62. Tan D., Zhang T. and Zheng Y., Delta-shock waves as limits of vanishing viscosity for hyperbolic systems of conservation laws. *J. Differential Equations* **112** (1994), 1-32.
63. Tartar L., Compensated compactness and applications to partial differential equations, In *Nonlinear Analysis and Mechanics, Heriot Watt Symposium, Vol. IV*, R.J. Knops, ed., Pitman Research Notes in Math., New York, 1979, pp. 136–192.
64. Truesdell C.A. and Noll W., *The Nonlinear Field Theories of Mechanics*, Handbuch der Physik III/3, Springer-Verlag, Berlin, 1965.

65. Tveito A. and Winther R., On the rate of convergence to equilibrium for a system of conservation laws with a relaxation term, *SIAM J. Math. Anal.* **28** (1997), 136-161.
66. Tzavaras A.E., Wave structure induced by fluid dynamic limits in the Broadwell model, *Arch. Rational Mech. Anal.* **127** (1994), 361-387.
67. Tzavaras A.E., Elastic as limit of viscoelastic response, in a context of self-similar viscous limits, *J. Diff. Equations*, **123** (1995), 305-341.
68. Tzavaras A.E., Wave interactions and variation estimates for self-similar zero-viscosity limits in systems of conservation laws, *Arch. Rational Mech. Anal.* **135** (1996), 1-60.
69. Tzavaras A.E., Materials with internal variables and relaxation to conservation laws, Arch. Rational Mech. Anal. (to appear).
70. Tupciev V.A., On the method of introducing viscosity in the study of problems involving the decay of a discontinuity, *Dokl. Akad. Nauk SSSR* **211** (1973), 55-58. English translation: *Soviet Math. Dokl.* **14** (1973), 978-982.
71. Volpert A.I., The space BV and quasilinear equations *Math. Sbornik* **73(115)** (2) (1967), 225–267.
72. Yong W.A., Singular perturbations of first-order hyperbolic systems, in "Nonlinear Hyperbolic Problems: Theoretical, Applied and Computational Aspects", Notes on Numerical Fluid Mechanics, Vol. 43, Vieweg, Braunschweig 1993; also Ph.D. Thesis, Univ. of heidelberg, 1992.

A Posteriori Error Analysis and Adaptivity for Finite Element Approximations of Hyperbolic Problems

Endre Süli

Oxford University Computing Laboratory,
Wolfson Building, Parks Road, Oxford OX1 3QD, U.K.

Abstract. The aim of this lecture series is to present an overview of recent developments in the area of *a posteriori* error estimation for finite element approximations of hyperbolic problems. The approach pursued here rests on the systematic use of hyperbolic duality arguments. We also discuss the question of computational implementation of the *a posteriori* error bounds into adaptive finite element algorithms.

Contents

1 Introduction

The numerical solution of hyperbolic conservation laws is of fundamental importance in several areas of applied science, particularly fluid dynamics and electromagnetics. Solutions to these partial differential equations frequently exhibit localised structures, such as propagating discontinuities and sharp transition layers whose reliable numerical approximation presents a challenging computational task. Indeed, in order to resolve such localised phenomena in an accurate and efficient way one has to use locally refined computational meshes. In computational fluid dynamics, at least, the traditional approach to the construction of such locally adapted meshes resorts to *ad hoc* criteria, usually justified on physical grounds, whose impact on the accuracy of the numerical solution is difficult to assess.

In contrast with such heuristic approaches, in these notes we shall be concerned with the question of quantitative error control for hyperbolic partial differential equations with the aim to achieve reliability, either in the sense that the numerical solution approximates the analytical solution in a given norm to within a given tolerance, or in the sense that physically relevant derived quantities, which can be thought of as functionals of the solution, are approximated to within a given tolerance. Reliability in the latter sense is particularly important in engineering applications; e.g. in fluid dynamics one may be concerned with calculating the lift and the drag coefficients of a body immersed into a viscous fluid whose flow is governed by the Navier-Stokes equations. The lift and drag coefficients are defined as integrals, over the boundary of the body, of the stress tensor components normal and tangential to the flow, respectively. Similarly, in elasticity theory, the quantities of prime interest, such as the stress intensity factor, or the moments of a shell or plate, are derived quantities. To achieve reliability in one sense or the other, we shall derive computable *a posteriori* error bounds in terms of the finite element residual which is obtained by inserting the computed solution into the partial differential equation under consideration. Such bounds represent the key ingredients of reliable adaptive finite element algorithms for hyperbolic problems: error control to within a given tolerance is achieved though a feed-back process where the *a posteriori* error bound plays the rôle of a stopping criterion.

The aim of the present paper is to discuss the construction and the practical implementation of *a posteriori* error bounds for finite element approximations of first-order hyperbolic equations. The derivation of the bounds rests on hyperbolic duality arguments; these will be exploited in a systematic manner throughout. In fact, following the paradigm of *a posteriori* error estimation outlined in [30] by Johnson, our analysis has two basic ingredients: the application of Galerkin orthogonality and the use of the strong stability of the dual (adjoint) problem; the rôle of these concepts in the error analysis will be highlighted below.

Over the last decade the *a posteriori* error analysis of finite element methods for partial differential equations has been an area of active research (see Ainsworth and Oden [2] Szabó and Babuška [60], and Verfürth [61]). Unfortunately, much of the interest has focussed on elliptic and parabolic equations and relatively little progress has been made on the *a posteriori* error analysis of finite element and finite volume approximations to hyperbolic and nearly-hyperbolic problems. For an overview of current activities in the latter area we refer to the articles [30] and [15]; see also, [29], [31], and references therein. The approach to *a posteriori* error estimation for hyperbolic problems pursued in those papers, particularly in the work of Johnson and Szepessy [32], and reviewed at the beginning of Section 4, rests on performing an elliptic or parabolic regularisation of the hyperbolic problem and exploiting the smoothing properties of the resulting adjoint problem in conjunction with Galerkin orthogonality to derive an *a posteriori* error bound in the L_2 norm; in Section 4 we present an alternative, more direct, process which avoids the need for regularisation of the hyperbolic operator, at the price of arriving at error bounds in weaker (negative Sobolev) norms. In the course of our analysis we shall require some basic results from the theory of function spaces; these are summarised in the next section, followed, in Section 3, by a brief overview of the theory of well-posedness of steady linear hyperbolic equations. Section 5 is devoted to the problem of error generation, error propagation and local error estimation in the context of *a posteriori* error analysis. We have already noted that *a posteriori* error estimation of linear functionals is of great practical importance; this topic is the subject of Section 6. Section 7 concerns the *a posteriori* error analysis of finite element approximations to unsteady hyperbolic problems; we shall also comment on the implementation of our error bounds into an adaptive algorithm. The final section discusses the theory for scalar nonlinear hyperbolic conservation laws; here we rely on the work of Tadmor [59] concerning the strong stability of the linearised dual problem (a backward linear transport equation with discontinuous coefficients) associated with the conservation law, in Lipschitz spaces. Thus we arrive at an *a posteriori* error bound in the dual Lipschitz (Lip') norm.

Acknowledgements: I wish to express my am gratitude to Bernardo Cockburn, Mike Giles, Claes Johnson, John Mackenzie, Rolf Rannacher, Thomas Sonar and Gerald Warnecke for helpful discussions on various aspects of *a posteriori* error analysis and adaptivity. I am particularly indebted to my colleague Paul Houston for performing the numerical experiments which appear in this paper.

2 Basic function spaces

In this section, we recall the definitions of some familiar function spaces, including those of continuously differentiable and Lebesgue integrable functions, and Sobolev spaces. For proofs and further details, we refer the reader to the monographs [1], [35] and [47].

2.1 Spaces of continuous functions

Let $\mathbb{N}$ denote the set of non-negative integers. An n-tuple $\alpha = (\alpha_1, \ldots, \alpha_n)$ in $\mathbb{N}^n$ is called a *multi-index*. The non-negative integer $|\alpha| = |\alpha_1| + \ldots + |\alpha_n|$ is called the length of α. We define $\partial^\alpha = \partial_1^{\alpha_1} \ldots \partial_n^{\alpha_n}$ where $\partial_j = \partial/\partial x_j$ for $j = 1, \ldots, n$.

Let Ω be an open set in $\mathbb{R}^n$. For $k \in \mathbb{N}$, we denote by $C^k(\Omega)$ the set of all continuous real-valued functions u, defined on Ω, such that $\partial^\alpha u$ is continuous on Ω for every multi-index α, $|\alpha| \leq k$. Further, we define $C^\infty(\Omega)$ as the intersection $\bigcap_{k \geq 0} C^k(\Omega)$. The notation $C^0(\Omega)$ is abbreviated to $C(\Omega)$.

For $k \in \mathbb{N}$, we denote by $C^k(\bar{\Omega})$ the set of all $u \in C^k(\Omega)$ such that $\partial^\alpha u$ can be continuously extended from Ω onto $\bar{\Omega}$, for every multi-index α, $|\alpha| \leq k$. Further, we define $C^\infty(\bar{\Omega})$ as the intersection $\bigcap_{k \geq 0} C^k(\bar{\Omega})$. The notation $C^0(\bar{\Omega})$ is abbreviated to $C(\bar{\Omega})$.

Assuming that Ω is a bounded open set in $\mathbb{R}^n$ and $k \in \mathbb{N}$, the linear space $C^k(\bar{\Omega})$ is a Banach space equipped with the norm

$$\|u\|_{C^k(\bar{\Omega})} = \max_{|\alpha| \leq k} \sup_{x \in \Omega} |\partial^\alpha u(x)|.$$

For $k \in \mathbb{N}$ and $0 < \lambda \leq 1$, we denote by $C^{k,\lambda}(\bar{\Omega})$ the set of all $u \in C^k(\bar{\Omega})$ such that the quantity

$$|u|_{C^{k,\lambda}(\bar{\Omega})} = \max_{|\alpha| = k} \sup_{x \neq y,\ x,y \in \Omega} \frac{|\partial^\alpha u(x) - \partial^\alpha u(y)|}{|x - y|^\lambda}$$

is finite. $C^{k,\lambda}(\bar{\Omega})$ is a Banach space with the norm

$$\|u\|_{C^{k,\lambda}(\bar{\Omega})} = \|u\|_{C^k(\bar{\Omega})} + |u|_{C^{k,\lambda}(\bar{\Omega})}.$$

When u belongs to $C^{0,1}(\bar{\Omega})$, it is said to be *Lipschitz continuous* on $\bar{\Omega}$.

The *support*, $\operatorname{supp} u$, of a continuous function u defined on an open set Ω is the closure in Ω of the set $\{x \in \Omega \ : \ u(x) \neq 0\}$; in other words, $\operatorname{supp} u$ is the smallest closed subset of Ω such that $u = 0$ on $\Omega \setminus \operatorname{supp} u$. For $k = 0, 1, \ldots, \infty$, $C_0^k(\Omega)$ denotes the set of all $u \in C^k(\Omega)$ whose support is a compact subset of Ω.

2.2 Spaces of integrable functions

For $p \geq 1$ and an open set $\Omega \subset \mathbb{R}^n$, let $L_p(\Omega)$ denote the set of all real-valued Lebesgue measurable functions u defined on Ω such that $|u|^p$ is integrable on Ω with respect to the Lebesgue measure $\mathrm{d}x = \mathrm{d}x_1 \ldots \mathrm{d}x_n$; we assume here that any two functions which are equal almost everywhere (i.e. equal, except maybe on a set of measure zero) are identified. $L_p(\Omega)$ is a Banach space with norm

$$\|u\|_{L_p(\Omega)} = \left(\int_\Omega |u(x)|^p \, \mathrm{d}x \right)^{1/p}.$$

In particular, for $p = 2$, $L_2(\Omega)$ is a Hilbert space with the inner product

$$(u, v) = \int_\Omega u(x) \, v(x) \, \mathrm{d}x.$$

$L_\infty(\Omega)$ denotes the set of all real-valued Lebesgue measurable functions u defined on Ω such that $|u|$ has finite essential supremum; the essential supremum of $|u|$ is defined as the infimum of the set of all positive real numbers M such that $|u| \leq M$ almost everywhere on Ω. Again, any two functions that are equal almost everywhere on Ω are identified. $L_\infty(\Omega)$ is a Banach space with norm

$$\|u\|_{L_\infty(\Omega)} = \text{ess.sup}_{x\in\Omega}|u(x)|.$$

Hölder's Inequality. Let $u \in L_p(\Omega)$ and $v \in L_q(\Omega)$, where $1/p + 1/q = 1$, $1 \leq p, q \leq \infty$. Then $uv \in L_1(\Omega)$ and

$$\left| \int_\Omega u(x) \, v(x) \, \mathrm{d}x \right| \leq \|u\|_{L_p(\Omega)} \|v\|_{L_q(\Omega)}.$$

For $p = q = 2$, this is referred to as the *Cauchy-Schwarz Inequality.*

2.3 Sobolev spaces

Suppose that Ω is an open set in $\mathbb{R}^n$. For a non-negative integer k and $1 \leq p \leq \infty$, we define

$$W_p^k(\Omega) = \{u \in L_p(\Omega) \, : \, \partial^\alpha u \in L_p(\Omega), \quad |\alpha| \leq k\}.$$

We equip $W_p^k(\Omega)$ with the Sobolev norm defined by

$$\|u\|_{W_p^k(\Omega)} = \left(\sum_{|\alpha|\leq k} \|\partial^\alpha u\|^p_{L_p(\Omega)} \right)^{1/p}$$

when $1 \leq p < \infty$, and by

$$\|u\|_{W_\infty^k(\Omega)} = \max_{|\alpha|\leq k} \|\partial^\alpha u\|_{L_\infty(\Omega)}$$

when $p = \infty$. The associated Sobolev seminorm is defined by

$$|u|_{W_p^k(\Omega)} = \left(\sum_{|\alpha|= k} \|\partial^\alpha u\|^p_{L_p(\Omega)} \right)^{1/p}$$

when $1 \le p < \infty$, and

$$|u|_{W^k_\infty(\Omega)} = \max_{|\alpha|=k} \|\partial^\alpha u\|_{L_\infty(\Omega)}$$

when $p = \infty$. In these definitions the derivatives are to be understood in the sense of distributions.

The Sobolev space $W^k_p(\Omega)$ can be shown to be a Banach space with the norm $\|\cdot\|_{W^k_p(\Omega)}$, $1 \le p \le \infty$, $k \ge 0$. A particularly important case occurs when $p = 2$; the normed linear space $W^k_2(\Omega)$ is a Hilbert space with the inner product

$$(u, v)_{W^k_2(\Omega)} = \sum_{|\alpha|\le k} (\partial^\alpha u, \partial^\alpha v),$$

where $(\cdot,\cdot)$ is the inner product in $L_2(\Omega)$.

In order to capture finer smoothness properties of integrable functions, we consider fractional-order Sobolev spaces defined in the following way: given that s is a positive real number, $s \notin \mathbb{N}$, let us write $s = m + \sigma$, where $0 < \sigma < 1$ and $m = [s]$ is the integer part of s. The fractional-order Sobolev space $W^s_p(\Omega)$, $1 \le p < \infty$, is the set of all $u \in W^m_p(\Omega)$ such that

$$|u|_{W^s_p(\Omega)} = \left\{ \sum_{|\alpha|=m} \int_\Omega \int_\Omega \frac{|D^\alpha u(x) - D^\alpha u(y)|^p}{|x-y|^{n+\sigma p}} \, dx \, dy \right\}^{1/p} < \infty,$$

with the usual modification when $p = \infty$. When equipped with the norm

$$\|u\|_{W^s_p(\Omega)} = \left\{ \|u\|^p_{W^m_p(\Omega)} + |u|^p_{W^s_p(\Omega)} \right\}^{1/p}, \qquad \text{if } 1 \le p < \infty,$$

or the norm

$$\|u\|_{W^s_\infty(\Omega)} = \|u\|_{W^m_\infty(\Omega)} + |u|_{W^s_\infty(\Omega)}, \qquad \text{if } p = \infty,$$

the Sobolev space $W^s_p(\Omega)$ is a Banach space.

When a boundary-value problem is considered for a partial differential equation on an open set Ω, it is convenient to incorporate the boundary condition on $\partial\Omega$, the boundary of Ω, into the definition of the function space in which a solution is sought. First, we characterise the smoothness of $\partial\Omega$.

Definition 1. Suppose that Ω is an open set in $\mathbb{R}^n$. The boundary $\partial\Omega$ of Ω is said to be Lipschitz continuous if, for every $x \in \partial\Omega$, there is an open set $\mathcal{O} \subset \mathbb{R}^n$ with $x \in \mathcal{O}$ and a local orthogonal coordinate system with coordinate $\zeta = (\zeta_1, \dots, \zeta_n) \equiv (\zeta', \zeta_n)$ and $a \in \mathbb{R}^n$, such that

$$\mathcal{O} = \{\zeta : -a_j < \zeta_j < a_j, \quad 1 \le j \le n\},$$

and there is a Lipschitz continuous function φ defined on

$$\mathcal{O}' = \{\zeta' \in \mathbb{R}^{n-1} : -a_j < \zeta_j < a_j, \quad 1 \le j \le n-1\},$$

with

$$|\varphi(\zeta')| \le a_n/2, \qquad \zeta' \in \mathcal{O}',$$

$$\Omega \cap \mathcal{O} = \{\zeta : \zeta_n < \varphi(\zeta'), \;\; \zeta' \in \mathcal{O}'\} \text{ and } \partial\Omega \cap \mathcal{O} = \{\zeta : \zeta_n = \varphi(\zeta'), \;\; \zeta' \in \mathcal{O}'\}.$$

A bounded open set with a Lipschitz continuous boundary is called a *Lipschitz domain.*

An important property of a Lipschitz domain Ω is that the unit outward normal to $\partial\Omega$ is defined almost everywhere with respect to the $(n-1)$-dimensional measure on $\partial\Omega$. A simple example of a Lipschitz domain is a bounded polyhedron in $\mathbb{R}^n$, $n \ge 2$.

Proposition 2. *Suppose that Ω is a Lipschitz domain contained in $\mathbb{R}^n$ and let $1 \le p < \infty$. Then $C^\infty(\bar{\Omega})$ is dense in $W_p^s(\Omega)$ for $s \ge 0$.*

We note that while $C^\infty(\bar{\Omega})$ is dense in $W_p^s(\Omega)$ for $s \ge 0$ and $1 \le p < \infty$, $C_0^\infty(\Omega)$ is not dense in $W_p^s(\Omega)$ for $s > 1/p$ (although it is dense in $W_p^s(\Omega)$, for $0 \le s < 1/p$, $1 \le p < \infty$).

We conclude this section with a brief discussion about Sobolev spaces on the boundary $\partial\Omega$ of a Lipschitz domain Ω. Let us begin by recalling from Definition 1 that for every x on $\partial\Omega$ there exists a Lipschitz continuous function $\varphi : \mathcal{O}' \subset \mathbb{R}^{n-1} \to \mathbb{R}$ such that, using the notation introduced in Definition 1,

$$\partial\Omega \cap \mathcal{O} = \{\zeta = (\zeta', \varphi(\zeta')) : \zeta' \in \mathcal{O}'\},$$

so that, locally, $\partial\Omega$ is an $(n-1)$-dimensional hypersurface in $\mathbb{R}^n$. We define the mapping ϕ by

$$\phi(\zeta') = (\zeta', \varphi(\zeta')).$$

Then ϕ^{-1} exists and it is Lipschitz continuous on $\phi(\mathcal{O}')$, which leads us to the following definition.

Definition 3. Let Ω be a Lipschitz domain in $\mathbb{R}^n$. For $0 \le s \le 1$ and $1 \le p < \infty$ we denote by $W_p^s(\partial\Omega)$ the set of all $u \in L_p(\partial\Omega)$ such that the composition $u \circ \phi$ belongs to $W_p^s(\mathcal{O}' \cap \phi^{-1}(\partial\Omega \cap \mathcal{O}))$ for all possible $\mathcal{O}'$ and φ satisfying the conditions of Definition 1, where $\phi(\zeta') = (\zeta', \varphi(\zeta'))$ for $\zeta' \in \mathcal{O}'$.

In order to equip $W_p^s(\partial\Omega)$ with a norm, we consider any *atlas* $(\mathcal{O}_j, \varphi_j)_{j=1}^J$ for $\partial\Omega$ such that $\mathcal{O}_j$ and φ_j, $j = 1, \dots, J$, satisfy the conditions of Definition 1. We define $\|\cdot\|_{W_p^s(\partial\Omega)}$ by

$$\|u\|_{W_p^s(\partial\Omega)} = \left(\sum_{j=1}^J \|u \circ \phi_j\|^p_{W_p^s(\mathcal{O}_j' \cap \phi_j^{-1}(\partial\Omega \cap \mathcal{O}_j))} \right)^{1/p},$$

where $\phi_j(\zeta') = (\zeta', \varphi_j(\zeta'))$ for $\zeta' \in \mathcal{O}_j'$, $j = 1, \ldots, J$.

In fact, for $0 < s < 1$ it can be shown that this is equivalent to the following norm

$$\|u\|_{W_p^s(\partial\Omega)} = \left(\int_{\partial\Omega} |u|^p \,\mathrm{d}\sigma + \int_{\partial\Omega} \int_{\partial\Omega} \frac{|u(x) - u(y)|^p}{|x - y|^{n-1+sp}} \,\mathrm{d}\sigma(x) \,\mathrm{d}\sigma(y) \right)^{1/p},$$

where $\mathrm{d}\sigma$ denotes the $(n-1)$-dimensional surface measure on $\partial\Omega$.

Finally, we recall the notion of trace of a function on the boundary $\partial\Omega$ of a Lipschitz domain $\Omega \subset \mathbb{R}^n$. If ψ belongs to $C^\infty(\bar{\Omega})$ then we put

$$\gamma_0(\psi) = \psi|_{\partial\Omega} \,. \tag{1}$$

The trace of a function u in $W_p^s(\Omega)$ is then defined by extending the operator γ_0 from the dense subspace $C^\infty(\bar{\Omega})$ to the whole of $W_p^s(\Omega)$.

Proposition 4. *Suppose that Ω is a Lipschitz domain in $\mathbb{R}^n$, and let $1 < p < \infty$. Assuming that $1/p < s \leq 1$, the mapping γ_0 defined on $C^\infty(\bar{\Omega})$ by (1) has a unique continuous extension to a linear operator, still denoted γ_0, from $W_p^s(\Omega)$ onto $W_p^{s-(1/p)}(\partial\Omega)$.*

We adopt the following notational convention: when $p = 2$ we shall write H^s in place of W_2^s to signify the fact that we are dealing with a Hilbert space. We define $H_0^1(\Omega)$ as the closure of $C_0^\infty(\Omega)$ in the norm of the Sobolev space $H^1(\Omega)$; when Ω is a Lipschitz domain, it can be shown that

$$H_0^1(\Omega) = \{u \in H^1(\Omega) \,:\, \gamma_0(u) = 0\}.$$

3 Steady hyperbolic problems

In this section we present a brief review of the theory of steady hyperbolic equations. In the first part of the section we focus on scalar hyperbolic equations, while the second part is concerned with symmetric hyperbolic systems.

3.1 Scalar hyperbolic equations

We consider the question of well-posedness of the hyperbolic boundary-value problem $\mathcal{P}(f)$:

$$\begin{aligned} \operatorname{div}(\mathbf{a}u) + c\,u = f \quad & \text{in } \Omega, \\ u = 0 \quad & \text{on } \partial_-\Omega, \end{aligned}$$

where Ω is a Lipschitz domain in $\mathbb{R}^n$, with *inflow boundary*

$$\partial_-\Omega = \{x \in \partial\Omega \,:\, \mathbf{a}(x) \cdot \nu(x) < 0\};$$

here $\nu(x)$ denotes the unit outward normal vector at $x \in \partial\Omega$ (whenever it is defined). The complement of $\partial_-\Omega$ with respect to $\partial\Omega$ will be denoted $\partial_+\Omega$ and will be referred to as the *outflow boundary*. For the sake of simplicity we shall suppose that $\Omega = (0,1)^n$, that $\mathbf{a} = (a_1, \ldots, a_n)$ is a real-valued continuously differentiable n-component vector function defined on $\bar{\Omega}$, c is a continuous real-valued function on $\bar{\Omega}$ and f is a real-valued square-integrable function on Ω.

We adopt the following additional hypothesis on $\mathbf{a}$.

Hypothesis 5. The components $a_1, \ldots, a_n$ of the vector field $\mathbf{a}$ belong to $C^1(\bar{\Omega})$ and are strictly positive functions on $\bar{\Omega}$.

This assumption ensures that $\partial_-\Omega$ is a non-characteristic hypersurface of dimension $(n-1)$ for the differential operator $v \mapsto \operatorname{div}(\mathbf{a}v) + cv$.

In order to set up the variational formulation of $\mathcal{P}(f)$, with $\mathbf{a}$ and c we associate the function space

$$H_-(\Omega) = \{v \in L_2(\Omega) \, : \, \operatorname{div}(\mathbf{a}v) + cv \in L_2(\Omega), \quad \gamma_\nu(\mathbf{a}v) = 0 \quad \text{on } \partial_-\Omega\}$$

in which the solution of the problem is sought. We note that the boundary condition is included into the definition of the space $H_-(\Omega)$; here $\gamma_\nu(\mathbf{a}v) = (\mathbf{a}v) \cdot \nu|_{\partial_-\Omega}$ signifies the normal trace of the vector field $\mathbf{a}v$ on $\partial_-\Omega$.

At this stage, the boundary condition should be understood formally: below we shall justify that the definition of $H_-(\Omega)$ is meaningful. To do so, we first recall from [20] (Chapter I, Theorem 2.5 and Corollary 2.8) that the normal trace operator, $\gamma_\nu(\cdot)$, is a continuous surjection of

$$H(\operatorname{div}, \Omega) = \{\mathbf{v} \in [L_2(\Omega)]^n \, : \, \operatorname{div} \mathbf{v} \in L_2(\Omega)\}$$

onto $H^{-1/2}(\partial\Omega)$, the latter being the dual space of the fractional-order Sobolev space $H^{1/2}(\partial\Omega) = W_2^{1/2}(\partial\Omega)$. Now, suppose that Γ is a connected relatively open subset of $\partial\Omega$ of positive $(n-1)$-dimensional measure (for our purposes, $\Gamma = \partial_-\Omega$). We denote by $H_0^1(\Gamma)$ the closure of $C_0^\infty(\Gamma)$ in the norm of the Sobolev space $H^1(\Gamma)$. Further, we define $H_{00}^{1/2}(\Gamma)$, using the K-method of function space interpolation (see Bergh and Löfström [8], for example), as the interpolation space 'halfway' between $L_2(\Gamma)$ and $H_0^1(\Gamma)$. Finally, we let $(H_{00}^{1/2}(\Gamma))'$ denote the dual space of $H_{00}^{1/2}(\Gamma)$. Since the trivial extension $\mathcal{E}_0$ is a continuous linear operator from $L_2(\Gamma)$ into $L_2(\partial\Omega)$ and from $H_0^1(\Gamma)$ into $H^1(\partial\Omega)$, we deduce by function space interpolation that it is also a continuous linear operator from $H_{00}^{1/2}(\Gamma)$ into $H^{1/2}(\partial\Omega)$. Thus, by applying the Transposition Theorem (see Theorem 4.1 in Baiocchi and Capelo [4]), we conclude that the transpose of the linear operator $\mathcal{E}_0 \, : \, H_{00}^{1/2}(\Gamma) \to H^{1/2}(\partial\Omega)$ is a continuous linear operator ${}^t\mathcal{E}_0$ from $(H^{1/2}(\partial\Omega))' = H^{-1/2}(\partial\Omega)$ into $(H_{00}^{1/2}(\Gamma))'$; ${}^t\mathcal{E}_0$ is called the *restriction* from $\partial\Omega$ to Γ.

Suppose that $v \in L_2(\Omega)$ and $\operatorname{div}(\mathbf{a}v) + cv \in L_2(\Omega)$. Then $\mathbf{a}v \in H(\operatorname{div}; \Omega)$, and it follows that $\gamma_\nu(\mathbf{a}v) \in H^{-1/2}(\partial\Omega)$. Hence the restriction of $\gamma_\nu(\mathbf{a}v)$ to

$\partial_-\Omega$ belongs to $(H_{00}^{1/2}(\partial_-\Omega))'$. The definition of $H_-(\Omega)$ is, therefore, meaningful; in fact, $H_-(\Omega)$ is a Hilbert space with the norm

$$\|v\|_{H_-(\Omega)} = (\|v\|^2_{L_2(\Omega)} + \|\mathcal{L}v\|^2_{L_2(\Omega)})^{1/2},$$

where $\mathcal{L} : H_-(\Omega) \to L_2(\Omega)$ denotes the linear operator defined by $\mathcal{L}v = \mathrm{div}(\mathbf{a}v) + c\,v$, $v \in H_-(\Omega)$.

With this notation, the boundary-value problem $\mathcal{P}(f)$, for $f \in L_2(\Omega)$, can be expressed as follows: find u in $H_-(\Omega)$ such that $\mathcal{L}u = f$. Alternatively, the variational formulation of $\mathcal{P}(f)$ is: find $u \in H_-(\Omega)$ satisfying

$$(\mathrm{div}(\mathbf{a}u) + c\,u, q) = (f, q) \qquad \forall q \in L_2(\Omega). \tag{2}$$

A solution of (2) can be thought of as a generalised solution of $\mathcal{P}(f)$, with the differential equation satisfied as an equality in $L_2(\Omega)$ and the boundary condition satisfied as an equality in $(H_{00}^{1/2}(\partial_-\Omega))'$.

Proposition 6. *Under Hypothesis 5 and given that $f \in L_2(\Omega)$ and $c \in C(\bar{\Omega})$, problem (2) has a unique solution u in $H_-(\Omega)$. In addition, the linear operator $\mathcal{L}$ is a continuous bijection of $H_-(\Omega)$ onto $L_2(\Omega)$ with a continuous inverse $\mathcal{L}^{-1} : L_2(\Omega) \to H_-(\Omega)$.*

Proof. The proof is based on Banach's Closed Range Theorem. The non-trivial step is to verify (4) below; to do so, let $C^1_-(\bar{\Omega})$ denote the set of all functions in $C^1(\bar{\Omega})$ which vanish on $\partial_-\Omega$. It is easily seen that, for an n-component real vector ξ,

$$\begin{aligned}(\mathrm{div}(\mathbf{a}v) + c\,v, \mathrm{e}^{-2\xi\cdot x}v) &= \left(c + \frac{1}{2}(\mathrm{div}\,\mathbf{a}) + \mathbf{a}\cdot\xi, |\mathrm{e}^{-\xi\cdot x}v|^2\right) \\ &\quad + \frac{1}{2}\int_{\partial_+\Omega} (\mathbf{a}\cdot\nu)|\mathrm{e}^{-\xi\cdot x}v|^2\,\mathrm{d}s \quad \forall v \in C^1_-(\overline{\Omega}).\end{aligned} \tag{3}$$

Let ξ be such that the constant

$$M_0 = \inf_{\Omega}\left(c + \frac{1}{2}(\mathrm{div}\,\mathbf{a}) + \mathbf{a}\cdot\xi\right)$$

is positive, and let M_1 and M_2 be two positive real numbers such that

$$M_1 \le \exp(-2\xi\cdot x) \le M_2 \qquad \forall x \in \bar{\Omega}.$$

Omitting the second term on the right-hand side of (3) and noting that $C^1_-(\bar{\Omega})$ is dense in $H_-(\Omega)$, it follows that

$$(\mathcal{L}v, \mathrm{e}^{-2\xi\cdot x}v) \ge M_0\|\mathrm{e}^{-\xi\cdot x}v\|^2_{L_2(\Omega)} \qquad \forall v \in H_-(\Omega),$$

and hence

$$\left(1+\left(\frac{M_2}{M_1 M_0}\right)^2\right)^{1/2} \|\mathcal{L}v\|_{L_2(\Omega)} \geq \|v\|_{H_-(\Omega)} \qquad \forall v \in H_-(\Omega). \qquad (4)$$

Inequality (4) implies that $\mathcal{L}$ is an injective operator from $H_-(\Omega)$ onto its range space $R(\mathcal{L})$, and that the inverse of $\mathcal{L}$ is continuous. Hence $\mathcal{L}$ is an isomorphism from $H_-(\Omega)$ onto $R(\mathcal{L})$; therefore $R(\mathcal{L})$ is a closed subspace of $L_2(\Omega)$. Exploiting the positivity assumption on the components of $\mathbf{a}$, it is easy to prove (using the method of characteristics, for example) that the transpose ${}^t\mathcal{L}$ of $\mathcal{L}$ has trivial kernel, i.e. $\mathrm{Ker}({}^t\mathcal{L}) = \{0\}$; the Closed Range Theorem then implies that $R(\mathcal{L}) = L_2(\Omega)$. Hence $\mathcal{L}$ is an isomorphism from $H_-(\Omega)$ onto $L_2(\Omega)$. In addition, (4) implies that

$$1 \leq \|\mathcal{L}^{-1}\|_{L_2 \to H_-} \leq \left(1+\left(\frac{M_2}{M_1 M_0}\right)^2\right)^{1/2}.$$

That completes the proof. □

We note that this existence and uniqueness result can be extended in several directions:

a) First, a theorem analogous to Proposition 6 holds for the boundary-value problem in the, so-called, *non-conservative form:*

$$\mathbf{a} \cdot \nabla u + c u = f \quad \text{in } \Omega, \qquad u = 0 \quad \text{on } \partial_- \Omega,$$

where $\mathbf{a}$, c, f and Ω are as in Proposition 6.

b) Second, a result analogous to Proposition 6 can be developed more generally, in L_p norms, $1 \leq p < \infty$. More precisely, suppose that in the definition of the solution space $H_-(\Omega)$, $L_2(\Omega)$ is replaced by $L_p(\Omega)$; then Proposition 6 holds with $L_2(\Omega)$ replaced by $L_p(\Omega)$ throughout.

c) Third, Hypothesis 5 can be relaxed by supposing that Ω is a Lipschitz domain in $\mathbb{R}^n$ such that $\partial\Omega$ is a non-characteristic hypersurface for $\mathcal{L}$, and that there exists a constant vector $\xi \in \mathbb{R}^n$ such that

$$M_0 = \inf_{\Omega} \left(c + \frac{1}{2}(\mathrm{div}\ \mathbf{a}) + \mathbf{a} \cdot \xi\right) > 0.$$

d) Finally, we note that under Hypothesis 5 the normal trace, $\gamma_\nu(\mathbf{a}v) \in H^{-1/2}(\partial_- \Omega)$ can be shown to belong to $L_2(\partial_- \Omega)$.

The next section extends the well-posedness results discussed here to symmetric positive hyperbolic systems on Lipschitz domains in $\mathbb{R}^n$.

3.2 Symmetric hyperbolic systems

The aim of this section is to give a brief account of the theory of well-posedness for a class of first-order systems of partial differential equations, usually referred to as Friedrichs systems or symmetric positive systems. These represent a natural generalisation of the scalar hyperbolic equation whose properties were discussed in the previous section. In fact, symmetric positive systems embrace a large class of partial differential equations irrespective of their type, allowing a unified treatment of certain elliptic and hyperbolic equations by means of a common tool, *energy analysis.* Historically, the main motivation for introducing this class of equations "was not the desire for a unified treatment of elliptic and hyperbolic equations, but the desire to handle equations which are partly elliptic, partly hyperbolic, such as the Tricomi equation" (see [16]):

$$\left(y\frac{\partial^2}{\partial x^2} - \frac{\partial^2}{\partial y^2}\right) u = 0,$$

which plays a basic rôle in the theory of transonic flow. Our presentation of the theory will closely follow the functional-analytic approach adopted in the work of Mackenzie, Süli and Warnecke [41] (see also [25]).

Suppose that Ω is a Lipschitz domain in $\mathbb{R}^n$. Let A_i, $i = 1, ..., n$, and C be (matrix-valued) mappings from $\bar{\Omega}$ into $\mathbb{R}^{m\times m}$, $m \geq 1$; we shall assume that, for each i, the entries of A_i are continuously differentiable on $\bar{\Omega}$ and the components of C are continuous on $\bar{\Omega}$. We consider the linear first-order system of partial differential equations

$$\mathcal{L}\mathbf{u} \equiv \sum_{i=1}^{n} \frac{\partial}{\partial x_i}(A_i\mathbf{u}) + C\mathbf{u} = \mathbf{f} \qquad \text{in } \Omega. \tag{5}$$

The system (5) is called *symmetric positive* if the following conditions hold:

(a) the matrices A_i, for $i = 1, ..., n$, are symmetric, i.e. $A_i = A_i^*$;
(b) there exists $\alpha \geq 0$ and a unit vector $\xi \in \mathbb{R}^n$, such that the symmetric part of the matrix

$$K_\xi = C + \frac{1}{2}\sum_{i=1}^{n} \frac{\partial A_i}{\partial x_i} + \alpha \sum_{i=1}^{n} \xi_i A_i$$

is positive definite, uniformly on $\bar{\Omega}$; namely, there exists a positive constant $C_0 = C_0(\Omega)$ such that

$$\frac{1}{2}(K_\xi(x) + K_\xi^*(x)) \geq C_0 I \tag{6}$$

for all x in $\bar{\Omega}$.

The positivity hypothesis (b) can be seen as a direct generalisation of the positivity condition imposed on the constant M_0 in the proof of Proposition 6; see also remark c) at the end of the previous section.

In practice a system of partial differential equations, such as (5), is rarely considered in isolation and will be accompanied with a boundary condition. In order to motivate the imposition of the boundary condition for symmetric positive systems, we note that for the scalar hyperbolic equation discussed in the previous section a well-posed boundary-value problem is arrived at by prescribing a boundary condition on the inflow part of $\partial\Omega$ only. By formal analogy, for the system, only the 'incoming components' of the solution vector should be imposed on the boundary, is accordance with the physical notion of causality. This will be made more precise below. To begin, let us consider the matrix

$$B = \nu_1 A_1 + \ldots + \nu_n A_n,$$

where $\nu = (\nu_1, \ldots, \nu_n)$ is the unit outward normal vector field on $\partial\Omega$. In order to simplify the exposition, we shall suppose that B is non-singular almost everywhere on $\partial\Omega$ (with respect to the $(n-1)$-dimensional measure on $\partial\Omega$); this is equivalent to requiring that the boundary of Ω is almost everywhere non-characteristic for $\mathcal{L}$. Since the matrix B is symmetric and of full rank it can be decomposed as $B = B^+ + B^-$, where B^+ is positive semi-definite and B^- is negative semi-definite.

Let us suppose that $\mathbf{g}$ is a (sufficiently smooth) real-valued vector function defined on Ω; then an 'admissible' boundary condition for (5) is of the form

$$B^-(\mathbf{u} - \mathbf{g})\big|_{\partial\Omega} = \mathbf{0}. \tag{7}$$

We shall also consider the homogeneous counterpart of this boundary condition, corresponding to $\mathbf{g} = \mathbf{0}$; namely,

$$B^-\mathbf{u}\big|_{\partial\Omega} = \mathbf{0}. \tag{8}$$

For the time being, these boundary conditions are to be understood formally. Below, following a similar approach as in the scalar case described in the previous section, we shall state a trace theorem which assigns a precise meaning to $B^-\mathbf{u}|_{\partial\Omega}$. First, however, we single out the class of functions for which such a trace will be considered. For this purpose, we introduce the *graph space* of the operator $\mathcal{L}$ as the linear space

$$H(\mathcal{L}, \Omega) = \{\mathbf{v} \in [L_2(\Omega)]^m \ : \ \mathcal{L}\mathbf{u} \in [L_2(\Omega)]^m\}.$$

It is a simple matter to verify that, when equipped with the norm

$$|||\mathbf{v}|||_{\Omega,\xi} = \left(\|\mathrm{e}^{-\alpha(\xi\cdot x)}\mathbf{v}\|^2_{[L_2(\Omega)]^m} + \|\mathrm{e}^{-\alpha(\xi\cdot x)}\mathcal{L}\mathbf{v}\|^2_{[L_2(\Omega)]^m}\right)^{\frac{1}{2}},$$

the graph space $H(\mathcal{L}, \Omega)$ is a Hilbert space. Many of the arguments that we shall use throughout this paper rely on the concept of duality, and we shall also require $\mathcal{L}^*$, the formal adjoint of $\mathcal{L}$, defined by

$$\mathcal{L}^*\mathbf{v} = -\sum_{i=1}^{n} A_i \frac{\partial \mathbf{v}}{\partial x_i} + C^*\mathbf{v};$$

the graph space $H(\mathcal{L}^*, \Omega)$ and graph norm $|||\cdot|||_{*,\Omega,\xi}$ associated with $\mathcal{L}^*$ are introduced is the same manner as for $\mathcal{L}$, but with the weight-function $\mathrm{e}^{-\alpha(\xi\cdot x)}$ replaced by $\mathrm{e}^{\alpha(\xi\cdot x)}$.

Let $\gamma_0 : [H^1(\Omega)]^m \to [H^{1/2}(\partial\Omega)]^m$ signify the usual trace operator, defined in Section 2.3, which to each element of $[H^1(\Omega)]^m$ assigns its restriction to $\partial\Omega$. We denote by $[H^{-1/2}(\partial\Omega)]^m$ the dual space of $[H^{1/2}(\partial\Omega)]^m$; the duality pairing between these two spaces will be labelled $\langle\cdot,\cdot\rangle$. The next proposition, stated and proved in [42] and [25], will play an important rôle in the rest of the paper.

Proposition 7. *Assuming that Ω is Lipschitz domain in $\mathbb{R}^n$, the mapping $\gamma_B : \mathbf{v} \mapsto B\gamma_0(\mathbf{v})$ defined on $[H^1(\Omega)]^m$ can be extended by continuity to a linear and continuous mapping, still denoted γ_B (and referred to as the* conormal trace operator*), from $H(\mathcal{L}, \Omega)$ into $[H^{-1/2}(\partial\Omega)]^m$. Moreover, for any $\mathbf{u} \in H(\mathcal{L}, \Omega)$ and $\mathbf{v} \in [H^1(\Omega)]^m$ the following Green's formula holds*

$$(\mathcal{L}\mathbf{u}, \mathbf{v}) - (\mathbf{u}, \mathcal{L}^*\mathbf{v}) = \langle \gamma_B(\mathbf{u}), \gamma_0(\mathbf{v})\rangle.$$

An analogous result holds for $H(\mathcal{L}^*, \Omega)$.

The proof of this result is identical to that of Theorem 18.6 in the monograph of Baiocchi and Capelo [4]. The interested reader may also wish to consult [42] and [25] for a detailed proof.

Proposition 7 assigns precise meaning to the conormal trace operator, by extending $B\gamma_0(\cdot)$ from $[H^1(\Omega)]^m$ to the graph space $H(\mathcal{L}, \Omega)$. However the 'admissible' boundary condition (7) involves B^- rather than B, so we need to introduce a trace operator based on B^-. To do so, we note that the splitting $B = B^+ + B^-$ induces a natural decomposition of γ_B which leads to the definition of the partial conormal trace operators $\gamma_{B^\pm}$. This can be seen by defining

$$\gamma_{B^\pm}(\mathbf{u}) = B^\pm\gamma_0(\mathbf{u}) \qquad \forall \mathbf{u} \in [H^1(\Omega)]^m,$$

and extending this definition from the dense subspace $[H^1(\Omega)]^m$ to $H(\mathcal{L}, \Omega)$ to arrive at continuous linear operators

$$\gamma_{B^\pm} : H(\mathcal{L}, \Omega) \to [H^{-1/2}(\partial\Omega)]^m$$

with $\gamma_B = \gamma_{B^+} + \gamma_{B^-}$. One can proceed in the same way for $H(\mathcal{L}^*, \Omega)$. Equipped with these definitions, in the case of the homogeneous boundary-value problem (5), (8) we can define the domains of the operators $\mathcal{L}$ and $\mathcal{L}^*$ as, respectively,

$$D(\mathcal{L}, \Omega) = \{\mathbf{u} \in H(\mathcal{L}, \Omega) : \gamma_{B^-}(\mathbf{u}) = \mathbf{0} \quad \text{on } \partial\Omega\},$$

$$D(\mathcal{L}^*, \Omega) = \{\mathbf{u} \in H(\mathcal{L}^*, \Omega) \; : \; \gamma_{B^+}(\mathbf{u}) = \mathbf{0} \quad \text{on } \partial\Omega\}.$$

When supplied with the associated graph-norms $|||\cdot|||_{\xi,\Omega}$, $|||\cdot|||_{*,\xi,\Omega}$, $D(\mathcal{L}, \Omega)$ and $D(\mathcal{L}^*, \Omega)$ are Hilbert subspaces of $H(\mathcal{L}, \Omega)$ and $H(\mathcal{L}^*, \Omega)$, respectively.

Now we are ready to define the concept of *solution*. Suppose that $\mathbf{f} \in [L_2(\Omega)]^m$; a function $\mathbf{u} \in [L_2(\Omega)]^m$, such that

$$(\mathbf{u}, \mathcal{L}^*\phi) = (\mathbf{f}, \phi) \qquad \forall \phi \in D(\mathcal{L}^*, \Omega) \cap [H^1(\Omega)]^m,$$

will be referred to as *weak solution* of the homogeneous boundary value problem (5), (8). If $\mathbf{u}$ is a weak solution of (5), (8) and $\mathbf{u}$ belongs to $H(\mathcal{L}, \Omega)$, we shall say that $\mathbf{u}$ is a *strong solution.* We note that the requirement that $\mathbf{u}$ be a strong solution does not preclude the possibility of $\mathbf{u}$ being discontinuous; indeed, since only $\mathcal{L}\mathbf{u} \in [L_2(\Omega)]^m$ is required for $\mathbf{u} \in [L_2(\Omega)]^m$ to be a strong solution (rather than $\mathbf{u} \in [H^1(\Omega)]^m$), discontinuities in a strong solution $\mathbf{u}$ may arise across characteristic hypersurfaces.

Proposition 8. *Let $\partial\Omega$ be a non-characteristic hypersurface for $\mathcal{L}$, and assume that $\mathbf{f} \in [L_2(\Omega)]^m$; then the homogeneous boundary-value problem (5), (8) has a unique strong solution $\mathbf{u} \in D(\mathcal{L}, \Omega)$. Further, the linear operator $\mathcal{L}$ is a continuous bijection from $D(\mathcal{L}, \Omega)$ onto $[L_2(\Omega)]^m$ with a continuous inverse $\mathcal{L}^{-1} : [L_2(\Omega)]^m \to D(\mathcal{L}, \Omega)$. An analogous result holds for $\mathcal{L}^*$.*

Proof. As in the scalar case discussed in the previous section, the proof is based on Banach's Closed Range Theorem. Let us suppose that $\mathbf{v} \in [H^1(\Omega)]^m$ and take the inner product of $\mathcal{L}\mathbf{v}$ with $\mathrm{e}^{-2\alpha(\xi\cdot x)}\mathbf{v}$; upon integrating by parts using Proposition 7 and splitting the conormal trace trace operator γ_B as $\gamma_{B^+} + \gamma_{B^-}$, we deduce that

$$\langle \mathrm{e}^{-\alpha(\xi\cdot x)}\gamma_{B^+}(\mathbf{v}), \mathrm{e}^{-\alpha(\xi\cdot x)}\gamma_0(\mathbf{v})\rangle + \left(\frac{1}{2}(K_\xi + K_\xi^*)\mathrm{e}^{-\alpha(\xi\cdot x)}\mathbf{v}, \mathrm{e}^{-\alpha(\xi\cdot x)}\mathbf{v}\right)$$

$$= -\langle \mathrm{e}^{-\alpha(\xi\cdot x)}\gamma_{B^-}(\mathbf{v}), \mathrm{e}^{-\alpha(\xi\cdot x)}\gamma_0(\mathbf{v})\rangle + (\mathrm{e}^{-\alpha(\xi\cdot x)}\mathcal{L}\mathbf{v}, \mathrm{e}^{-\alpha(\xi\cdot x)}\mathbf{v}).$$

Noting (6) of hypothesis (b), we arrive at the following Gårding inequality:

$$C_0\|\, \mathrm{e}^{-\alpha(\xi\cdot x)}\mathbf{v}\|_{[L_2(\Omega)]^m} \leq \|\, \mathrm{e}^{-\alpha(\xi\cdot x)}\mathcal{L}\mathbf{v}\|_{[L_2(\Omega)]^m} \quad \forall \mathbf{v} \in [H^1(\Omega)]^m \cap D(\mathcal{L}, \Omega). \tag{9}$$

As $[H^1(\Omega)]^m \cap D(\mathcal{L}, \Omega)$ is dense in $D(\mathcal{L}, \Omega)$, it follows that

$$C_0'\|\mathrm{e}^{-\alpha(\xi\cdot x)}\mathcal{L}\mathbf{v}\|_{[L_2(\Omega)]^m} \geq |||\mathbf{v}|||_{\xi,\Omega} \qquad \forall \mathbf{v} \in D(\mathcal{L}, \Omega), \tag{10}$$

where $C_0' = (1 + C_0^{-2})^{1/2}$. The rest of the proof is identical to the final part of the proof of Proposition 6, with $D(\mathcal{L}, \Omega)$ and $[L_2(\Omega)]^m$ replacing $H_-(\Omega)$ and $L_2(\Omega)$, respectively. □

We deduce from the proof of this theorem that the strong solution to the homogeneous boundary-value problem (5), (8) obeys the *stability estimate*

$$\|\mathbf{u}\|_{[L_2(\Omega)]^m} \leq \frac{1}{C_0} e^{2\alpha D} \|\mathbf{f}\|_{[L_2(\Omega)]^m},$$

where $D = \mathrm{diam}(\Omega)$.

More generally, consider the non-homogeneous boundary-value problem (5), (7), where $\mathbf{f} \in [L_2(\Omega)]^m$ and $\mathbf{g} \in H(\mathcal{L}, \Omega)$. A function $\mathbf{u} \in [L_2(\Omega)]^m$ satisfying

$$(\mathbf{u}, \mathcal{L}^* \phi) + \langle \gamma_{B^-}(\mathbf{g}), \phi \rangle = (\mathbf{f}, \phi) \qquad \forall \phi \in D(\mathcal{L}^*, \Omega) \cap [H^1(\Omega)]^m$$

is called a *weak solution* of the boundary-value problem. A weak solution $\mathbf{u}$ to the non-homogeneous problem (5), (7) which belongs to $H(\mathcal{L}, \Omega)$ is called a *strong solution.* Lax and Phillips [37] proved, under the assumption that $\partial\Omega$ is a non-characteristic hypersurface for $\mathcal{L}$, that every weak solution is a strong solution. Further, if the data are more regular then so is the solution; specifically, if $\mathbf{f} \in [H_0^1(\Omega)]^m$, the entries of the A_i are in $C^2(\bar{\Omega})$, and the entries of C are continuously differentiable on $\bar{\Omega}$, then the following inequality holds in the case of a homogeneous boundary condition (corresponding to $\mathbf{g} = \mathbf{0}$); see, [58]:

$$\|\mathbf{u}\|_{[H^1(\Omega)]^m} \leq C_1 \|\mathbf{f}\|_{[H^1(\Omega)]^m}.$$

An analogous bound is valid for the dual problem

$$\begin{aligned} \mathcal{L}^* \phi &= \mu \qquad && \text{in } \Omega, \\ \gamma_{B^+}(\phi) &= \mathbf{0} \qquad && \text{on } \partial\Omega, \end{aligned}$$

with corresponding stability constant C_1'; namely,

$$\|\phi\|_{[H^1(\Omega)]^m} \leq C_1' \|\mu\|_{[H^1(\Omega)]^m}, \tag{11}$$

provided that $\mu \in [H_0^1(\Omega)]^m$. We shall refer to the last inequality as *strong stability of the dual problem.*

4 A posteriori error analysis for steady problems

In this section we present the basic theory of *a posteriori* error estimation for linear hyperbolic problems.

4.1 A posteriori error analysis á la Johnson

In order to motivate the approach followed in this paper, we present a brief review of the general theoretical framework of *a posteriori* error analysis, in the context of linear first-order hyperbolic systems, pursued by Johnson and his co-workers; for a detailed account, see [30] and [15].

Let us suppose that Y is a Hilbert space with inner product $(\cdot,\cdot)$ and norm $\|\cdot\|$, and let $\mathcal{L} : Y \to Y$ be a linear operator on Y with domain $D(\mathcal{L}) \subset Y$; in our case, $\mathcal{L}$ is a symmetric positive system of linear first-order hyperbolic differential operators on $Y = [L_2(\Omega)]^m$ with domain $D(\mathcal{L}) = D(\mathcal{L}, \Omega)$. Given that $\mathbf{f} \in Y$, we consider the problem of finding $\mathbf{u} \in D(\mathcal{L})$ such that

$$\mathcal{L}\mathbf{u} = \mathbf{f}.$$

Next we consider a Galerkin finite element approximation to this problem. We select a sequence of finite-dimensional spaces $\{X_h\}$, parametrised by the positive discretisation parameter h; for the sake of simplicity we shall suppose that we are dealing with a conforming approximation in the sense that $X_h \subset D(\mathcal{L})$ for each h. Simultaneously, we consider a sequence of finite-dimensional spaces $\{Y_h\}$, with Y_h contained in Y for each h. For the present purposes, X_h and Y_h can be thought of as standard finite element spaces consisting of piecewise polynomial functions on a partition, of granularity h, of the computational domain Ω; X_h is called the *trial space* while Y_h is referred to as the *test space*. Let Π_h denote the orthogonal projector in Y onto Y_h. The Galerkin finite element method can then be formulated as follows: find an approximation $\mathbf{u}_h$ to $\mathbf{u}$ in X_h such that

$$\Pi_h \mathcal{L}\mathbf{u}_h = \Pi_h \mathbf{f}.$$

Equivalently, we can write this as follows: find $\mathbf{u}_h$ in X_h such that

$$(\mathcal{L}\mathbf{u}_h, \mathbf{v}_h) = (\mathbf{f}, \mathbf{v}_h) \qquad \forall \mathbf{v}_h \in Y_h.$$

In order to obtain a computable bound on the *global error* $\mathbf{e}_h = \mathbf{u} - \mathbf{u}_h$ in terms of the *finite element residual* $\mathbf{r}_h$, defined by

$$\mathbf{r}_h = \mathbf{f} - \mathcal{L}\mathbf{u}_h,$$

we note the *Galerkin orthogonality* property

$$(\mathbf{r}_h, \mathbf{v}_h) = 0 \qquad \forall \mathbf{v}_h \in Y_h$$

which will play a crucial rôle in the analysis. Further, denoting by $\mathcal{L}^*$ the adjoint of $\mathcal{L}$, we consider the following auxiliary problem, referred to as the *dual problem*: find $\mathbf{z} \in D(\mathcal{L}^*) = D(\mathcal{L}^*, \Omega)$ such that

$$\mathcal{L}^*\mathbf{z} = \mathbf{u} - \mathbf{u}_h.$$

As already indicated in the introduction, the *a posteriori* error analysis is based on a duality argument. The first step is to derive a representation of the global error in terms of the residual; this is achieved as follows:

$$\begin{aligned} \|\mathbf{u} - \mathbf{u}_h\|^2 &= (\mathbf{u} - \mathbf{u}_h, \mathbf{u} - \mathbf{u}_h) = (\mathbf{u} - \mathbf{u}_h, \mathcal{L}^*\mathbf{z}) \\ &= (\mathcal{L}(\mathbf{u} - \mathbf{u}_h), \mathbf{z}) = (\mathcal{L}\mathbf{u} - \mathcal{L}\mathbf{u}_h, \mathbf{z}) \\ &= (\mathbf{f} - \mathcal{L}\mathbf{u}_h, \mathbf{z}) = (\mathbf{r}_h, \mathbf{z}), \end{aligned}$$

where the Green's identity stated in Proposition 7 has been used in the transition from line one to line two. Exploiting the Galerkin orthogonality property, namely that $(\mathbf{r}_h, \mathbf{z}_h) = 0$ for any $\mathbf{z}_h \in Y_h$, we deduce that

$$\|\mathbf{u} - \mathbf{u}_h\|^2 = (\mathbf{r}_h, \mathbf{z} - \mathbf{z}_h) = (h^s \mathbf{r}_h, h^{-s}(\mathbf{z} - \mathbf{z}_h)),$$

where s is a non-negative real number, to be chosen below. According to the Cauchy-Schwarz inequality,

$$\|\mathbf{u} - \mathbf{u}_h\|^2 \le \|h^s \mathbf{r}_h\| \, \|h^{-s}(\mathbf{z} - \mathbf{z}_h)\|.$$

The first term on the right-hand side of this inequality is of the desired form, involving the (computable) residual $\mathbf{r}_h$ multiplied by an appropriate power of the discretisation parameter, while the second term incorporates $\mathbf{z}$, the solution to the dual problem. Since the dual-problem has the (unknown) global error as data, $\mathbf{z}$ is unknown and has to be eliminated from the analysis by relating it to $\mathbf{u} - \mathbf{u}_h$; moreover, an appropriate choice of $\mathbf{z}_h$ has to be made. The details of these steps are described below.

Let us suppose that $\{W_\sigma\}_{\sigma \ge 0}$ is a scale of Hilbert spaces, with corresponding norms $\|\cdot\|_\sigma$, such that $W_0 = Y$ and W_{σ_2} is continuously embedded into W_{σ_1} when $\sigma_2 \ge \sigma_1$. We hypothesise the following approximation property: for each $\mathbf{z} \in W_s$ there exist $\mathbf{z}_h \in Y_h$ and a positive constant C_{appr} such that

$$\|h^{-s}(\mathbf{z} - \mathbf{z}_h)\| \le C_{appr} \|\mathbf{z}\|_s.$$

For finite element methods, this hypothesis is easily fulfilled by choosing $W_s = [H^s(\Omega)]^m$, for an appropriate $s = s_{appr} > 0$, and referring to standard approximation properties of piecewise polynomial functions in Sobolev spaces, with $\mathbf{z}_h \in Y_h$ taken as the interpolant, the quasi-interpolant or the projection of $\mathbf{z}$. Thus we arrive at the bound

$$\|\mathbf{u} - \mathbf{u}_h\|^2 \le C_{appr} \|h^s \mathbf{r}_h\| \, \|\mathbf{z}\|_s.$$

Now we have reached the final and most subtle step in the *a posteriori* error analysis. The norm $\|\mathbf{z}\|_s$ appearing on the right-hand side of the last inequality has to be eliminated in terms of $\mathbf{u} - \mathbf{u}_h$ by noting the relationship between $\mathbf{z}$ and $\mathbf{u} - \mathbf{u}_h$, namely that $\mathcal{L}^* \mathbf{z} = \mathbf{u} - \mathbf{u}_h$. In order to proceed, we shall suppose that $\mathcal{L}^*$ is invertible and that $(\mathcal{L}^*)^{-1}$ is a bounded linear operator from Y into W_s for some $s \in [0, s_{appr}]$; thus,

$$\|\mathbf{z}\|_s = \|(\mathcal{L}^*)^{-1}(\mathbf{u} - \mathbf{u}_h)\|_s \le C_{stab} \|\mathbf{u} - \mathbf{u}_h\|,$$

where C_{stab} is a positive constant (referred to as the stability constant of the dual problem), greater than or equal to the norm of $(\mathcal{L}^*)^{-1}$. Upon combining the last two bounds we deduce the desired *a posteriori* bound on the global error $\mathbf{e}_h = \mathbf{u} - \mathbf{u}_h$ in terms of the finite element residual $\mathbf{r}_h$:

$$\|\mathbf{u} - \mathbf{u}_h\| \le C_{appr} \, C_{stab} \|h^s \mathbf{r}_h\|.$$

Once the numerical solution $\mathbf{u}_h$ has been determined, the finite element residual $\mathbf{r}_h$ and $\|h^s \mathbf{r}_h\|$ are easy to compute. However the last inequality will only be of practical use if the constants C_{appr} and C_{stab} are also available. Estimating C_{appr} is a relatively simple task, using readily available results from approximation theory (see, for example, Exercise 3.1.2 in Ciarlet's monograph [12] for an explicit formula for C_{appr} in the case of standard finite element spaces consisting of continuous piecewise polynomials on simplices, or the work of Handscomb [23] for sharper estimates of C_{appr} for piecewise linear finite elements on triangles). On the other hand, providing a numerical value for C_{stab} is much harder, involving the study of the well-posedness of the dual problem. Since any value of C_{stab} which is arrived at through the use of general analytical (worst-case-scenario) arguments is bound to be a considerable overestimate of the ratio $\|\mathbf{z}\|_s/\|\mathbf{u}-\mathbf{u}_h\|$, in practice the stability constant C_{stab} is determined computationally for the specific problem at hand, as part of the process of *a posteriori* error estimation or by other computational means (see, for example, the Thesis of Sandboge [51], where strong stability constants of dual problems are predicted by means of statistical analysis).

Finally, we have to determine s, the exponent of h in the error bound. Ideally, one would like the *a posteriori* bound to reflect the approximation property of the test space Y_h to its full extent; consequently, one would wish to choose s as large as possible, and pick $s = s_{appr}$. Unfortunately, for first-order hyperbolic systems the weak smoothing properties of $(\mathcal{L}^*)^{-1}$ (which is a bounded linear operator from $Y = [L_2(\Omega)]^m$ into the anisotropic space $H(\mathcal{L}^*, \Omega)$, but not into $W_1 = [H^1(\Omega)]^m$) pose an unsurmountable limitation on the choice of s. Indeed, since we have restricted ourselves to operating within the realm of standard isotropic Sobolev spaces, such as $W_s = [H^s(\Omega)]^m$, where approximation theory by piecewise polynomial functions is well developed, it follows that the strongest statement that we can make (in terms of these spaces) is that $(L^*)^{-1}$ is a bounded operator from Y into $Y = W_0$, only; consequently, s cannot exceed 0 and we end up with the error bound

$$\|\mathbf{u}-\mathbf{u}_h\| \le C_{appr}\, C_{stab}\|\mathbf{r}_h\|.$$

In fact, we note that when $s = 0$ we do not benefit from the application of Galerkin orthogonality, and we may simply take $\mathbf{z}_h = 0$ in our argument to simplify this bound to

$$\|\mathbf{u}-\mathbf{u}_h\| \le C_{stab}\|\mathbf{r}_h\|.$$

Either way, we see that in the case of a first-order hyperbolic system the *a posteriori* error bound that we arrive at on the basis of the reasoning outlined above is unsatisfactory in that it fails to display explicitly, in terms of powers of h, the approximation properties of the test space Y_h. Worse still, when the data are discontinuous, linear hyperbolic equations may possess solutions that are discontinuous across characteristic hypersurfaces and, under mesh refinement, the associated residual norm $\|\mathbf{r}_h\|$ will then converge to 0 very slowly, if at all; consequently, in the absence of the compensating factor h^s, any adaptive algorithm driven by this error bound is likely to be inefficient.

A possible approach to rectifying the problem is based on perturbing the first-order hyperbolic operator $\mathcal{L}^*$ (or, indeed, both $\mathcal{L}$ and $\mathcal{L}^*$) through the addition of a second-order elliptic term with a small coefficient (see [32]), which then provides additional isotropic regularity; in favourable circumstances the inverse of the perturbed adjoint operator is bounded from $W_0 = [L_2(\Omega)]^m$ into $W_2 = [H^2(\Omega)]^m$ which then allows one to take $s = 2$. This approach, however, is associated with undesirable complications when applied in bounded domains, related to the fact that artificial boundary conditions have to be supplied for the resulting second-order operator in such a way that the features of the solution to the hyperbolic system are retained, particularly in the vicinity of the boundary; a further difficulty with elliptic regularisation of non-dissipative hyperbolic problems, such as the Maxwell system of electro-magnetism, is that it may introduce a physically unacceptable level of damping into the model. Our aim in these notes is to pursue a direct approach to the *a posteriori* error analysis of finite element methods for first-order hyperbolic problems, namely one that does not require elliptic regularisation of the hyperbolic operator. The next section is devoted to some simple examples which illustrate the technique.

4.2 Petrov-Galerkin finite element methods

In this section we shall present a general framework of *a posteriori* error estimation for Galerkin finite element approximations of hyperbolic problems. Suppose that X and Y are two real Banach spaces, equipped with norms $\|\cdot\|_X$ and $\|\cdot\|_Y$, respectively. In the context of the problems discussed here, we can think of two natural choices:

α) When the boundary condition $\gamma_{B^-}(\mathbf{u}) = \mathbf{0}$ is enforced *strongly*, we take $X = D(\mathcal{L}, \Omega)$ equipped with the graph norm $\|\cdot\|_X = |||\cdot|||_{\xi,\Omega}$ and $Y = [L_2(\Omega)]^m$ with the norm $\|\cdot\|_Y = \|\cdot\|_{[L_2(\Omega)]^m}$;

β) When the boundary condition $\gamma_{B^-}(\mathbf{u}-\mathbf{g}) = \mathbf{0}$ is enforced *weakly* (through the definition of the bilinear functional in the variational formulation, rather than through the definition of the solution space), then we take $X = H(\mathcal{L}, \Omega)$ equipped with the graph norm $\|\cdot\|_X = |||\cdot|||_{\xi,\Omega}$ and $Y = [H^1(\Omega)]^m$ with the norm $\|\cdot\|_Y = \|\cdot\|_{[H^1(\Omega)]^m}$.

Suppose further that $a(\cdot,\cdot)$ is a bilinear functional on $X \times Y$, and let $l(\cdot)$ be a linear functional on Y; in the context of symmetric positive systems we adopt the following definitions, corresponding to cases α) and β).

α) In the case of strongly imposed boundary condition,

$$a(\mathbf{w},\mathbf{v}) = (\mathcal{L}\mathbf{w},\mathbf{v}), \qquad \mathbf{w} \in X = D(\mathcal{L},\Omega), \quad \mathbf{v} \in Y = [L_2(\Omega)]^m$$

and

$$l(\mathbf{v}) = (\mathbf{f},\mathbf{v}), \qquad \mathbf{v} \in Y = [L_2(\Omega)]^m.$$

β) In the case of weakly imposed boundary condition,

$$a(\mathbf{w},\mathbf{v}) = (\mathcal{L}\mathbf{w},\mathbf{v}) - \langle \gamma_{B^-}(\mathbf{w}), \gamma_0(\mathbf{v})\rangle, \qquad \mathbf{w} \in X = H(\mathcal{L},\Omega), \mathbf{v} \in Y = [H^1(\Omega)]^m$$

and

$$l(\mathbf{v}) = (\mathbf{f},\mathbf{v}) - \langle \gamma_{B^-}(\mathbf{g}), \gamma_0(\mathbf{v})\rangle, \quad \mathbf{v} \in Y = [H^1(\Omega)]^m.$$

Either way, we arrive at a variational problem of the following form: find $\mathbf{u}$ in X such that

$$a(\mathbf{u},\mathbf{v}) = l(\mathbf{v}) \quad \forall \mathbf{v} \in Y. \tag{12}$$

The existence of a unique solution to this problem in the case of a strongly imposed homogeneous boundary condition has been shown in the previous section (see Proposition 8). The problem with the weakly imposed non-homogeneous boundary condition is easily seen to be equivalent to the problem with strongly imposed non-homogeneous boundary condition, considered at the end of Section 3.2: $\mathbf{u}$ is a solution to one problem if and only if it is a solution to the other. While in the case of $\mathbf{g} = \mathbf{0}$ the formulations α) and β) are entirely equivalent, this is not necessarily true of the associated Galerkin discretisations; this will be discussed in more detail below.

Let us consider the Galerkin finite element discretisation of problem (12). Given that the computational domain Ω is a Lipschitz domain in $\mathbb{R}^n$, we consider a *partition* of Ω; namely, we select a finite collection $\mathcal{T}_h = \{\kappa_i\}$ of Lipschitz subdomains κ_i of Ω such that:

(1) $\kappa_i \cap \kappa_j$ is an empty set if $i \neq j$, and
(2) $\cup_i \bar{\kappa}_i = \bar{\Omega}$.

Furthermore, in order to avoid the presence of "hanging nodes", a partition will be assumed to have an additional property which, for the sake of simplicity, we only formulate in the two-dimensional case using triangular elements; an analogous assumption is adopted for higher dimensions and other types of elements:

(3) No vertex of any triangle lies in the interior of an edge of another triangle.

We choose a family of finite element spaces X_h, parametrised by h, $0 < h \leq h_0$ (typically, h is taken to be a piecewise constant function whose value on element κ is equal to the diameter of κ), consisting of piecewise polynomial functions on the partition $\mathcal{T}_h = \{\kappa_i\}$ of Ω, such that $X_h \subset X$ for all $h > 0$. Analogously, we suppose that Y_h is a finite element space contained in Y. It will be assumed that the spaces X_h and Y_h are equipped with the norms of X and Y, respectively. We consider the following approximation of problem (12): find $\mathbf{u}_h$ in X_h such that

$$a(\mathbf{u}_h,\mathbf{v}_h) = l(\mathbf{v}_h) \quad \forall \mathbf{v}_h \in Y_h. \tag{13}$$

The only hypothesis that we shall adopt throughout is that the spaces X_h and Y_h have been chosen so that *(13) has a unique solution* $\mathbf{u}_h$, *for each* $h \in (0, h_0]$; we note in passing that this can be ensured by satisfying the conditions of the next proposition (see [3]).

Proposition 9. *Suppose that the bilinear functional* $a(\cdot,\cdot)$ *is bounded on* $X_h \times Y_h$ *and that the linear functional* $l(\cdot)$ *is bounded on* Y_h*; namely, there exists a positive constant* M_1 *such that*

$$|a(\mathbf{w}_h, \mathbf{v}_h)| \leq M_1 \|\mathbf{w}_h\|_X \|\mathbf{v}_h\|_Y$$

for all $\mathbf{w}_h$ *in* X_h *and all* $\mathbf{v}_h$ *in* Y_h*, and a positive constant* M_2 *such that*

$$|l(\mathbf{v}_h)| \leq M_2 \|\mathbf{v}_h\|_Y$$

for all $\mathbf{v}_h$ *in* Y_h*. Suppose further that* $a(\cdot,\cdot)$ *satisfies the following* inf-sup condition*: there exists a positive constant* M_0 *such that*

$$\inf_{0 \neq \mathbf{w}_h \in X_h} \sup_{0 \neq \mathbf{v}_h \in Y_h} \frac{a(\mathbf{w}_h, \mathbf{v}_h)}{\|\mathbf{w}_h\|_X \|\mathbf{v}_h\|_Y} \geq M_0;$$

and

$$\sup_{\mathbf{w}_h \in X_h} a(\mathbf{w}_h, \mathbf{v}_h) > 0 \quad \forall \mathbf{v}_h \in Y_h.$$

Then there exists a unique $\mathbf{u}_h$ *in* X_h *such that*

$$a(\mathbf{u}_h, \mathbf{v}_h) = l(\mathbf{v}_h) \quad \forall \mathbf{v}_h \in Y_h.$$

Furthermore,

$$\|\mathbf{u}_h\|_X \leq \frac{1}{M_0} \|l\|_{Y'}.$$

Of the conditions listed in Proposition 9 the boundedness of the bilinear functional $a(\cdot,\cdot)$ on $X_h \times Y_h$ and the boundedness of the functional $l(\cdot)$ on Y_h follow automatically from the boundedness of these functionals on $X \times Y$ and Y, respectively. On the other hand the verification of the inf-sup condition on $X_h \times Y_h$ can be a non-trivial exercise, depending on the choice of the spaces X_h and Y_h. As the precise structure of the conditions which guarantee the existence and uniqueness of $\mathbf{u}_h$ is of no relevance in the *a posteriori* error analysis that we wish to pursue, we shall simply suppose that (13) possesses a unique solution for each $h \in (0, h_0]$; no structural conditions on $a(\cdot,\cdot)$ and $l(\cdot)$ of the kind appearing in Proposition 9 will be made.

Next we derive an *a posteriori* bound on the global error $\mathbf{e}_h = \mathbf{u} - \mathbf{u}_h$. Suppose that $\psi \in [C_0^\infty(\Omega)]^m$, and consider the auxiliary (dual) problem: find $\mathbf{z}$ in Y such that

$$a(\mathbf{w}, \mathbf{z}) = (\mathbf{w}, \psi) \quad \forall \mathbf{w} \in X, \tag{14}$$

where $(\cdot,\cdot)$ denotes the inner product of $[L_2(\Omega)]^m$. Thus,

$$(\mathbf{e}_h,\psi) = a(\mathbf{e}_h,\mathbf{z}) = a(\mathbf{u}-\mathbf{u}_h,\mathbf{z}) = a(\mathbf{u}-\mathbf{u}_h,\mathbf{z}-\mathbf{z}_h),$$

where $\mathbf{z}_h$ is any element in Y_h. Hence

$$(\mathbf{e}_h,\psi) = l(\mathbf{z}-\mathbf{z}_h) - a(\mathbf{u}_h,\mathbf{z}-\mathbf{z}_h).$$

We write the right-hand side as $\langle \mathbf{r}_h, \mathbf{z}-\mathbf{z}_h\rangle$, where $\langle\cdot,\cdot\rangle$ is the duality pairing between Y', the dual space of Y, and Y. The quantity $\mathbf{r}_h \in Y'$ is referred to as the finite element residual. Consequently,

$$(\mathbf{u}-\mathbf{u}_h,\psi) = \langle \mathbf{r}_h, \mathbf{z}-\mathbf{z}_h\rangle.$$

Our aim is to derive a bound on the global error in terms of the finite element residual, based on this error representation formula.

From here on, we shall distinguish between the two cases, labelled α) and β), which were formulated earlier on, corresponding to strongly imposed and weakly imposed boundary conditions, respectively. We begin by developing an error analysis for case α).

α) In this case $Y = [L_2(\Omega)]^m$, and therefore

$$(\mathbf{u}-\mathbf{u}_h,\psi) = (\mathbf{r}_h, \mathbf{z}-\mathbf{z}_h). \tag{15}$$

We shall suppose that $\mathbf{f} \in [L_2(\Omega)]^m$, that $\mathcal{T}_h$ is a finite element partition of Ω into elements κ, and we adopt the following standard approximation property for the test space Y_h:

(c) There exists a positive constant C_2, independent of h, such that for each $\mathbf{v} \in [H^1(\Omega)]^m$ there is $\mathbf{v}_h \in Y_h$ with

$$\|h^{-1}(\mathbf{v}-\mathbf{v}_h)\|_{[L_2(\Omega)]^m} \leq C_2\|\mathbf{v}\|_{[H^1(\Omega)]^m}.$$

Theorem 10. *Suppose that hypotheses (a), (b) and (c) hold, and that the entries of the matrices A_i, $i=1,\dots,n$, are in $C^2(\bar{\Omega})$ and those of C are continuously differentiable on $\bar{\Omega}$. Then,*

$$\|\mathbf{u}-\mathbf{u}_h\|_{[H^{-1}(\Omega)]^m} \leq C_1' C_2 \|h\mathbf{r}_h\|_{[L_2(\Omega)]^m},$$

where $\|\cdot\|_{[H^{-1}(\Omega)]^m}$ denotes the norm of the dual space of $[H_0^1(\Omega)]^m$.

Proof. Given that $\psi \in [C_0^\infty(\Omega)]^m$, it follows from (15) that

$$\begin{aligned}(\mathbf{u}-\mathbf{u}_h,\psi) &= (\mathbf{r}_h,\mathbf{z}-\mathbf{z}_h)\\ &\leq \|h\mathbf{r}_h\|_{[L_2(\Omega)]^m}\|h^{-1}(\mathbf{z}-\mathbf{z}_h)\|_{[L_2(\Omega)]^m}\\ &\leq C_2\|h\mathbf{r}_h\|_{[L_2(\Omega)]^m}\|\mathbf{z}\|_{[H^1(\Omega)]^m}.\end{aligned} \tag{16}$$

According to the strong stability result (11) with $\mu = \psi$,

$$\|\mathbf{z}\|_{[H^1(\Omega)]^m} \leq C_1' \|\psi\|_{[H^1(\Omega)]^m}.$$

Substituting this into (16), dividing both sides by $\|\psi\|_{[H^1(\Omega)]^m}$, taking the supremum over all $\psi \in [C_0^\infty(\Omega)]^m$ and noting that $[C_0^\infty(\Omega)]^m$ is dense in $[H_0^1(\Omega)]^m$, we obtain the desired error bound. □

Next, we carry out a similar analysis for case β) corresponding to weakly imposed non-homogeneous boundary condition, with $\mathbf{g} \in [H^1(\Omega)]^m$.

β) Arguing in the same way as in case α), we deduce that

$$\begin{aligned}(\mathbf{u} - \mathbf{u}_h, \psi) &= (\mathbf{r}_h, \mathbf{z} - \mathbf{z}_h) \\ &\quad + \langle \gamma_{B^-}(\mathbf{u}_h - \mathbf{g}), \gamma_0(\mathbf{z} - \mathbf{z}_h)\rangle \equiv I + II.\end{aligned}$$

To proceed, replace c) with the following hypothesis:

(c') There exists a positive constant C_2, independent of h, such that for each $\mathbf{v} \in [H^1(\Omega)]^m$ there is $\mathbf{v}_h \in Y_h$ with

$$\|h^{-1}(\mathbf{v} - \mathbf{v}_h)\|_{[L_2(\Omega)]^m} + \|h^{-1/2}\gamma_0(\mathbf{v} - \mathbf{v}_h)\|_{[L_2(\partial\Omega)]^m} \leq C_2 \|\mathbf{v}\|_{[H^1(\Omega)]^m}.$$

Term I is dealt with as in case α) to deduce that

$$I \leq C_2 \|h\mathbf{r}_h\|_{[L_2(\Omega)]^m} \|\mathbf{z}\|_{[H^1(\Omega)]^m}.$$

To estimate II, we note that

$$\gamma_{B^-}(\mathbf{u}_h - \mathbf{g}) = B^- \gamma_0(\mathbf{u}_h - \mathbf{g}) \in [L_2(\partial\Omega)]^m,$$

so that

$$II \leq \|h^{1/2} B^- \gamma_0(\mathbf{u}_h - \mathbf{g})\|_{[L_2(\partial\Omega)]^m} \|h^{-1/2}(\mathbf{z} - \mathbf{z}_h)\|_{[L_2(\partial\Omega)]^m}$$

and therefore

$$II \leq C_2 \|h^{1/2} B^- \gamma_0(\mathbf{u}_h - \mathbf{g})\|_{[L_2(\partial\Omega)]^m} \|\mathbf{z}\|_{[H^1(\Omega)]^m}.$$

Thus, appealing to the strong stability result (11) for the adjoint (dual) problem, we deduce the following theorem.

Theorem 11. *Suppose that hypotheses (a), (b) and (c') hold, and that the entries of the matrices A_i, $i = 1, \ldots, n$, are in $C^2(\bar{\Omega})$ and those of C are continuously differentiable on $\bar{\Omega}$. Then,*

$$\|\mathbf{u} - \mathbf{u}_h\|_{[H^{-1}(\Omega)]^m} \leq C_1' C_2 \left(\|h\mathbf{r}_h\|_{[L_2(\Omega)]^m} + \|h^{1/2}\mathbf{r}_h^-\|_{[L_2(\partial\Omega)]^m} \right),$$

where $\|\cdot\|_{[H^{-1}(\Omega)]^m}$ denotes the norm of the dual space of $[H_0^1(\Omega)]^m$, $\mathbf{r}_h = \mathbf{f} - \mathcal{L}\mathbf{u}_h$ denotes the interior residual, *and $\mathbf{r}_h^- = B^-\gamma_0(\mathbf{g} - \mathbf{u}_h)$ signifies the* boundary residual.

The interior and boundary residuals measure the extent to which $\mathbf{u}_h$ fails to satisfy the partial differential equation and the boundary condition, respectively. In Theorem 10 the boundary residual term did not arise since there we had $X_h \subset X = D(\mathcal{L}, \Omega)$, so the boundary condition was satisfied exactly by all elements of the finite element trial space, including $\mathbf{u}_h$.

4.3 The streamline diffusion method

For Petrov-Galerkin approximations of symmetric positive systems ensuring stability is a non-trivial matter. We have already touched on this issue in the previous section when we commented on Proposition 9: to prove stability, one has to verify the inf-sup condition, with a constant M_0 (preferably, independent of h), for the particular choice of test and trial space. For examples of stable Petrov-Galerkin methods for hyperbolic systems, we refer to [38], [39] and [62].

As an alternative to these techniques, in this section we consider a family of methods which use the same test and trial space and a bilinear functional $a_\delta(\cdot,\cdot)$ which is a consistent perturbation of the bilinear functional $a(\cdot,\cdot)$ such that the resulting Galerkin method is stable. The stabilising perturbation term acts along the characteristic hyperplanes of the differential operator $\mathcal{L}$ and can be thought of physically as a numerical diffusion term in the direction of the streamlines; hence the name of the resulting discretisation technique: the *streamline diffusion finite element method.*

In order to highlight the key issues concerning the *a posteriori* error analysis of the streamline diffusion finite element method we consider the symmetric positive system

$$\mathcal{L}\mathbf{u} = \mathbf{f} \quad \text{in } \Omega, \qquad \gamma_{B^-}(\mathbf{u} - \mathbf{g}) = \mathbf{0} \quad \text{on } \partial\Omega, \tag{17}$$

where $\mathbf{f} \in [L_2(\Omega)]^m$ and $\mathbf{g} \in [H^1(\Omega)]^m$.

Let us suppose that $\bar{\Omega}$ has been subdivided by a finite element partition $\mathcal{T}_h = \{\kappa_i\}$; here h is a piecewise constant mesh function with $h(x) = \mathrm{diam}(\kappa)$ when x is in element κ. On this partition we consider the finite element space X_h, $X_h \subset [H^1(\Omega)]^m$, consisting of continuous piecewise polynomials of fixed degree k, $k \geq 1$. It will be assumed that X_h possesses the following standard approximation property:

(c") Given that $\mathbf{v} \in [H^1(\Omega)]^m$, there exists $\mathbf{v}_h \in X_h$ and a constant C_2, independent of $\mathbf{v}$ and h, such that

$$\begin{aligned} \|h^{-1}(\mathbf{v} - \mathbf{v}_h)\|_{[L_2(\Omega)]^m} + \|h^{-1/2}\gamma_0(\mathbf{v} - \mathbf{v}_h)\|_{[L_2(\partial\Omega)]^m} \\ + \|\mathbf{v}_h\|_{[H^1(\Omega)]^m} \leq C_2\|\mathbf{v}\|_{[H^1(\Omega)]^m}. \end{aligned} \tag{18}$$

Given a function $\mathbf{v}$ and an associated $\mathbf{v}_h$ satisfying (18), we shall write $\mathbf{v}_h = P_h\mathbf{v}$ to denote that $\mathbf{v}_h$ is assigned to $\mathbf{v}$.

Next, we introduce the streamline diffusion finite element approximation of (17); to do so, we define the *streamline diffusion parameter* δ as a piecewise constant function on $\bar{\Omega}$ whose value on $\kappa \in \mathcal{T}_h$ is

$$\delta|_\kappa = K_0 \,\mathrm{diam}(\kappa), \quad \kappa \in \mathcal{T}_h, \tag{19}$$

where K_0 is a fixed positive constant. Further, we consider the bilinear form $a_\delta(\cdot,\cdot)$ defined by

$$a_\delta(\mathbf{w}, \mathbf{v}) = (\mathcal{L}\mathbf{w}, \mathbf{v} + \delta\mathcal{L}\mathbf{v}) - \langle\gamma_{B^-}(\mathbf{w}), \gamma_0(\mathbf{v})\rangle,$$

and the linear functional

$$l_\delta(\mathbf{v}) = (\mathbf{f}, \mathbf{v} + \delta\mathcal{L}\mathbf{v}) - \langle \gamma_{B^-}(\mathbf{g}), \gamma_0(\mathbf{v})\rangle.$$

In these definitions $(\cdot,\cdot)$ denotes the inner product of $[L_2(\Omega)]^m$.

Streamline diffusion method: Find $\mathbf{u}_h \in X_h$ such that

$$a_\delta(\mathbf{u}_h, \mathbf{v}_h) = l_\delta(\mathbf{v}_h) \quad \forall \mathbf{v}_h \in X_h. \tag{20}$$

Assuming that X_h has been equipped with the norm of $H(\mathcal{L},\Omega)$ and (6) holds for $\alpha = 0$, it is a simple matter to show that a_δ and l_δ satisfy the hypotheses of Theorem 9 on X_h with $X = Y = H(\mathcal{L},\Omega)$ and $Y_h = X_h$, and thereby (20) has a unique solution $\mathbf{u}_h$ in X_h. Formally, (20) can be thought of as a perturbation of the standard Galerkin method corresponding to $\delta \equiv 0$.

In order to illustrate the qualitative improvement over the standard Galerkin finite element method offered by the streamline diffusion method, we show in Figure 1 the results of a numerical experiment. Our model problem is

$$\nabla \cdot (\mathbf{a}u) = 0 \quad \text{on } \Omega = (0,1)^2, \qquad u = g \quad \text{on } \partial_-\Omega, \tag{21}$$

where $\mathbf{a} = (2,1)$, $g(0,y) = 1$ for $0 \le y \le 1$ and $g(x,0) = 0$ for $0 < x \le 1$. On Ω we considered a triangulation which arises from a 21×21 uniform mesh by connecting the bottom-left corner of each mesh-square with its top-right corner; Figure 1 shows the numerical solution obtained by using: (a) the standard Galerkin finite element method with continuous piecewise linear trial and test functions, and (b) the numerical solution given by the streamline diffusion finite element method with the same trial and test spaces, and stabilisation parameter $K_0 = 0.5/\|\mathbf{a}\|$, where $\|\mathbf{a}\|$ denotes the Euclidean norm of $\mathbf{a}$.

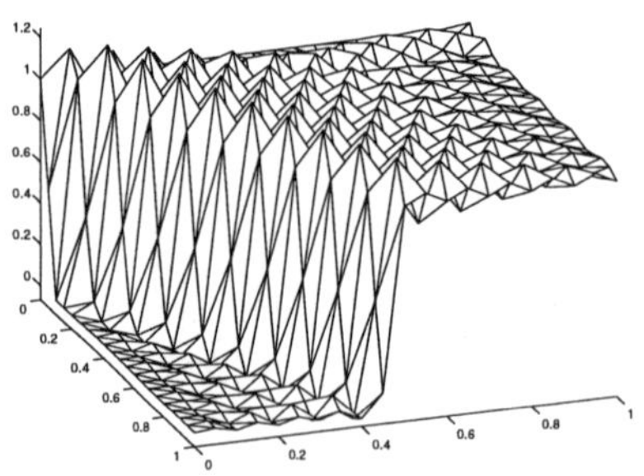

(a)

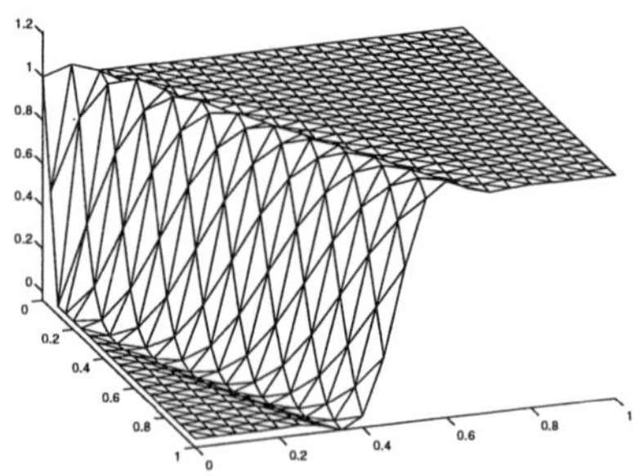

(b)

Fig. 1. Corner discontinuity problem: (a) Standard Galerkin finite element method; (b) Streamline diffusion method with $K_0 = 0.5/\|\mathbf{a}\|$.

Here we shall be concerned with the *a posteriori* error analysis of the streamline diffusion method. The analysis relies on the following *Galerkin orthogonality* property:

$$a_\delta(\mathbf{u} - \mathbf{u}_h, \mathbf{v}_h) = 0 \quad \forall \mathbf{v}_h \in X_h. \tag{22}$$

The equation (22) is easily seen to hold by noting (20) and that

$$a_\delta(\mathbf{u}, \mathbf{v}_h) = l_\delta(\mathbf{v}_h) \quad \forall \mathbf{v}_h \in X_h.$$

The starting point in the argument is the following *dual problem*: given that $\psi \in [C_0^\infty(\Omega)]^m$, find $\mathbf{z}$ in $H(\mathcal{L}^*, \Omega)$ such that

$$\mathcal{L}^*\mathbf{z} = \psi \quad \text{in } \Omega, \qquad \gamma_{B^+}(\mathbf{z}) = \mathbf{0} \quad \text{on } \partial\Omega. \tag{23}$$

The *a posteriori* error bound that we shall state below will be expressed in terms of the finite element residual

$$\mathbf{r}_h = \mathbf{f} - \mathcal{L}\mathbf{u}_h$$

which measures the extent to which $\mathbf{u}_h$ fails to satisfy the differential equation in Ω; thus, as in the previous section, we shall refer to it as the *internal residual*. Also, since $\mathbf{u}_h$ satisfies the boundary condition only approximately rather then in the pointwise sense, the difference $B^-\gamma_0(\mathbf{g} - \mathbf{u}_h)$ is not necessarily zero and will be seen to enter the *a posteriori* error bound; thus, we also define the *boundary residual*

$$\mathbf{r}_h^- = B^-\gamma_0(\mathbf{g} - \mathbf{u}_h).$$

With these definitions we have the following result.

Theorem 12. *Let us suppose that the entries of the matrices A_i, $i = 1, \ldots, n$, are in $C^2(\bar{\Omega})$ and those of C are continuously differentiable on $\bar{\Omega}$; then the following a posteriori error bound holds:*

$$\|\mathbf{u} - \mathbf{u}_h\|_{[H^{-1}(\Omega)]^m} \leq C_3 \|h\mathbf{r}_h\|_{[L_2(\Omega)]^m} + C_1' C_2 \|h^{1/2}\mathbf{r}_h^-\|_{[L_2(\partial\Omega)]^m},$$

where $C_3 = C_1'(C_2 + K_0 C_4)$ with C_1' the strong stability constant for the dual problem appearing in (11), C_2 and K_0 the constants from conditions (18) and (19) and C_4 defined in (25) below.

Proof. Let $\psi \in [C_0^\infty(\Omega)]^m$. Recalling the dual problem (23), integrating by parts, and appealing to the Galerkin orthogonality property (22), we deduce that, for any $\mathbf{z}_h \in X_h$,

$$\begin{aligned}(\mathbf{u} - \mathbf{u}_h, \psi) &= (\mathbf{r}_h, \mathbf{z} - \mathbf{z}_h) - \langle \gamma_{B^-}(\mathbf{g} - \mathbf{u}_h), \gamma_0(\mathbf{z} - \mathbf{z}_h)\rangle + (\delta\mathbf{r}_h, \mathcal{L}\mathbf{z}_h) \\ &\equiv I + II + III.\end{aligned} \tag{24}$$

Next, we make a specific choice of $\mathbf{z}_h$: we take $\mathbf{z}_h = P_h\mathbf{z}$, where P_h is defined in hypothesis (c"). Terms I and II are dealt with as in case β) of the previous section; thus, noting (18), we have that

$$\begin{aligned}|\,I\,| &\leq \|h\mathbf{r}_h\|_{[L_2(\Omega)]^m}\|h^{-1}(\mathbf{z}-\mathbf{z}_h)\|_{[L_2(\Omega)]^m}\\ &\leq C_2\|h\mathbf{r}_h\|_{[L_2(\Omega)]^m}\|\mathbf{z}\|_{[H^1(\Omega)]^m},\end{aligned}$$

and applying (18) and noting that $\gamma_{B^-}(\mathbf{g}-\mathbf{u}_h) = B^-\gamma_0(\mathbf{g}-\mathbf{u}_h)$ yields

$$|\,II\,| \leq C_2\|h^{1/2}\mathbf{r}_h^-\|_{[L_2(\partial\Omega)]^m}\,\|\mathbf{z}\|_{[H^1(\Omega)]^m}.$$

Further,

$$\begin{aligned}|\,III\,| &\leq \|\delta\mathbf{r}_h\|_{[L_2(\Omega)]^m}\|\mathcal{L}\mathbf{z}_h\|_{[L_2(\Omega)]^m}\\ &\leq K_0C_4\|h\mathbf{r}_h\|_{[L_2(\Omega)]^m}\|\mathbf{z}\|_{[H^1(\Omega)]^m},\end{aligned}$$

where

$$C_4 = C_2\left(\sum_{i=1}^{n}\|A_i\|^2_{[C(\bar{\Omega})]^{m\times m}} + \|C\|^2_{[C(\bar{\Omega})]^{m\times m}}\right)^{1/2}. \tag{25}$$

Upon collecting the bounds on I, II and III, and inserting them into (24), we deduce that

$$\begin{aligned}|(\mathbf{u}-\mathbf{u}_h,\psi)| \leq\; & C_2\|h\mathbf{r}_h\|_{[L_2(\Omega)]^m}\|\mathbf{z}\|_{[H^1(\Omega)]^m}\\ &+C_2\|h^{1/2}\mathbf{r}_h^-\|_{[L_2(\partial\Omega)]^m}\,\|\mathbf{z}\|_{[H^1(\Omega)]^m}\\ &+K_0C_4\|h\mathbf{r}_h\|_{[L_2(\Omega)]^m}\|\mathbf{z}\|_{[H^1(\Omega)]^m}.\end{aligned} \tag{26}$$

Finally, recalling the strong stability result for the dual problem, the last inequality implies the desired *a posteriori* error bound. □

When we formally set $K_0 = 0$, the streamline diffusion finite element method reduces to the standard Galerkin finite element method; similarly, the bound derived in Theorem 12 collapses to that in Theorem 11, as expected.

4.4 The cell vertex finite volume method

In the previous section we outlined the error analysis of the streamline diffusion method and we saw that perturbing the bilinear form $a(\cdot,\cdot)$ to $a_\delta(\cdot,\cdot)$ did not affect the *a posteriori* error bound in the H^{-1} norm (except, perhaps, altering the constant in the error bound). In this section, we consider a different perturbation of the basic Galerkin framework by applying numerical quadrature to a Petrov-Galerkin finite element method; as an illustration of the effects of such a "non-Galerkin" perturbation, we discuss the *a posteriori* error analysis of the cell vertex finite volume method.

For the sake of simplicity, we suppose in this subsection that Ω is the unit square $(0,1)^2$. We consider the symmetric positive system in conservation form subject to a non-homogeneous boundary condition:

$$\mathcal{L}\mathbf{u} = \sum_{i=1}^{2} \frac{\partial}{\partial x_i}(A_i \mathbf{u}) + C\mathbf{u} = \mathbf{f} \quad \text{in } \Omega, \tag{27}$$

$$\gamma_{B^-}(\mathbf{u} - \mathbf{g}) = \mathbf{0} \quad \text{on } \partial\Omega, \tag{28}$$

where $\mathbf{f} \in [L_2(\Omega)]^m$ and $\mathbf{g} \in [H^1(\Omega)]^m$. In order to formulate the cell vertex finite volume discretisation of this problem, we subdivide Ω by a structured mesh consisting of convex quadrilaterals. Here the word *structured* signifies the fact that the partition is topologically equivalent to a uniform square mesh on Ω. The cell vertex finite volume approximation of this boundary value problem is obtained by integrating the system of partial differential equations (27) over each quadrilateral in the partition, and exploiting the fact that the equations are in divergence form: Gauss' theorem is applied to convert integrals over quadrilaterals into contour integrals over the boundaries of quadrilaterals; these contour integrals are then approximated by means of the trapezium rule. This process provides a four-point finite difference scheme referred to as the cell vertex finite volume method, given that the unknowns are carried at the vertices of the cells – the quadrilateral elements in the partition.

For the purposes of the present paper it is useful to note that the construction of the cell vertex scheme can be also described in the language of finite element methods. Thus, let $\mathcal{F} = \{\mathcal{T}_h\}$, $h > 0$, be a regular family of structured partitions $\mathcal{T}_h$ of $\Omega = (0,1)^2$ into convex quadrilateral elements κ_{ij}. In order to introduce the relevant finite element spaces, we define the reference square $\hat{\kappa} = (-1,1)^2$, and denote by $F_{\kappa_{ij}}$ the bilinear function that maps $\hat{\kappa}$ onto the 'finite volume' κ_{ij}. Let $\mathcal{Q}_1(\hat{\kappa})$ be the set of bilinear functions on $\hat{\kappa}$, and $\mathcal{Q}_0(\hat{\kappa})$ the set of constant functions on $\hat{\kappa}$. We define

$$Y_h = \left\{ \mathbf{v} \in [L_2(\Omega)]^m \,:\, \mathbf{v} = \hat{\mathbf{v}} \circ F_{\kappa_{ij}}^{-1},\ \hat{\mathbf{v}} \in [\mathcal{Q}_0(\hat{\kappa})]^m,\ \kappa_{ij} \in \mathcal{T}_h \right\},$$

$$X_h = \left\{ \mathbf{w} \in [H^1(\Omega)]^m \,:\, \mathbf{w} = \hat{\mathbf{w}} \circ F_{\kappa_{ij}}^{-1},\ \hat{\mathbf{w}} \in [\mathcal{Q}_1(\hat{\kappa})]^m,\ \kappa_{ij} \in \mathcal{T}_h \right\}.$$

Let us denote by $\Pi_h : [L_2(\Omega)]^m \to Y_h$ the orthogonal projector in $[L_2(\Omega)]^m$ onto the linear subspace Y_h, and by

$$\mathcal{I}_h : (H(L,\Omega) \cap [C(\bar{\Omega})]^m)^2 \to X_h \times X_h$$

the interpolation projector onto $X_h \times X_h$. With this notation, we put

$$a_h(\mathbf{w}_h, \mathbf{v}_h) = (\operatorname{div} \mathcal{I}_h(\mathcal{A}\mathbf{w}_h), \mathbf{v}_h) + (C\mathbf{w}_h, \mathbf{v}_h) - \langle \gamma_{B^-}(\mathbf{w}_h), \gamma_0(\mathbf{v}_h) \rangle,$$

and

$$l(\mathbf{v}_h) = (\mathbf{f}, \mathbf{v}_h) - \langle \gamma_{B^-}(\mathbf{g}), \gamma_0(\mathbf{v}_h) \rangle.$$

The cell vertex finite volume approximation of the boundary-value problem (27), (28) is now defined as follows: find $\mathbf{u}_h \in X_h$ such that

$$a_h(\mathbf{u}_h, \mathbf{v}_h) = l(\mathbf{v}_h) \qquad \forall \mathbf{v}_h \in Y_h. \tag{29}$$

In order to proceed we shall suppose that (29) has a unique solution $\mathbf{u}_h$; for the discussion of conditions which are sufficient to ensure that this is the case, and for theoretical results concerning the stability and convergence of the cell vertex scheme we refer to [5], [44], [45], [53], [54] and [55].

Theorem 13. *Suppose that the entries of the matrices A_i, $i = 1, 2$, are in $C^2(\bar{\Omega})$ and those of C belong to $C^1(\bar{\Omega})$. Then, the cell vertex scheme obeys the following* a posteriori *error bound:*

$$\|\mathbf{u}-\mathbf{u}_h\|_{[H^{-1}(\Omega)]^m} \le C_1'\Big[C_2\Big(\|h\mathbf{r}_h\|_{[L_2(\Omega)]^m} + \|h^{1/2}\mathbf{r}_h^-\|_{[L_2(\partial\Omega)]^m}\Big) + \|\hat{\mathbf{r}}_h\|_{[L_2(\Omega)]^m}\Big]$$

where C_1' is the constant from the strong stability estimate (11), C_2 is the constant from the approximation property (c'), $\mathbf{r}_h$ and $\mathbf{r}_h^-$ are the interior and boundary residual, respectively, and $\hat{\mathbf{r}}_h = \Pi_h \mathrm{div}\,(\mathcal{I}_h(\mathcal{A}\mathbf{u}_h) - \mathcal{A}\mathbf{u}_h)$.

Proof. Suppose that $\psi \in [C_0^\infty(\Omega)]^m$ and consider the adjoint problem

$$\mathcal{L}^*\mathbf{z} = \psi \qquad \text{on } \Omega, \qquad\qquad \gamma_{B^+}(\phi) = \mathbf{0}.$$

Then,

$$\begin{aligned}(\mathbf{u}-\mathbf{u}_h, \psi) &= (\mathbf{u}-\mathbf{u}_h, \mathcal{L}^*\mathbf{z}) \\ &= (\mathcal{L}(\mathbf{u}-\mathbf{u}_h), \mathbf{z}) - \langle \gamma_{B^-}(\mathbf{u}-\mathbf{u}_h), \mathbf{z}\rangle \\ &= (\mathbf{r}_h, \mathbf{z}) - \langle \mathbf{r}_h^-, \mathbf{z}\rangle \\ &= (\mathbf{r}_h, \mathbf{z}-\mathbf{z}_h) - \langle \mathbf{r}_h^-, \mathbf{z}-\mathbf{z}_h\rangle + (\hat{\mathbf{r}}_h, \mathbf{z}_h) \\ &= I + II + III.\end{aligned}$$

Thus, choosing $\mathbf{z}_h = \Pi_h\mathbf{z}$ and exploiting the fact that with this choice of $\mathbf{z}_h$ the approximation property (c') holds, we have that

$$|III| \le \|\hat{\mathbf{r}}_h\|_{[L_2(\Omega)]^m}\, \|\mathbf{z}\|_{[L_2(\Omega)]^m},$$

and

$$|I + II| \le C_2\left(\|h\mathbf{r}_h\|_{[L_2(\Omega)]^m} + \|h^{1/2}\mathbf{r}_h^-\|_{[L_2(\partial\Omega)]^m}\right)\|\mathbf{z}\|_{[H^1(\Omega)]^m},$$

as in the proof of Theorem 11. Adding up the bounds on I, II and III, and recalling the strong stability estimate

$$\|\mathbf{z}\|_{[H^1(\Omega)]^m} \le C_1'\|\psi\|_{[H^1(\Omega)]^m},$$

we obtain the desired result. □

The third term on the right-hand side of this *a posteriori* error bound can be rewritten as

$$\sup_{\mathbf{v}_h \in Y_h} \frac{|a(\mathbf{u}_h, \mathbf{v}_h) - a_h(\mathbf{u}_h, \mathbf{v}_h)|}{\|\mathbf{v}_h\|_{[L_2(\Omega)]^m}},$$

and can be thought of as the consistency error between the bilinear functional $a(\cdot,\cdot)$ and its discretisation $a_h(\cdot,\cdot)$, resulting from the non-Galerkin-type error committed by applying a numerical integration rule.

4.5 Reliable quantitative error control and adaptivity

We have established a number of *a posteriori* error bounds on the global error $\mathbf{u} - \mathbf{u}_h$ for finite element and finite volume approximations of symmetric positive systems. These bounds are of the following generic form:

$$\|\mathbf{u} - \mathbf{u}_h\|_{[H^{-1}(\Omega)]^m} \leq C_* \left(\sum_{\kappa \in \mathcal{T}_h} |\eta_\kappa(\mathbf{u}_h)|^2 \right)^{1/2}, \tag{30}$$

where C_* is a 'computable' constant and $\eta_\kappa(\mathbf{u}_h)$ is a *local error indicator* on element κ involving the numerical solution $\mathbf{u}_h$; in particular, in Theorem 10

$$\eta_\kappa(\mathbf{u}_h) = \|h\mathbf{r}_h\|_{[L_2(\kappa)]^m},$$

in Theorems 11 and 12

$$\eta_\kappa(\mathbf{u}_h) = \left(\|h\mathbf{r}_h\|^2_{[L_2(\kappa)]^m} + \|h^{1/2}\mathbf{r}_h^-\|^2_{[L_2(\partial\kappa \cap \partial\Omega)]^m} \right)^{1/2},$$

and in Theorem 13 we have that

$$\eta_\kappa(\mathbf{u}_h) = \left(\|h\mathbf{r}_h\|^2_{[L_2(\kappa)]^m} + \|h^{1/2}\mathbf{r}_h^-\|^2_{[L_2(\partial\kappa \cap \partial\Omega)]^m} + \|\hat{\mathbf{r}}_h\|^2_{[L_2(\kappa)]^m} \right)^{1/2}.$$

The right-hand side in the error bound (30) can be evaluated once the finite element solution $\mathbf{u}_h$ has been computed and can be used to estimate the size of the global error in the norm of $[H^{-1}(\Omega)]^m$. Moreover, exploiting the *a posteriori* error bound it is possible to adaptively control the global error to a desired tolerance level by suitably refining the partition. In order to achieve *reliability* in the sense that

$$\|\mathbf{u} - \mathbf{u}_h\|_{[H^{-1}(\Omega)]^m} \leq \mathrm{TOL},$$

where TOL is the prescribed *error tolerance*, it suffices to ensure that

$$\eta_\Omega(\mathbf{u}_h) \equiv C_* \left(\sum_{\kappa \in \mathcal{T}_h} |\eta_\kappa(\mathbf{u}_h)|^2 \right)^{1/2} \leq \mathrm{TOL}.$$

In addition to reliability we are also concerned with *efficiency*, which means that among all possible partitions which yield an approximation with this accuracy we want to determine (the) one that has the smallest number of degrees of freedom. Constructing a partition that is optimal in this sense is a difficult nonlinear optimisation problem whose solution is rarely attempted in practice. The usual approach to constructing a partition which does not contain an excessively large number of elements is to proceed iteratively: we start with a coarse mesh and refine it successively based on the size of the *a posteriori* error estimate, and in the course of doing so we try to keep the number of elements as small as possible. The last inequality can be thought of as a stopping criterion in this iterative processes. In fact, one can adopt various strategies to generate a sequence of partitions from an initial coarse mesh; here we mention only three of the most popular approaches, following Rannacher and Suttmeier [49].

Let an error tolerance TOL or a maximal number of elements $N_{\max}$ be given. Starting from some initial coarse partition, the refinement criteria are chosen in terms of the local error indicators $\eta_\kappa(\mathbf{u}_h)$.

1. *Error-per-cell strategy.* In this approach the mesh generation aims to equilibrate the local error indicators by refining or coarsening the elements κ in the current partition $\mathcal{T}_h$ according to the criterion
$$\eta_\kappa(\mathbf{u}_h) \approx \frac{TOL}{C_* \sqrt{N}},$$
where N is the (predicted) number of elements in the resulting new partition. Since N depends on the result of the refinement decision, this strategy is implicit and requires an iterative implementation. It is common practice to work with a varying value of N on each refinement level, with N successively updated according to the outcome of the refinement process. This strategy will deliver a partition on which $\eta_\Omega(\mathbf{u}_h) \approx TOL$, provided that N_{max} is not exceeded.
2. *Fixed-fraction strategy.* In each refinement step, the elements are ordered according to the size of the local error indicator $\eta_\kappa(\mathbf{u}_h)$, and then a fixed portion (in two dimensions, typically 30%) of the elements κ with largest $\eta_\kappa(\mathbf{u}_h)$ is refined (resulting in about doubling the number of elements). This process is repeated until the stopping criterion $\eta_\Omega(\mathbf{u}_h) \leq TOL$ is satisfied, or $N_{\max}$ is exceeded.
3. *Fixed-reduction strategy.* Here one works with a variable tolerance TOL_{var}. Supposing that on a partition the approximate solution $\mathbf{u}_h$ has been obtained, the tolerance is set to $TOL_{var} = \sigma\,\eta(\mathbf{u}_h)$, where $\sigma \in (0,1)$ is a fixed reduction factor (e.g. $\sigma = 0.5$). In the next step one (or several) cycles of the *error-per-cell* strategy are performed with tolerance TOL_{var}; this provides a new mesh $\mathcal{T}_h^{new}$ and a new solution $\mathbf{u}_h^{new}$ with associated error estimator $\eta(\mathbf{u}_h^{new})$. Then the tolerance is reduced again by setting $TOL_{var} = \sigma\,\eta(\mathbf{u}_h^{new})$ and a new refinement cycle begins. This iterative process is repeated until $TOL_{var} \leq TOL$, or N_{max} is exceeded.

In each of the three strategies we repeat mesh modification followed by solution on the new partition until the tolerance is satisfied, or the prescribed maximum number of elements is exceeded.

We conclude this section by showing that the adaptive algorithms outlined above will terminate in a finite number of steps. We shall suppose, for simplicity, that only mesh refinements are carried out and no derefinements are done. It is clear that if reaching a prescribed maximum number is taken as stopping criterion, then the mesh refinement algorithm will terminate after a finite number of steps. If an error tolerance is given instead as stopping criterion, then termination of the refinement algorithm can be ensured by proving that the finite element method satisfies the *a priori* error bound

$$\|\mathbf{u}-\mathbf{u}_h\|_{[L_2(\Omega)]^m}+|h|\,|\mathbf{u}-\mathbf{u}_h|_{[H^1(\Omega)]^m}\leq C(\mathbf{u})|h|^{1-\epsilon}, \tag{31}$$

where $C(\mathbf{u})$ depends on $\mathbf{u}$ (and its Sobolev smoothness),

$$|h|=\max\{h_\kappa\,:\,\kappa\in\mathcal{T}_h\},$$

and ϵ is a fixed real number in the interval $[0,1)$.

Concerning finite element approximations of the kind mentioned in Theorems 10 – 12, an *a priori* error bound of the type (31) can be derived under suitable assumptions on the smoothness of $\mathbf{u}$, the choice of the trial and test space, and the regularity of the partition. Then, noting that $\mathbf{r}_h=\mathcal{L}(\mathbf{u}-\mathbf{u}_h)$, it follows that

$$\|h\mathbf{r}_h\|_{[L_2(\Omega)]^m}\leq Const.|h|\|\mathbf{u}-\mathbf{u}_h\|_{[H^1(\Omega)]^m}\leq Const.|h|^{1-\epsilon},$$

and similarly,

$$\begin{aligned}\|h^{1/2}\mathbf{r}_h^-\|_{[L_2(\partial\Omega)]^m}&\leq Const.|h|^{1/2}\|\gamma_0(\mathbf{u}-\mathbf{u}_h)\|_{[L_2(\partial\Omega)]^m}\\&\leq Const.|h|^{1/2}\|\mathbf{u}-\mathbf{u}_h\|^{1/2}_{[L_2(\Omega)]^m}\|\mathbf{u}-\mathbf{u}_h\|^{1/2}_{[H^1(\Omega)]^m}\\&\leq Const.|h|^{1-\epsilon}.\end{aligned}$$

Thus, considering Theorems 10 – 12, it is a simple matter to show that, under the same assumptions as are required to ensure that the *a priori* error bound (31) holds, we have that

$$\left(\sum_{\kappa\in\mathcal{T}_h}|\eta_\kappa(\mathbf{u}_h)|^2\right)^{1/2}\to 0\qquad\text{as } |h|\to 0,$$

and, therefore, the stopping criterion will be satisfied eventually as the mesh is refined.

Concerning Theorem 13, it can be shown that the cell vertex finite volume method satisfies an *a priori* error bound of the type (31). More precisely, suppose that $\mathbf{u}\in[H^s(\Omega)]^m$ with $s>1$, that the components of the matrices

A_i, $i = 1, \dots, n$, and C belong to $C^{[s]+1}(\bar{\Omega})$, that the matrices A_i are positive definite, uniformly on $\bar{\Omega}$, and that the family of structured partitions $\{\mathcal{T}_h\}$ is quasi-parallel (namely, there exists a fixed positive constant c_* independent of $|h|$ such that, for each κ in $\mathcal{T}_h$, the distance between the midpoints of the two diagonals is bounded by $c_*|h|^2$); then

$$\begin{aligned}\|\mathbf{u} - \mathbf{u}_h\|_{[L_2(\Omega)]^m} &+ \|\Pi_h \mathrm{div}(\mathcal{A}(\mathbf{u} - \mathbf{u}_h))\|_{[L_2(\Omega)]^m} \\ &+|h||\mathbf{u} - \mathbf{u}_h|_{[H^1(\Omega)]^m} \leq Const.|h|^{r-1}\|\mathbf{u}\|_{[H^r(\Omega)]^m},\end{aligned}$$

for $1 < r \leq \min(s, 3)$. In the scalar case ($m = 1$) and uniform square meshes this has been proved in [5]; the extension of the error analysis presented in [5] to the case of $m > 1$ is straightforward, while quasi-parallel meshes can be dealt with by following the analysis in [54]. At any rate, under these hypotheses and taking $r = 2 - \epsilon$, $\epsilon \in [0, 1)$, it follows that that

$$\left(\sum_{\kappa \in \mathcal{T}_h} |\eta_\kappa(\mathbf{u}_h)|^2\right)^{1/2} \to 0 \qquad \text{as } |h| \to 0,$$

and, therefore, the stopping criterion will be satisfied eventually as the mesh is refined.

The use of an *a priori* error bound to prove that the refinement algorithm terminates after a finite number of steps presupposes that the hypotheses under which the *a priori* error bound has been established are valid; in practice, this may be a restrictive requirement, and it is likely that termination will occur in circumstances which are less demanding than those in *a priori* error analysis.

5 Local considerations for steady problems

In the previous section we derived various *a posteriori* error bounds of the general form

$$\|\mathbf{u} - \mathbf{u}_h\|_{[H^{-1}(\Omega)]^m} \leq \text{Const.} \left(\sum_{\kappa \in \mathcal{T}_h} |\eta_\kappa(\mathbf{u}_h)|^2\right)^{1/2},$$

and we showed how reliable quantitative error control, to within a given tolerance, can be achieved through a feed-back process based on mesh adaptation. We also showed that, under mesh refinement, the local error indicator $\eta_\kappa(\mathbf{u}_h)$ must converge to zero. However, it is not clear from these global considerations *to what extent the reduction of the local error indicator on element κ contributes to the reduction of the global error $\mathbf{e}_h = \mathbf{u} - \mathbf{u}_h$ restricted to element κ?* In this section we shall approach this question from two different viewpoints:

1. We argue that, due to error propagation phenomena, the local error indicator $\eta_\kappa(\mathbf{u}_h)$ on a particular element κ controls only a portion of $\mathbf{e}_h|_\kappa$, namely a local quantity $\mathbf{e}_\kappa^{cell}$ called the *cell error*; error propagation is a non-local process and the complementary portion of the error, $\mathbf{e}_\kappa^{trans} = \mathbf{e}_h - \mathbf{e}_\kappa^{cell}$, called the *transmitted error*, does not obey a local bound.
2. Having shown that the local error indicator puts a bound only on part of the global error on each element, we prove that, at least for scalar problems, $\mathbf{e}_h$ restricted to element κ_0 can be bounded by the sum of local error indicators $\eta_\kappa(\mathbf{u}_h)$ over all κ that intersect the *domain of dependence* of κ_0.

The first viewpoint, analysed in Section 5.1 below, highlights the fact that by reducing the local residual $\mathbf{r}_h|_\kappa$ we reduce only the part of the global error which has been 'created' in κ, but not the part which has been 'transported' into κ. The second viewpoint, discussed in Section 5.2, shows that in order to reduce the whole of the global error in an element κ_0, we have to reduce the residual in each element κ whose *domain of influence* intersects κ_0.

5.1 What is controlled by the local residual?

This section is based on the papers [25], [42] and [56]. Here we shall restrict ourselves to an overview of the main results; the reader is referred to these papers for further details.

Let us suppose that κ is a Lipschitz subdomain of Ω whose boundary $\partial\kappa$ is a non-characteristic hypersurface for the operator $\mathcal{L}$. The domain κ can be an element in the finite element partition of Ω or a union of neighbouring elements; we shall refer to κ as *cell.* Throughout this section $(\cdot,\cdot)_\kappa$ will denote the inner product of the Hilbert space $[L_2(\kappa)]^m$, and $(\cdot,\cdot)_{\partial\kappa}$ will signify the inner product of $[L_2(\partial\kappa)]^m$.

On κ we consider the local boundary-value problem

$$\mathcal{L}\tilde{\mathbf{u}}_h = \mathbf{f} \quad \text{on } \kappa, \qquad \gamma_{B^-}(\tilde{\mathbf{u}}_h - \mathbf{u}_h) = \mathbf{0} \quad \text{on } \partial\kappa.$$

According to the theory outlined in Section 3.2, this problem has a unique strong solution $\tilde{\mathbf{u}}_h$; in fact, $\tilde{\mathbf{u}}_h$ can be thought of as a local solution of the partial differential equation (12) subject to a boundary condition whose data is a distortion of the correct local boundary data $B^-\mathbf{u}|_{\partial\kappa}$, due to the numerical error that has been 'created' outside the cell and is being advected into κ through the boundary $\partial\kappa$. We shall refer to the quantity

$$\mathbf{e}_\kappa^{cell} = \tilde{\mathbf{u}}_h - \mathbf{u}_h$$

as the *cell error*; clearly, $\mathbf{e}_\kappa^{cell}$ belongs to $D(\mathcal{L},\kappa)$ and it is the solution of the local boundary-value problem:

$$\mathcal{L}\mathbf{e}_\kappa^{cell} = \mathbf{r}_h \quad \text{on } \kappa, \qquad \gamma_{B^-}(\mathbf{e}_\kappa^{cell}) = \mathbf{0} \quad \text{on } \partial\kappa. \tag{32}$$

Thus $\mathbf{e}_\kappa^{cell}$ is governed by $\mathbf{r}_h|_\kappa$ and is not influenced by numerical effects which occur outside κ. The complementary quantity

$$\mathbf{e}_\kappa^{trans} = \mathbf{u} - \tilde{\mathbf{u}}_h,$$

called the *transmitted error*, represents the component of the global error $\mathbf{e}_h$ which has been created upwind of the cell κ and is merely advected into it. The transmitted error restricted to κ is the solution of the local problem

$$\mathcal{L}\mathbf{e}_\kappa^{trans} = \mathbf{0} \quad \text{on } \kappa, \qquad \gamma_{B^-}(\mathbf{e}_\kappa^{trans} - \mathbf{e}_h) = \mathbf{0} \quad \text{on } \partial\kappa.$$

We note that the concept of cell error is analogous to the notion of *local error* arising in the theory of numerical approximations to ordinary differential equations (see Hairer, Norsett and Wanner [22]).

With these definitions, we have the following decomposition of the global error:

$$\mathbf{e}_h\big|_\kappa = \mathbf{e}_\kappa^{cell} + \mathbf{e}_\kappa^{trans}.$$

Equation (32) shows that the local residual $\mathbf{r}_h\big|_\kappa$ is directly related to the 'locally created' part of the global error, $\mathbf{e}_\kappa^{cell}$, on cell κ.

Next we shall state sharp two-sided bounds on the cell error in terms of the cell residual; these will show that it is reasonable to attempt to improve the accuracy of the numerical solution by reducing the size of the residual on those cells where it is largest. In order to simplify the presentation, we shall suppose that the centre of the coordinate system is the centroid (centre of gravity) of cell κ; if this is not the case, then the local exponential weight function $\exp\{-\alpha(\xi\cdot x)\}$ in Theorems 14 and 15 below should be replaced by $\exp\{-\alpha(\xi\cdot(x-x_c))\}$, where x_c is the centroid of cell κ; this ensures that the local weight function is close to 1 when $h = \mathrm{diam}(\kappa) << 1$.

Theorem 14. *We have the two-sided local error bound:*

$$\min_{x\in\kappa} w(x)\|\mathbf{r}_h\|_{[L_2(\kappa)]^m} \le |||\mathbf{e}_\kappa^{cell}|||_{\kappa,\xi} \le c_0'(\kappa)\max_{x\in\kappa} w(x)\|\mathbf{r}_h\|_{[L_2(\kappa)]^m}, \quad (33)$$

where $w(x) = \exp\{-\alpha(\xi\cdot x)\}$, $c_0'(\kappa) = (1 + 1/c_0(\kappa)^2)^{1/2}$, *and* $c_0(\kappa)$ *is the constant from condition (b) applied on the cell* κ *(clearly,* $c_0(\kappa) \ge C_0(\Omega)$ *for all* $\kappa \subset \Omega$*).*

Proof. Recalling that $\mathbf{r}_h = \mathcal{L}\mathbf{e}_\kappa^{cell}$ on cell κ, the first inequality is a straightforward consequence of the definition of the norm $|||\cdot|||_{\kappa,\xi}$. The second inequality follows from the Gårding inequality (10) with Ω replaced by κ and $\mathbf{v} = \mathbf{e}_\kappa^{cell}$. □

It is intuitively clear that the global error $\mathbf{e}_h$ is non-local in character, and an error committed in certain part of the computational domain (say, near an inflow boundary, for a scalar hyperbolic boundary-value problem) will be also felt in other parts of the domain. Thus, it is unreasonable to expect that the restriction of the global error to a cell is controllable merely in terms of the residual on that cell. This is, indeed, the case: in contrast with the local two-sided estimate obeyed by the cell error, the transmitted error satisfies only a *non-local* one-sided error bound; namely,

$$c_0(\kappa)\|w\mathbf{e}_\kappa^{trans}\|^2_{[L_2(\kappa)]^m}+(B^+w\mathbf{e}_\kappa^{trans},w\mathbf{e}_\kappa^{trans})_{\partial\kappa}\leq(-B^-w\mathbf{e}_\kappa^{trans},w\mathbf{e}_\kappa^{trans})_{\partial\kappa},$$

where $w(x)$ is as in the previous theorem. We say that the bound is non-local because it involves $B^-\mathbf{e}_\kappa^{trans}|_{\partial\kappa}$, the 'incoming components' of $\mathbf{e}_\kappa^{trans}$ which have been created outside κ and are being transported into κ; from the point of view of an observer sitting in cell κ these are pollution effects from outside κ. The proof of this error bound is based on taking the L_2 inner product on κ of the equality $\mathcal{L}\mathbf{e}_\kappa^{trans}=\mathbf{0}$ with $\mathbf{e}_\kappa^{trans}$, integrating by parts, and splitting the conormal trace operator into partial conormal traces.

Using a duality argument, it is possible to derive a local two-sided bound on the L_2 norm of the cell error in terms of the dual graph norm of the residual. This is stated in the next theorem.

Theorem 15. *Suppose that* $\mathbf{e}_\kappa^{cell}\in[H^1(\kappa)]^m$*; then, we have the two-sided local error bound:*

$$\min_{x\in\kappa}\hat{w}(x)|||\mathbf{r}_h|||'_{*,\kappa,\xi}\leq\|\mathbf{e}_\kappa^{cell}\|_{[L_2(\kappa)]^m}\leq c_0'(\kappa)\max_{x\in\kappa}\hat{w}(x)|||\mathbf{r}_h|||'_{*,\kappa,\xi},\qquad(34)$$

where $\hat{w}(x)=1/w(x)$*, and* $w(x)$ *and* $c_0'(\kappa)$ *are as in the previous theorem.*

Proof. Recalling the definition of the dual graph norm $|||\cdot|||'_{*,\kappa}$, we have that

$$|||\mathbf{r}_h|||'_{*,\kappa,\xi}=|||\mathcal{L}\mathbf{e}_\kappa^{cell}|||'_{*,\kappa,\xi}=\sup_{\phi\in D(\mathcal{L}^*,\kappa)}\frac{(\mathcal{L}\mathbf{e}_\kappa^{cell},\phi)_\kappa}{|||\phi|||_{*,\kappa,\xi}}=\sup_{\phi\in D(\mathcal{L}^*,\kappa)}\frac{(\mathbf{e}_\kappa^{cell},\mathcal{L}^*\phi)_\kappa}{|||\phi|||_{*,\kappa,\xi}}$$

$$\leq\sup_{\phi\in D(\mathcal{L}^*,\kappa)}\frac{\|\mathrm{e}^{-\alpha(\xi\cdot x)}\mathbf{e}_\kappa^{cell}\|_{[L_2(\kappa)]^m}\|\mathrm{e}^{\alpha(\xi\cdot x)}\mathcal{L}^*\phi\|_{[L_2(\kappa)]^m}}{|||\phi|||_{*,\kappa,\xi}}$$

$$\leq\|\mathrm{e}^{-\alpha(\xi\cdot x)}\mathbf{e}_\kappa^{cell}\|_{[L_2(\kappa)]^m}\leq\max_{x\in\kappa}w(x)\|\mathbf{e}_\kappa^{cell}\|_{[L_2(\kappa)]^m}.$$

Hence the first inequality. To prove the second inequality, we consider the local adjoint boundary-value problem

$$\mathcal{L}^*\varphi=\mathrm{e}^{-2\alpha(\xi\cdot x)}\mathbf{e}_\kappa^{cell}\quad\text{on }\kappa,\qquad B^+\varphi=\mathbf{0}\quad\text{on }\partial\kappa,$$

and we note that the corresponding (unique) solution φ belongs to $D(L^*,\kappa)$. Now since $\mathbf{e}_\kappa^{cell}\in D(L,\kappa)\cap[H^1(\kappa)]^m$, upon integration by parts we have that

$$|||\mathbf{r}_h|||'_{*,\kappa,\xi}=\sup_{\phi\in D(\mathcal{L}^*,\kappa)}\frac{(\mathcal{L}\mathbf{e}_\kappa^{cell},\phi)_\kappa}{|||\phi|||_{*,\kappa,\xi}}\geq\frac{(\mathcal{L}\mathbf{e}_\kappa^{cell},\varphi)_\kappa}{|||\varphi|||_{*,\kappa,\xi}}=\frac{(\mathbf{e}_\kappa^{cell},\mathcal{L}^*\varphi)_\kappa}{|||\varphi|||_{*,\kappa,\xi}}$$

$$=\frac{(\mathrm{e}^{\alpha(\xi\cdot x)}\mathcal{L}^*\varphi,\mathrm{e}^{\alpha(\xi\cdot x)}\mathcal{L}^*\varphi)_\kappa}{|||\varphi|||_{*,\kappa,\xi}}=\frac{\|\mathrm{e}^{\alpha(\xi\cdot x)}\mathcal{L}^*\varphi\|^2_{[L_2(\kappa)]^m}}{|||\varphi|||_{*,\kappa},\xi}$$

$$\geq\frac{1}{c_0'(\kappa)}\|\mathrm{e}^{\alpha(\xi\cdot x)}\mathcal{L}^*\varphi\|_{[L_2(\kappa)]^m}=\frac{1}{c_0'(\kappa)}\|\mathrm{e}^{-\alpha(\xi\cdot x)}\mathbf{e}_\kappa^{cell}\|_{[L_2(\kappa)]^m}.$$

That completes the proof. □

The *a posteriori* bounds (33) and (34) provide sharp estimates of the cell error; unfortunately, the dual graph norm of the residual is difficult to compute in practice since its definition involves a supremum over the infinite set $D(\mathcal{L}^*, \kappa)$ (although, in [52] the dual graph norm $|||\cdot|||'_{*,\kappa,\xi}$ was approximated by partitioning the cell κ and considering the supremum over a finite dimensional subspace of $D(\mathcal{L}^*, \kappa)$ consisting of piecewise linear functions on such micro-partitions; this approximation was then successfully implemented into an adaptive finite volume algorithm for the numerical solution of the Euler equations of compressible gas dynamics in two space dimensions). In addition to the fact that the dual graph norm is unattractive from the computational point of view, it is not clear at this stage how (34) relates to the *a posteriori* error bounds established in the previous section which involved $\|h\mathbf{r}_h\|_{[L_2(\kappa)]^m}$ instead of $|||\mathbf{r}_h|||'_{*,\kappa,\xi}$. Our aim now is to resolve these issues by showing that $|||\mathbf{r}_h|||'_{*,\kappa,\xi}$ can be further bounded above by a constant multiple of $\|h\mathbf{r}_h\|_{[L_2(\kappa)]^m}$, a quantity that is simple and cheap to compute; we shall also prove that there is a similar lower bound. Thus we shall obtain a local two-sided bound on the L_2 norm of the cell error in terms of the L_2 norm of the finite element residual $\mathbf{r}_h$ scaled by the local mesh size. These results will establish a connection between the global *a posteriori* bounds of Section 4 and the local estimates on the cell error stated earlier in this section.

The theory of error estimation that we described so far is valid for any symmetric positive system, irrespective of its type. In order to proceed, we shall replace the positivity condition (b) by a stronger hypothesis, thereby restricting ourselves to symmetric hyperbolic systems. Namely, we shall suppose the following:

(b') There exists $\xi \in \mathbb{R}^n$ such that the matrix $\sum_{i=1}^n \xi_i A_i$ is positive definite, uniformly on $\bar{\Omega}$; i.e. there is a positive constant $c_0 = c_0(\Omega)$, such that

$$\sum_{i=1}^{n} \xi_i A_i(x) \geq c_0 I \qquad \text{for all } x \in \bar{\Omega}.$$

This condition is referred to as *hyperbolicity in the sense of Lax* (see [36]). In the rest of this section we shall assume that (b') holds instead of (b).

Theorem 16. *Suppose that $\mathbf{e}_\kappa^{cell} \in [H^1(\kappa)]^m$; then we have the following one-sided* a posteriori *error bound on the cell error:*

$$\|\mathbf{e}_\kappa^{cell}\|_{[L_2(\kappa)]^m} \leq c_3(\kappa)\|h\mathbf{r}_h\|_{[L_2(\kappa)]^m},$$

where

$$c_3(\kappa) = c_2(\kappa)(1 + 1/c_0(\kappa)^2)^{1/2}\exp(\alpha(1+h)|\xi|),$$
$$c_2(\kappa) = (h^2 + c_0(\kappa)^2/4)^{-1/2}\exp(\alpha|\xi|),$$

$c_0(\kappa)$ is the constant from condition (b') applied on the cell κ and h denotes the diameter of κ.

Proof. Let us choose $\zeta \in \mathbb{R}^n$ (to be fixed later on) and let κ be any element in the partition of Ω. In order to simplify the presentation we shall assume that the origin of the coordinate system is the centroid (centre of gravity) of κ; if this is not the case then the local exponential weight-function $\exp(\alpha(\zeta \cdot x))$ in the expressions below should be replaced by $\exp(\alpha(\zeta \cdot (x - x_c)))$ where x_c is the centroid of κ, so that the local weight function remains bounded on κ as $h = \operatorname{diam}(\kappa)$ converges to zero. The proof consists of two parts. First we prove the local Gårding inequality stated in (39) below. In the second part of the proof, we use this inequality to show that the dual graph norm of the residual, $|||\mathbf{r}_h|||'_{*,\kappa,\xi}$, is bounded above by a constant multiple of $||h\mathbf{r}_h||_{[L_2(\kappa)]^m}$; this, together with the second inequality in (34) will yield the desired result.

Part 1: Given that ϕ is an element of $D(\mathcal{L}^*, \kappa) \cap [H^1(\kappa)]^m$, we have that

$$\int_\kappa \mathrm{e}^{2\alpha(\zeta\cdot x)} \left(-\sum_{i=1}^n A_i \frac{\partial \phi}{\partial x_i} + C^*\phi \right) \cdot \phi \, \mathrm{d}x = -\frac{1}{2} \int_{\partial\kappa} \mathrm{e}^{2\alpha(\zeta\cdot x)} \phi \cdot (B\phi) \, \mathrm{d}s$$
$$+ \int_\kappa \mathrm{e}^{2\alpha(\zeta\cdot x)} \phi \cdot \left(C + \alpha \sum_{i=1}^n \zeta_i A_i + \frac{1}{2} \sum_{i=1}^n \frac{\partial A_i}{\partial x_i} \right) \phi \, \mathrm{d}x. \quad (35)$$

Since we are dealing with real-valued functions,

$$(\mathrm{e}^{2\alpha(\zeta\cdot x)}\phi, C\phi)_\kappa = (\mathrm{e}^{2\alpha(\zeta\cdot x)} C^*\phi, \phi)_\kappa = (\mathrm{e}^{2\alpha(\zeta\cdot x)}\phi, C^*\phi)_\kappa.$$

Applying this identity in the second integral on the right-hand side of (35) gives

$$\int_\kappa \mathrm{e}^{2\alpha(\zeta\cdot x)} \left(-\sum_{i=1}^n A_i \frac{\partial \phi}{\partial x_i} + C^*\phi \right) \cdot \phi \, \mathrm{d}x = -\frac{1}{2} \int_{\partial\kappa} \mathrm{e}^{2\alpha(\zeta\cdot x)} \phi \cdot (B\phi) \, \mathrm{d}s$$
$$+ \int_\kappa \mathrm{e}^{2\alpha(\zeta\cdot x)} \phi \cdot \left(C^* + \alpha \sum_{i=1}^n \zeta_i A_i + \frac{1}{2} \sum_{i=1}^n \frac{\partial A_i}{\partial x_i} \right) \phi \, \mathrm{d}x. \quad (36)$$

Adding (35) and (36), and noting that $B = B^+ + B^-$ with $B^+\phi = 0$ on $\partial\kappa$ and $-\phi \cdot (B^-\phi) \geq 0$ on $\partial\kappa$, we deduce that

$$\int_\kappa \mathrm{e}^{2\alpha(\zeta\cdot x)} \mathcal{L}^*\phi \cdot \phi \, \mathrm{d}x \geq \int_\kappa \mathrm{e}^{2\alpha(\zeta\cdot x)} \phi \cdot \frac{1}{2}(K_\zeta(x) + K_\zeta^*(x))\phi \, \mathrm{d}x. \quad (37)$$

The remainder of Part 1 of the proof is devoted to showing that the matrix $K_\zeta(x) + K_\zeta^*(x)$ is positive definite, uniformly on κ, and that the inequality (38) below holds. Substituting (38) into (37) will then yield the Gårding inequality (39). Let us therefore consider

$$\frac{1}{2}(K_\zeta(x) + K_\zeta^*(x)) = \frac{1}{2}(C + C^*) + \frac{1}{2} \sum_{i=1}^n \frac{\partial A_i}{\partial x_i} + \alpha \sum_{i=1}^n \zeta_i A_i.$$

So far ζ has been an arbitrary vector from $\mathbb{R}^n$; now (as promised at the beginning of the proof) we fix its value and take

$$\zeta_i = h^{-1}\xi_i, \quad i = 1, \dots, n, \qquad \text{where } h = \text{diam}(\kappa).$$

Applying hypothesis (b') we have that

$$\frac{1}{2}(K_\zeta(x) + K_\zeta^*(x)) \geq h^{-1}\left(\alpha c_0(\kappa) I + h\left(\frac{1}{2}(C + C^*) + \frac{1}{2}\sum_{i=1}^{n}\frac{\partial A_i}{\partial x_i}\right)\right).$$

Since, by assumption, the entries of C and $\partial A_i/\partial x_i$ belong to $C(\overline{\Omega})$, it follows that, for h sufficiently small and all $x \in \kappa$,

$$\frac{1}{2}(K_\zeta(x) + K_\zeta^*(x)) \geq \frac{\alpha c_0(\kappa)}{2h} I. \tag{38}$$

Substituting (38) into (37) gives the local Gårding inequality

$$\int_\kappa e^{2\alpha(\zeta\cdot x)} \mathcal{L}^*\phi \cdot \phi \, dx \geq \frac{\alpha c_0(\kappa)}{2h} \int_\kappa e^{2\alpha(\zeta\cdot x)} |\phi|^2 \, dx \tag{39}$$

for all $\phi \in D(\mathcal{L}^*, \kappa) \cap [H^1(\Omega)]^m$, and by density also for all $\phi \in D(\mathcal{L}^*, \kappa)$. Further, applying the Cauchy-Schwarz inequality to the left-hand side of (39), it follows that

$$\left(\int_\kappa e^{2\alpha(\zeta\cdot x)} |\phi|^2 \, dx\right)^{\frac{1}{2}} \leq \frac{2h}{\alpha c_0(\kappa)} \left(\int_\kappa e^{2\alpha(\zeta\cdot x)} |\mathcal{L}^*\phi|^2 \, dx\right)^{\frac{1}{2}} \tag{40}$$

for all ϕ in $D(\mathcal{L}^*, \kappa)$. This completes the first part of the proof.

Part 2: Now we use (40) to show that the dual graph norm of $\mathbf{r}_h$ is bounded above by a constant multiple of $\|h\mathbf{r}_h\|_{[L_2(\kappa)]^m}$; indeed,

$$\begin{aligned}
|||\mathbf{r}_h|||'_{*,\kappa,\zeta} &= \sup_{\phi \in D(\mathcal{L}^*,\kappa)} \frac{|(\mathbf{r}_h, \phi)_\kappa|}{(\|e^{\alpha(\zeta\cdot x)}\phi\|^2_{[L_2(\kappa)]^m} + \|e^{\alpha(\zeta\cdot x)}\mathcal{L}^*\phi\|^2_{[L_2(\kappa)]^m})^{\frac{1}{2}}} \\
&\leq \sup_{\phi \in D(\mathcal{L}^*,\kappa)} \frac{\|e^{-\alpha(\zeta\cdot x)}\mathbf{r}_h\|_{[L_2(\kappa)]^m} \|e^{\alpha(\zeta\cdot x)}\phi\|_{[L_2(\kappa)]^m}}{(\|e^{\alpha(\zeta\cdot x)}\phi\|^2_{[L_2(\kappa)]^m} + \|e^{\alpha(\zeta\cdot x)}\mathcal{L}^*\phi\|^2_{[L_2(\kappa)]^m})^{\frac{1}{2}}} \\
&\leq h(h^2 + \alpha^2 c_0(\kappa)^2/4)^{-1/2} \|e^{-\alpha(\zeta\cdot x)}\mathbf{r}_h\|_{[L_2(\kappa)]^m}.
\end{aligned} \tag{41}$$

This is essentially the desired bound on the dual graph norm of the residual, except that the left-hand side includes $|||\mathbf{r}_h|||'_{*,\kappa,\zeta}$ instead of $|||\mathbf{r}_h|||'_{*,\kappa,\xi}$, and an exponential term appears under the norm sign on the right. The concluding part of the proof shows that this exponential term is bounded independent of h and that the norms $|||\cdot|||'_{*,\kappa,\zeta}$ and $|||\cdot|||'_{*,\kappa,\xi}$ are equivalent.

As $|\alpha(\zeta \cdot x)| = h^{-1}\alpha|\xi \cdot x|$ with $h(< 1)$ denoting the diameter of κ, upon recalling that the origin of the coordinate system is at the centroid of κ, it follows that

$$|\alpha(\zeta \cdot x)| \leq \alpha|\xi|. \tag{42}$$

Substituting (42) into (41) gives

$$|||\mathbf{r}_h|||'_{*,\kappa,\zeta} \leq c_2(\kappa)||h\mathbf{r}_h||_{[L_2(\kappa)]^m}, \tag{43}$$

where $c_2(\kappa) = \mathrm{e}^{\alpha|\xi|}(h^2 + \alpha^2 c_0(\kappa)^2/4)^{-1/2}$. Now let us note that, with

$$|||\phi|||_{*,\kappa,\xi} = (||\mathrm{e}^{\alpha(\xi\cdot x)}\phi||^2_{[L_2(\kappa)]^m} + ||\mathrm{e}^{\alpha(\xi\cdot x)}\mathcal{L}^*\phi||^2_{[L_2(\kappa)]^m})^{\frac{1}{2}}$$

and

$$\mathrm{e}^{\alpha(\xi\cdot x)} = \mathrm{e}^{\alpha(\zeta\cdot x)}\,\mathrm{e}^{-\alpha(\zeta-\xi)\cdot x} = \mathrm{e}^{\alpha(\zeta\cdot x)}\,\mathrm{e}^{-\alpha\xi(1-h)\cdot(x/h)},$$

we have that

$$\mathrm{e}^{-\alpha|\xi|}|||\phi|||_{*,\kappa,\zeta} \leq |||\phi|||_{*,\kappa,\xi}.$$

Consequently,

$$|||\mathbf{r}_h|||'_{*,\kappa,\xi} = \sup_{\phi\in D(\mathcal{L}^*,\kappa)} \frac{|(\mathbf{r}_h,\phi)_\kappa|}{|||\phi|||_{*,\kappa,\xi}} \leq \mathrm{e}^{\alpha|\xi|}|||\mathbf{r}_h|||'_{*,\kappa,\zeta}. \tag{44}$$

Finally, we recall the second inequality in (34),

$$||\mathbf{e}^{cell}_\kappa||_{[L_2(\kappa)]^m} \leq c_0'(\kappa)\max_{x\in\kappa}\mathrm{e}^{\alpha(\xi\cdot x)}|||\mathbf{r}_h|||'_{*,\kappa,\xi},$$

and combine this with (43) and (44) to deduce that

$$||\mathbf{e}^{cell}_\kappa||_{[L_2(\kappa)]^m} \leq c_0'(\kappa)c_2(\kappa)\mathrm{e}^{\alpha(1+h)|\xi|}||h\mathbf{r}_h||_{[L_2(\kappa)]^m},$$

and hence the desired bound. □

Theorem 16 provides an upper bound on the L_2 norm of the cell error, analogous to the second inequality in (34). Now we prove a lower bound on the L_2 norm of the cell error, similar to the first inequality in (34). To do so, we consider a uniformly regular family of micro-partitions of the cell κ, and let $S_{\hat{h}} = S_{\hat{h}}(\kappa)$ be a finite element subspace of $D(\mathcal{L}^*,\kappa)$ on such a micro-partition; here $\hat{h} = \hat{h}(\kappa)$ denotes the maximum diameter of elements in the micro-partition. We denote by $P_{\hat{h}}$ the orthogonal projector in $[L_2(\kappa)]^m$ onto the finite element space $S_{\hat{h}}$.

Theorem 17. *We have the following* a posteriori *lower bound on the cell error:*

$$c_4(\kappa)||\hat{h}P_{\hat{h}}\mathbf{r}_h||_{[L_2(\kappa)]^m} \leq ||e^{cell}_\kappa||_{[L_2(\kappa)]^m},$$

where $c_4(\kappa)$ is a computable constant.

Proof. According to the definition of the dual graph norm, we have that

$$\begin{aligned}\|\mathbf{r}_h\|'_{*,\kappa,\xi} &= \sup_{\phi\in D(\mathcal{L}^*,\kappa)} \frac{|(\mathbf{r}_h,\phi)_\kappa|}{(\|e^{\alpha(\xi\cdot x)}\phi\|^2_{[L_2(\kappa)]^m} + \|e^{\alpha(\xi\cdot x)}\mathcal{L}^*\phi\|^2_{[L_2(\kappa)]^m})^{\frac{1}{2}}} \\ &\geq \sup_{\phi_{\hat{h}}\in S_{\hat{h}}} \frac{|(\mathbf{r}_h,\phi_{\hat{h}})_\kappa|}{(\|e^{\alpha(\xi\cdot x)}\phi_{\hat{h}}\|^2_{[L_2(\kappa)]^m} + \|e^{\alpha(\xi\cdot x)}\mathcal{L}^*\phi_{\hat{h}}\|^2_{[L_2(\kappa)]^m})^{\frac{1}{2}}},\end{aligned} \tag{45}$$

where we made use of the fact that $D(\mathcal{L}^*,\kappa) \supset S_{\hat{h}}(\kappa)$. Recalling that the family of micro-partitions of κ has been assumed uniformly regular, we can apply the standard inverse inequality (see [12])

$$\left(\sum_{i=1}^n \|e^{\alpha(\xi\cdot x)}\frac{\partial\phi_{\hat{h}}}{\partial x_i}\|^2_{[L_2(\kappa)]^m}\right)^{\frac{1}{2}} \leq \frac{c_5(\kappa)}{\hat{h}}\|e^{\alpha(\xi\cdot x)}\phi_{\hat{h}}\|_{[L_2(\kappa)]^m}, \qquad \phi_{\hat{h}}\in S_{\hat{h}},$$

to deduce that

$$\|e^{\alpha(\xi\cdot x)}\mathcal{L}^*\phi_{\hat{h}}\|_{[L_2(\kappa)]^m} \leq \hat{h}^{-1}c_6(\kappa)\|e^{\alpha(\xi\cdot x)}\phi_{\hat{h}}\|_{[L_2(\kappa)]^m},$$

where

$$c_6(\kappa) = \hat{h}\|C\|_{[L_\infty(\kappa)]^{m\times m}} + c_5(\kappa)\left(\sum_{i=1}^n \|A_i\|^2_{[L_\infty(\kappa)]^{m\times m}}\right)^{\frac{1}{2}}.$$

Therefore,

$$|||\phi_{\hat{h}}|||_{*,\kappa,\xi} \leq \hat{h}^{-1}(\hat{h}^2 + c_6(\kappa)^2)^{1/2}\|e^{\alpha(\xi\cdot x)}\phi_{\hat{h}}\|_{[L_2(\kappa)]^m}$$

for all $\phi_{\hat{h}}\in S_{\hat{h}}$. Substituting this inequality into (45) gives

$$\begin{aligned}\|\mathbf{r}_h\|'_{*,\kappa,\xi} &\geq \hat{h}(\hat{h}^2 + c_6(\kappa)^2)^{-1/2} \sup_{\phi_{\hat{h}}\in S_{\hat{h}}} \frac{|(\mathbf{r}_h,\phi_{\hat{h}})_\kappa|}{\|e^{\alpha(\xi\cdot x)}\phi_{\hat{h}}\|_{[L_2(\kappa)]^m}} \\ &= e^{-\alpha h|\xi|}(\hat{h}^2 + c_6(\kappa)^2)^{-\frac{1}{2}}\|\hat{h}P_{\hat{h}}\mathbf{r}_h\|_{[L_2(\kappa)]^m}.\end{aligned}$$

Combining this result with the first inequality of (34) we obtain the desired lower bound on the cell error. □

The upper bound stated in Theorem 16 and the lower bound from Theorem 17 can be coupled into a single two-sided bound on the L_2 norm of the cell error; namely,

$$c_4(\kappa)\|\hat{h}P_{\hat{h}}\mathbf{r}_h\|_{[L_2(\kappa)]^m} \leq \|\mathbf{e}_\kappa^{cell}\|_{[L_2(\kappa)]^m} \leq c_3(\kappa)\|h\mathbf{r}_h\|_{[L_2(\kappa)]^m}, \tag{46}$$

where $c_3(\kappa)$ and $c_4(\kappa)$ are computable constants. We note here that unlike the sharp two-sided bound on the L_2 norm of the cell error in terms of the dual graph norm of the finite element residual given in Theorem 15, the two-sided bound (46) is not asymptotically sharp because of the mismatch between the expressions under the norm signs in the lower and the upper estimate. Nevertheless, (46) is 'almost sharp' in the following sense: in practice $\hat{h}$ can be chosen to be a *fixed fraction* of h such that

$$\|\mathbf{r}_h - P_{\hat{h}}\mathbf{r}_h\|_{[L_2(\kappa)]^m} \leq \epsilon \|\mathbf{r}_h\|_{[L_2(\kappa)]^m},$$

where $\epsilon \in (0,1)$ is a fixed real number; then, by Pythagoras' Theorem,

$$\|\hat{h}P_{\hat{h}}\mathbf{r}_h\|_{[L_2(\kappa)]^m} \geq (1-\epsilon^2)^{1/2}\|\hat{h}\mathbf{r}_h\|_{[L_2(\kappa)]^m}.$$

This indicates that the lower bound in inequality (46) is at least 'comparable' in form, if not in size, with the upper bound.

To conclude, inequality (46) shows that in adaptive mesh refinement processes driven by residual-based error indicators, such as the ones listed at the beginning of Section 4.5, only cells with large cell error will be flagged for refinement (ignoring boundary conditions); the complementary part of the global error on a cell cannot be detected by measuring the residual on that cell only; to achieve local error control, a more global bound is required. This is the theme of the next section.

5.2 What controls the local size of the global error?

In the previous section we showed that, due to pollution effects, the finite element residual restricted to a cell puts a bound only on part of the global error restricted to that cell, and we derived local bounds on this part of the global error in terms of the residual. Here we show that in order to bound the whole of the global error restricted to cell κ we have to involve residuals over all the cells which intersect the domain of dependence of cell κ: the resulting *a posteriori* error bound is non-local in nature. The results presented in this section are based on the paper [28]. For the sake of simplicity we focus on scalar hyperbolic equations, corresponding to $m = 1$, and consider the following boundary-value problem:

$$\mathcal{L}u \equiv \nabla \cdot (\mathbf{a}u) + cu = f \quad \text{in } \Omega, \qquad\qquad (\nu \cdot \mathbf{a})^- u\big|_{\partial\Omega} = 0,$$

in a convex Lipschitz domain $\Omega \subset \mathbb{R}^n$, where $(x)^- = \min(x,0)$ denotes the negative part of the number x. We shall suppose that $\mathbf{a} = (a_1, \ldots, a_n)$ has its components in $C^2(\bar{\Omega})$ and that $c \in C^1(\bar{\Omega})$. The function f will be assumed to be in $L_2(\Omega)$. Recalling the definition of the inflow boundary $\partial_-\Omega$ from Section 3.1, we can restate the problem as follows:

$$\nabla \cdot (\mathbf{a}u) + cu = f \quad \text{in } \Omega, \qquad\qquad u = 0 \quad \text{on } \partial_-\Omega.$$

The adjoint operator $\mathcal{L}^*$ of $\mathcal{L}$ on Ω is defined by

$$\mathcal{L}^* z \equiv -\mathbf{a} \cdot \nabla z + cz.$$

Suppose that κ is an element in the partition of Ω. Let $D(\kappa)$ denote the union of all forward (with respect to $\partial_+\Omega$) characteristics of $\mathcal{L}^*$ contained in Ω which emanate from κ. Equivalently, $D(\kappa)$ can be described as the union of all backward (with respect to $\partial_-\Omega$) characteristics of $\mathcal{L}$ contained in Ω which emanate from κ. Further, we denote by $D_h(\kappa)$ the set of all elements in the partition which intersect $D(\kappa)$.

Similarly as in Section 4.2, we shall consider two situations, referred to symbolically as α) and β), corresponding to a Galerkin finite element method with strongly and weakly imposed boundary condition, respectively.

α) In this case, we have the following result (note that since we are dealing with the scalar case hypothesis (a) is redundant).

Theorem 18. *Suppose that (b) and (c) hold with $m = 1$, that the entries of* $\mathbf{a}$ *are in $C^2(\bar{\Omega})$ and that $c \in C^1(\bar{\Omega})$; then*

$$\|u - u_h\|_{H^{-1}(\kappa)} \leq C_1' C_2 \|h r_h\|_{L_2(D_h(\kappa))},$$

where $r_h = f - \mathcal{L}u_h$ denotes the residual corresponding to the finite element approximation u_h to u.

Proof. For $\psi \in C_0^\infty(\kappa)$ consider the adjoint (dual) problem

$$\mathcal{L}^* z = \psi \qquad \text{on } \Omega, \qquad\qquad (\nu \cdot \mathbf{a})^+ z\big|_{\partial\Omega} = 0,$$

where $(x)^+ = \max(x, 0)$ denotes the positive part of the number x. Since ψ has compact support in κ, characteristic theory implies that the support of z is contained in $D(\kappa)$. Let $z_h \in Y_h$ denote the quasi-interpolant of z (see [9], [11]); then, by Galerkin orthogonality, we have that

$$(u - u_h, \psi) = (r_h, z - z_h).$$

Further, since the support of z_h is contained in $D_h(\kappa)$ and $D_h(\kappa) \supset D(\kappa)$, it follows that

$$(u - u_h, \psi) = (r_h, z - z_h)_{D_h(\kappa)}.$$

Applying the approximation property (c) (with $m = 1$) and noting that $D_h(\kappa) \subset \Omega$ gives

$$|(u - u_h, \psi)| \leq C_2 \|h r_h\|_{L_2(D_h(\kappa))} \|z\|_{H^1(\Omega)}.$$

Now, recalling the strong stability of the dual problem,

$$\|z\|_{H^1(\Omega)} \leq C_1' \|\psi\|_{H^1(\Omega)} = C_1' \|\psi\|_{H^1(\kappa)}.$$

Hence, for any $\psi \in C_0^\infty(\kappa)$,

$$|(u - u_h, \psi)| \leq C_1' C_2 \|h r_h\|_{L_2(D_h(\kappa))} \|\psi\|_{H^1(\kappa)}.$$

Dividing both sides of this inequality by $\|\psi\|_{H^1(\kappa)}$ and taking the supremum over all ψ in $C_0^\infty(\kappa)$, upon noting that $C_0^\infty(\kappa)$ is dense in $H_0^1(\kappa)$ and recalling the definition of the negative Sobolev norm $\|\cdot\|_{H^{-1}(\kappa)}$, we arrive at the desired *a posteriori* error bound. □

In the previous section we decomposed the global error e, restricted to cell κ, as

$$e|_\kappa = e_\kappa^{cell} + e_\kappa^{trans},$$

and we gave, in Theorem 16, a bound on the L_2 norm of the cell error in terms of the local finite element residual, which in the case of a scalar hyperbolic equation ($m = 1$) has the following form:

$$\|e_\kappa^{cell}\|_{L_2(\kappa)} \leq c_3(\kappa)\|hr_h\|_{L_2(\kappa)},$$

We also noted that the L_2 norm of the transmitted error on κ can be bounded by the L_2 norm of the incoming component of the transmitted error on $\partial\kappa$. Here we derive a further bound on the transmitted error which is closer in spirit to that in Theorem 18.

Theorem 19. *Suppose that (b) and (c) hold with $m = 1$, that the entries of* $\mathbf{a}$ *belong to $C^2(\bar{\Omega})$ and that $c \in C^1(\bar{\Omega})$; also suppose that $e_\kappa^{cell} \in H^1(\kappa)$. Then*

$$\|e_\kappa^{trans}\|_{H^{-1}(\kappa)} \leq (C_1' C_2 + c_3(\kappa))\|hr_h\|_{L_2(D_h(\kappa))}.$$

Proof. Writing $e_\kappa^{trans} = e|_\kappa - e_\kappa^{cell}$ and applying the triangle inequality for the $\|\cdot\|_{H^{-1}(\kappa)}$ norm, we have that

$$\|e_\kappa^{trans}\|_{H^{-1}(\kappa)} \leq \|e\|_{H^{-1}(\kappa)} + \|e_\kappa^{cell}\|_{H^{-1}(\kappa)}. \tag{47}$$

According to Theorem 18,

$$\|e\|_{H^{-1}(\kappa)} \leq C_1' C_2 \|hr_h\|_{L_2(D_h(\kappa))}. \tag{48}$$

Further, by the definition of the $H^{-1}(\kappa)$ norm and recalling Theorem 16,

$$\|e_\kappa^{cell}\|_{H^{-1}(\kappa)} \leq \|e_\kappa^{cell}\|_{L_2(\kappa)} \leq c_3(\kappa)\|hr_h\|_{L_2(\kappa)}. \tag{49}$$

Substituting (48) and (49) into (47) and noting that $\kappa \subset D_h(\kappa)$ we complete the proof. □

β) We consider the scalar hyperbolic equation

$$\nabla \cdot (\mathbf{a}u) + cu = f \qquad \text{in } \Omega,$$

subject to the non-homogeneous boundary condition

$$(\mathbf{a} \cdot \nu)^-(u - g) = 0 \qquad \text{on } \partial\Omega,$$

with the same hypotheses on $\mathbf{a}$, c, f and Ω as in case α) above; further, we suppose that $g|_{\partial\Omega} \in L_2(\partial\Omega)$. In the case of a Petrov-Galerkin finite element approximation with weakly imposed boundary condition we have the following *a posteriori* error bound on the global error restricted to element κ in terms of the internal and boundary residual.

Theorem 20. *Suppose that hypotheses (b) and (c') hold, that the entries of* $\mathbf{a}$ *are in* $C^2(\bar{\Omega})$ *and that* c *belongs to* $C^1(\bar{\Omega})$. *Then,*

$$\|u - u_h\|_{H^{-1}(\kappa)} \leq C_1' C_2 \left(\|h r_h\|_{L_2(D_h(\kappa))} + \|h^{1/2} r_h^-\|_{L_2(\partial\Omega \cap D_h(\kappa))} \right),$$

where $r_h = f - \mathcal{L}u_h$ *denotes the interior residual, and* $r_h^- = (\mathbf{a}\cdot\nu)^-(g - u_h)|_{\partial\Omega}$ *signifies the boundary residual.*

We shall omit the proof, as it can be easily reconstructed from the proofs of Theorems 11 and 18. A similar result can be shown for the streamline diffusion finite element approximation of the scalar hyperbolic problem, and a bound on the transmitted error akin to that in Theorem 19 can be established.

Figure 2 shows the qualitative behaviour of the different error components in the numerical solution of the model problem (21) using the streamline diffusion method on an unstructured triangular mesh consisting of 504 nodes and 926 elements. In the figure caption $\mathcal{E}_1(u_h, h)$ and $\mathcal{E}_2(u_h, h)$ denote the right-hand sides of the error bounds in Theorems 16 and 20, respectively.

6 A posteriori error estimation for functionals

It is frequently the case in engineering problems that the main quantity of concern is not the solution of a partial differential equation, but a derived quantity which can be though of as a functional of the solution. In such instances it is unlikely that *a posteriori* error bounds of the kind stated in Section 4 will be of use in the design of efficient adaptive algorithms.

Our aim in this section is to propose an approach to the derivation of *a posteriori* bounds on the error in linear functionals directly, without attempting to obtain an upper bound on the error in a *norm* in which the functional is bounded. In order to illustrate the key ideas we begin by discussing some specific examples: in the next subsection we consider the *a posteriori* error analysis for a particular linear functional, the normal flux through the boundary of the computational domain; in the second subsection, we present a similar analysis for the local weighted average of the solution. In the third subsection, we approach the problem of *a posteriori* error estimation for linear functionals from a general viewpoint, following the ideas of Becker and Rannacher [7].

6.1 Estimation of the normal flux through the boundary

Let us consider the non-homogeneous boundary-value problem (12), (14), where $\mathbf{f} \in [L_2(\Omega)]^m$ and $\mathbf{g} \in [H^1(\Omega)]^m$. For the sake of simplicity, we assume that the restriction of $\mathbf{g}$ to $\partial\Omega$ belongs to the restriction of the trial space $X_h \subset H(\mathcal{L}, \Omega)$ to $\partial\Omega$, and the test space Y_h is contained in $[L_2(\Omega)]^m$. We

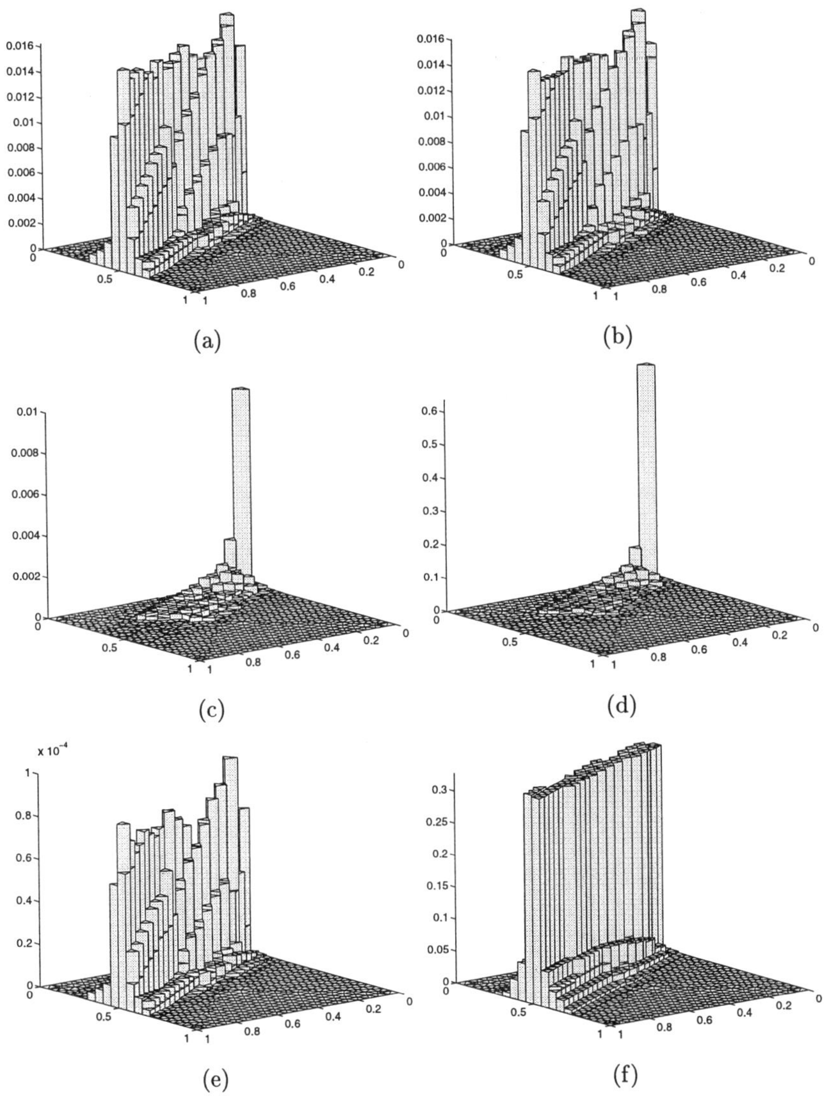

Fig. 2. Corner discontinuity problem: (a) $\|e\|_{L^2(\kappa)}$; (b) $\|e^{trans}\|_{L^2(\kappa)}$; (c) $\|e^{cell}\|_{L^2(\kappa)}$; (d) $\mathcal{E}_1(u_h, h)$; (e) $\|e\|_{H^{-1}(\kappa)}$; (f) $\mathcal{E}_2(u_h, h)$.

approximate the non-homogeneous problem by the following method: find $\mathbf{u}_h$ in X_h such that

$$\begin{aligned} \Pi_h \mathcal{L}\mathbf{u}_h &= \Pi_h \mathbf{f} && \text{in } \Omega, \\ B^-(\mathbf{u}_h - \mathbf{g})\big|_{\partial\Omega} &= \mathbf{0} && \text{on } \partial\Omega. \end{aligned}$$

Here Π_h denotes the orthogonal projector in $[L_2(\Omega)]^m$ onto Y_h.

Next we define the functional that represents the normal outflow flux through the boundary of the domain Ω. Given any $\psi \in [H^1(\Omega)]^m$, we consider the linear functional

$$N_\psi(\mathbf{v}) = \int_{\partial\Omega} (B^+\mathbf{v}) \cdot \psi \, \mathrm{d}s, \qquad \mathbf{v} \in H(\mathcal{L}, \Omega);$$

here ψ plays the rôle of a weight function that can be chosen freely (e.g. $\psi|_{\partial\Omega}$ can be taken to have its support contained in a subset of $\partial\Omega$, etc.). We seek to approximate the normal flux $N_\psi(\mathbf{u})$ by $N_\psi(\mathbf{u}_h)$. In order to avoid technical details concerning the regularity of the strong solution to a non-homogeneous boundary-value problem for a symmetric positive system, we make the following hypothesis on the data (see, [58]):

(d) Suppose that the entries of the matrices A_i, $i = 1, \dots, n$, and C and the vector ψ are so smooth that the strong solution to the dual problem

$$\mathcal{L}^*\mathbf{z} = \mathbf{0} \quad \text{in } \Omega, \qquad \gamma_{B^+}(\mathbf{z} - \psi) = \mathbf{0} \quad \text{on } \partial\Omega \tag{50}$$

belongs to $H^1(\Omega)$, and there is constant C_1', independent of ψ, such that

$$\|\mathbf{z}\|_{[H^1(\Omega)]^m} \leq C_1' \|\psi\|_{[H^s(\Omega)]^m}$$

for some $s \geq 1$.

Theorem 21. *Suppose that hypotheses (a), (b), (c) and (d) hold. Then, we have that*

$$|N_\psi(\mathbf{u}) - N_\psi(\mathbf{u}_h)| \leq C_1' C_2 \left(\sum_\kappa \|h\mathbf{r}_h\|^2_{L_2(\kappa)} \right)^{1/2} \|\psi\|_{H^s(\Omega)}.$$

Proof. Consider the dual problem (50). Since $\mathbf{u} - \mathbf{u}_h$ is in $D(\mathcal{L}, \Omega)$ and $\mathbf{z} \in [H^1(\Omega)]^m$, Green's formula gives

$$0 = (\mathbf{u} - \mathbf{u}_h, \mathcal{L}^*\mathbf{z}) = (\mathcal{L}(\mathbf{u} - \mathbf{u}_h), \mathbf{z}) - N_\psi(\mathbf{u} - \mathbf{u}_h).$$

Consequently, for any $\mathbf{z}_h \in Y_h$,

$$N_\psi(\mathbf{u}) - N_\psi(\mathbf{u}_h) = (\mathbf{r}_h, \mathbf{z} - \mathbf{z}_h).$$

Exploiting hypothesis (c), it follows that

$$|N_\psi(\mathbf{u}) - N_\psi(\mathbf{u}_h)| \leq C_2 \left(\sum_\kappa \|h\mathbf{r}_h\|^2_{L_2(\kappa)} \right)^{1/2} \|\mathbf{z}\|_{H^1(\Omega)},$$

and therefore, by hypothesis (d),

$$|N_\psi(\mathbf{u}) - N_\psi(\mathbf{u}_h)| \leq C_1' C_2 \left(\sum_\kappa \|h\mathbf{r}_h\|^2_{L_2(\kappa)} \right)^{1/2} \|\psi\|_{H^s(\Omega)}.$$

□

From the practical point of view, a particularly relevant situation concerns the estimation of the flux through a relatively open subset $\gamma \subset \partial\Omega$. In this case it is tempting to choose ψ such that the restriction of each of its components to $\partial\Omega$ is equal to the characteristic function of γ; unfortunately such ψ does not belong to $[H^1(\Omega)]^m$ (since the characteristic function of γ is in $H^{1/2-\varepsilon}(\partial\Omega)$, for all $\varepsilon > 0$, but not in $H^{1/2}(\partial\Omega)$), so Theorem 21 does not apply. The reader is referred to [56] for details of a slightly more refined analysis in the case of a scalar hyperbolic equation which carries over to this case. Nevertheless, we note that the choice of a smoother cut-off function ψ, with $\psi|_{\partial\Omega} \in [H^{s-1/2}(\partial\Omega)]^m$, $s \geq 1$, is covered by the hypotheses of Theorem 21, and this may suffice in practice.

6.2 Estimation of the local mean value

In this section we consider the *a posteriori* error analysis of finite element approximations to the local mean value

$$M_\psi(\mathbf{u}) = \int_\Omega \mathbf{u}(x) \cdot \psi(x)\, \mathrm{d}x,$$

where $\psi \in [C_0^\infty(\Omega)]^m$, and $\mathbf{u}$ is the solution of the non-homogeneous boundary value problem (12), (14) with $\mathbf{f} \in [L_2(\Omega)]^m$ and $\mathbf{g} \in [H^1(\Omega)]^m$. Again, for the sake of simplicity, we assume that the restriction of $\mathbf{g}$ to $\partial\Omega$ belongs to the restriction of the trial space $X_h \subset H(\mathcal{L}, \Omega)$ to the boundary, and we approximate the non-homogeneous problem by the following method: find $\mathbf{u}_h$ in X_h such that

$$\begin{aligned} \Pi_h \mathcal{L}\mathbf{u}_h &= \Pi_h \mathbf{f} && \text{in } \Omega, \\ B^-(\mathbf{u}_h - \mathbf{g})\big|_{\partial\Omega} &= \mathbf{0} && \text{on } \partial\Omega. \end{aligned}$$

Here, as in the previous section, Π_h denotes the orthogonal projection in $[L_2(\Omega)]^m$ onto the finite element test space $Y_h \subset [L_2(\Omega)]^m$. We have the following *a posteriori* bound on the error between $M_\psi(\mathbf{u})$ and its approximation $M_\psi(\mathbf{u}_h)$.

Theorem 22. *Suppose that hypotheses (a), (b) and (c) hold, the entries of the matrices A_i, $i = 1, \ldots, n$, belong to $C^2(\bar{\Omega})$ and those of C are continuously differentiable on $\bar{\Omega}$. Then, for each $\psi \in [C_0^\infty(\Omega)]^m$, we have that*

$$|M_\psi(\mathbf{u}) - M_\psi(\mathbf{u}_h)| \leq C_1' C_2 \left(\sum_\kappa \|h\mathbf{r}_h\|^2_{L_2(\kappa)} \right)^{1/2} \|\psi\|_{H^1(\Omega)}.$$

Proof. In contrast with the previous section, here the appropriate dual problem is

$$\begin{aligned} \mathcal{L}^*\mathbf{z} &= \psi \qquad \text{in } \Omega, \\ \gamma_{B^+}(\mathbf{z}) &= \mathbf{0} \qquad \text{on } \partial\Omega. \end{aligned}$$

Since $\mathbf{u} - \mathbf{u}_h$ is in $D(\mathcal{L}, \Omega)$ and $\mathbf{z} \in [H^1(\Omega)]^m$, Green's formula gives

$$M_\psi(\mathbf{u}) - M_\psi(\mathbf{u}_h) = (\mathbf{u} - \mathbf{u}_h, \psi) = (\mathbf{u} - \mathbf{u}_h, \mathcal{L}^*\mathbf{z}) = (\mathcal{L}(\mathbf{u} - \mathbf{u}_h), \mathbf{z}).$$

Consequently, for any $\mathbf{z}_h \in Y_h$,

$$M_\psi(\mathbf{u}) - M_\psi(\mathbf{u}_h) = (\mathbf{r}_h, \mathbf{z} - \mathbf{z}_h).$$

Exploiting hypothesis (c), it follows that

$$|M_\psi(\mathbf{u}) - M_\psi(\mathbf{u}_h)| \leq C_2 \left(\sum_\kappa \|h\mathbf{r}_h\|^2_{L_2(\kappa)} \right)^{1/2} \|\mathbf{z}\|_{H^1(\Omega)},$$

and therefore, by the strong stability of the dual problem,

$$|M_\psi(\mathbf{u}) - M_\psi(\mathbf{u}_h)| \leq C_1' C_2 \left(\sum_\kappa \|h\mathbf{r}_h\|^2_{L_2(\kappa)} \right)^{1/2} \|\psi\|_{H^1(\Omega)}.$$

□

Comparing this analysis with the one performed in the previous subsection for the boundary flux, one quickly recognises the similarities. Indeed, the question arises, whether it is possible to provide a general approach to the *a posteriori* error analysis of functionals. This is the theme of the next subsection.

6.3 A general duality argument

We give a brief overview of a general duality argument due to Becker and Rannacher (see [7]) for the *a posteriori* error estimation of functionals. For an alternative perspective on duality arguments, we refer to the work of Giles

[18] and, in a slightly different context, the articles of Peraire, Paraschivoiu and Patera [48] and Giles, Larson, Levenstam and Süli [19].

Suppose that X and Y are two reflexive Banach spaces equipped with their norms $\|\cdot\|_X$ and $\|\cdot\|_Y$, respectively, that $a(\cdot,\cdot)$ is a continuous bilinear functional on $X \times Y$ and $l(\cdot)$ is a continuous linear functional on Y. In the framework of symmetric positive systems, appropriate choices of X and Y and of $a(\cdot,\cdot)$ and $l(\cdot)$ are given at the beginning of Section 4.2, in α) and β).

We consider the variational problem: find $\mathbf{u}$ in X such that

$$a(\mathbf{u},\mathbf{v}) = l(\mathbf{v}) \qquad \text{for all } \mathbf{v} \text{ in } Y.$$

This problem is approximated by a Galerkin finite element method using a sequence of trial spaces $X_h \subset X$ and test spaces $Y_h \subset Y$ parametrised by a discretisation parameter h. The discrete problem reads: find $\mathbf{u}_h$ in X_h such that

$$a(\mathbf{u}_h,\mathbf{v}_h) = l(\mathbf{v}_h) \qquad \text{for all } \mathbf{v}_h \text{ in } Y_h.$$

Letting $\mathbf{e}_h = \mathbf{u} - \mathbf{u}_h$ denote the global error, we observe the Galerkin orthogonality property

$$a(\mathbf{e}_h,\mathbf{v}_h) = l(\mathbf{v}_h) \qquad \text{for all } \mathbf{v}_h \text{ in } Y_h.$$

Now suppose that $M(\cdot)$ is a continuous linear functional on X. In order to derive an *a posteriori* bound on the error between $M(\mathbf{u})$ and its approximation $M(\mathbf{u}_h)$, we introduce the following dual problem: find $\mathbf{z}$ in Y such that

$$a(\mathbf{w},\mathbf{z}) = M(\mathbf{w}) \qquad \text{for all } \mathbf{w} \text{ in } X.$$

Assuming that $a(\cdot,\cdot)$ satisfies the hypotheses of Theorem 9 on $X \times Y$ (rather than $X_h \times Y_h$) with X and Y interchanged[1], we deduce that the dual problem has a unique solution $\mathbf{z}$ in Y. Next, we see that

$$M(\mathbf{u}) - M(\mathbf{u}_h) = M(\mathbf{e}_h) = a(\mathbf{e}_h,\mathbf{z}) = a(\mathbf{e}_h,\mathbf{z}-\mathbf{z}_h)$$

for all $\mathbf{z}_h$ in Y_h. Equivalently, we can write this identity as

$$M(\mathbf{u}) - M(\mathbf{u}_h) = l(\mathbf{z}-\mathbf{z}_h) - a(\mathbf{u}_h,\mathbf{z}-\mathbf{z}_h).$$

Further, noting that $l(\cdot) - a(\mathbf{u}_h,\cdot)$ is a continuous linear functional on Y, we can rewrite the right-hand side as $\langle \mathbf{r}_h, \mathbf{z}-\mathbf{z}_h\rangle$ where $\mathbf{r}_h$ is the residual and $\langle\cdot,\cdot\rangle$ denotes the duality pairing between Y', the dual space of Y, and Y. This leads to the error representation

$$M(\mathbf{u}) - M(\mathbf{u}_h) = \langle \mathbf{r}_h, \mathbf{z}-\mathbf{z}_h\rangle,$$

[1] It can be shown that the inf-sup condition with X and Y interchanged is, in fact, equivalent to the inf-sup condition in its usual form; for a proof, we refer to Proposition A.2 in the paper of Melenk and Schwab [43].

for all $\mathbf{z}_h$ in Y_h. At this point we are at the same stage in the analysis as in Section 4.2 in the paragraph preceding Theorem 1. Following the same reasoning as there, one can derive an *a posteriori* bound on the error in the functional, $M(\mathbf{u}) - M(\mathbf{u}_h)$. We refer to [7] for further details, in the context of elliptic boundary-value problems.

To conclude this section, we note that this approach allows one to derive *a posteriori* bounds on the global error in norms stronger than $\|\cdot\|_{[H^{-1}(\Omega)]^m}$. Indeed, to obtain an *a posteriori* bound on the global error $\mathbf{e}_h = \mathbf{u} - \mathbf{u}_h$ in the norm of $[L_p(\Omega)]^m$, $1 \le p < \infty$, we consider the functional

$$M_p(\mathbf{w}) = \int_\Omega \frac{|\mathbf{e}_h|^{p-1}}{\|\mathbf{e}_h\|^{p-1}_{[L_p(\Omega)]^m}} \operatorname{sgn}(\mathbf{e}_h) \cdot \mathbf{w}\, dx,$$

where $\operatorname{sgn}(\mathbf{v})$ is a vector whose jth entry is the sign of the jth entry of $\mathbf{v}$. Clearly (remembering that we are dealing with real-valued functions!),

$$\|\mathbf{u} - \mathbf{u}_h\|_{[L_p(\Omega)]^m} = M_p(\mathbf{u}) - M_p(\mathbf{u}_h),$$

and thereby,

$$\|\mathbf{u} - \mathbf{u}_h\|_{[L_p(\Omega)]^m} = \langle \mathbf{r}_h, \mathbf{z} - \mathbf{z}_h \rangle, \tag{51}$$

for any $\mathbf{z}_h$ in the test space; here $\mathbf{z}$ denotes the solution to the dual problem

$$a(\mathbf{w}, \mathbf{z}) = M_p(\mathbf{w}) \qquad \text{for all } \mathbf{w} \text{ in } X. \tag{52}$$

We see that the right-hand side of (51) is of the usual form; so, at least formally, one can proceed with the *a posteriori* error analysis as before. However, there is a fundamental difference between (52) and the dual problems which occurred in Subsections 6.1 and 6.2: while in those problems ψ was a given function, so the data for the dual was known, here the dual problem involves the (unknown) global error $\mathbf{e}_h$. From the point of view of implementation a possible approach might be to compute the numerical solution of two (or more) successively refined meshes, and use their difference as an approximation to $\mathbf{e}_h$ in the functional $M_p(\cdot)$ to fix the right-hand side of the dual problem, and repeat this process in the course of the adaptive mesh refinement driven by the resulting *a posteriori* error bound. More analysis is required to quantify the effects of this additional approximation.

7 A posteriori analysis for unsteady problems

So far, we have been concerned with the *a posteriori* error analysis of finite element approximations to steady hyperbolic problems. In the present section we consider similar questions for unsteady problems.

In the next subsection we discuss a general class of (semi-discrete in time) Petrov-Galerkin methods for strictly hyperbolic systems, while in the second subsection we restrict ourselves to (fully-discrete) evolution Galerkin methods for scalar hyperbolic equations, although the basic steps in the error analysis would be identical for fully-discrete finite element approximations multi-dimensional hyperbolic systems.

7.1 A posteriori error analysis for strictly hyperbolic systems

In this section we shall consider the *a posteriori* error analysis of finite element approximations to the system of partial differential equations

$$\frac{\partial \mathbf{u}}{\partial t} = \sum_{i=1}^{n} A_i(x,t)\frac{\partial \mathbf{u}}{\partial x_i} + C(x,t)\mathbf{u} + \mathbf{f}(x,t), \tag{53}$$

where A_i, $i = 1, \ldots, n$, and C are smooth $m \times m$ matrix-valued functions, constant outside a compact subset of $\Omega \times \mathrm{I\!R}$ with

$$\begin{aligned}\Omega &= \{x \in \mathrm{I\!R}^n \, : \, x_1 > 0\},\\ \partial\Omega &= \{x \in \mathrm{I\!R}^n \, : \, x_1 = 0\},\\ x &= (x_1, \ldots, x_n) = (x_1, x').\end{aligned}$$

In physical applications modelled by this system, the variable t plays the rôle of time, and $x = (x_1, \ldots, x_n)$ represent the spatial independent variables.

It will be assumed that the differential operator is strictly hyperbolic; in other words, we shall suppose that the matrix $\sum_{i=1}^{n} A_i(x,t)\xi_i$ has m distinct real eigenvalues for all $\xi \in \mathrm{I\!R}^n \setminus \{\mathbf{0}\}$ and $(x,t) \in \bar{\Omega} \times \mathrm{I\!R}$. Furthermore, we shall require that the boundary $\partial\Omega \times \mathrm{I\!R}$ is a non-characteristic hypersurface for the differential operator, namely, $\det(A_1) \neq 0$ when $x_1 = 0$.

Equation (53) is solved in tandem with an initial condition at $t = 0$, and a boundary condition on $\partial\Omega \times [0,T]$ which is imposed by considering the boundary operator $B(x',t)$, a smooth $l \times m$ matrix-valued function, independent of (x',t) for $|x'| + t$ large, such that $\mathrm{rank}(B) = l$, where l is the number of negative eigenvalues of A_1. Then it is known that, for any $T > 0$, $\mathbf{f} \in L_2([0,T] \times \Omega)$ and $\mathbf{u}_0 \in L_2(\Omega)$, there is a unique strong solution $\mathbf{u}$ of (53) subject to the initial condition

$$\mathbf{u}(x,0) = \mathbf{u}_0(x) \qquad \text{for } x \in \Omega \tag{54}$$

and the boundary condition

$$B\mathbf{u} = \mathbf{0} \qquad \text{on } \partial\Omega \times [0,T], \tag{55}$$

provided that the latter is admissible in a sense that will be made precise below.

By admissibility of the boundary condition (55) we mean the following. Let us note that, upon a smooth change of coordinates, A_1 can be written in the form

$$A_1 = \begin{bmatrix} A_1^I & 0 \\ 0 & A_2^{II} \end{bmatrix},$$

where A_1^I is negative definite and A_2^{II} is positive definite; by splitting the vector $\mathbf{u}$ in a similar manner into $\mathbf{u}^I = (u_1, \ldots, u_l)$ and $\mathbf{u}^{II} = (u_{l+1}, \ldots, u_m)$, the boundary condition $B\mathbf{u} = \mathbf{0}$, upon decomposing B correspondingly, can be restated as $S_I\mathbf{u}^I - S_{II}\mathbf{u}^{II} = \mathbf{0}$. We shall say that the boundary condition is *admissible* if S_I is invertible; in that case, (55) can be written in the equivalent form

$$\mathbf{u}^I - S\mathbf{u}^{II} = \mathbf{0},$$

where $S = S_I^{-1}S_{II}$.

Now we consider a finite element approximation to this problem. Suppose that we have chosen a finite element trial space $X_h \subset [H^1(\Omega)]^m$ consisting of continuous piecewise polynomial m-component vector functions on a partition $\mathcal{T}_h = \{\kappa\}$ of Ω which satisfy the boundary condition (55), and a finite element test space $Y_h \subset [L_2(\Omega)]^m$ on the same partition. Then, we approximate (53) – (55) by a semi-discrete Petrov-Galerkin finite element method of the following form: find $\mathbf{u}_h(t) \in X_h$, $0 < t \leq T$, such that, for all $\mathbf{q}_h \in Y_h$,

$$\left(\frac{\partial \mathbf{u}_h}{\partial t}, \mathbf{q}_h\right) = \sum_{i=1}^{n} \left(A_i(\cdot, t)\frac{\partial \mathbf{u}_h}{\partial x_i}, \mathbf{q}_h\right) + (C(\cdot, t)\mathbf{u}_h, \mathbf{q}_h) + (\mathbf{f}(\cdot, t), \mathbf{q}_h),$$

$$(\mathbf{u}_h(\cdot, 0), \mathbf{q}_h) = (\mathbf{u}_0(\cdot), \mathbf{q}_h).$$

We note in passing that the inclusion of the boundary condition into the definition of the trial space is unreasonable from the practical point of view (and implausible from the theoretical point of view, unless S is a constant matrix); we have adopted this assumption only to simplify the error analysis. A practical method would involve a weakly imposed boundary condition, in the same spirit as in the steady case discussed earlier on. Moreover, in practice, a time-discretisation is required; as long as the latter is also a Galerkin-type method, the error analysis that we provide below in the semi-discrete case is easily modified to include the effects of the time discretisation.

We define the finite element residual

$$\mathbf{r}_h(x, t) = \mathbf{f}(x, t) - \frac{\partial \mathbf{u}_h}{\partial t} + \sum_{i=1}^{n} A_i(x, t)\frac{\partial \mathbf{u}_h}{\partial x_i} + C(x, t)\mathbf{u}_h.$$

It is in terms of this quantity that we wish to obtain a bound on the discretisation error in the spatial H^{-1} norm at time T.

Theorem 23. *Suppose that hypothesis (c) holds. Then, there exists a computable constant C_5 such that*

$$\|\mathbf{u}(\cdot, T) - \mathbf{u}_h(\cdot, T)\|_{H^{-1}(\Omega)} \leq C_5 \left(\|h\mathbf{r}_h\|^2_{L_2(0,T;L_2(\Omega))} + \|h\mathbf{r}_h^0\|^2_{L_2(\Omega)}\right)^{\frac{1}{2}},$$

where $\mathbf{r}_h^0(x) = \mathbf{u}_0(x) - \mathbf{u}_h(x, 0)$.

Proof. Suppose that $\psi \in [C_0^\infty(\Omega)]^m$ and consider the dual problem:

$$\frac{\partial \mathbf{z}}{\partial t} = \sum_{i=1}^{n} \frac{\partial}{\partial x_i}(A_i^* \mathbf{z}) - C^* \mathbf{z} \qquad \text{in } \Omega \times [0,T),$$
$$\mathbf{z}(x,T) = \psi(x) \qquad \text{for } x \in \Omega,$$

subject to the boundary condition

$$\mathbf{z}^{II} - \hat{S}\mathbf{z}^{I} = \mathbf{0} \qquad \text{on } \partial\Omega \times [0,T],$$

where $\hat{S} = -A_1^{II^{-*}} S^* A_1^{I^*}$. Then

$$(\mathbf{e}_h, \mathbf{z})|_0^T = \int_0^T \frac{d}{dt}(\mathbf{e}_h, \mathbf{z})\ dt = \int_0^T (\frac{\partial \mathbf{e}_h}{\partial t}, \mathbf{z}) + (\mathbf{e}_h, \frac{\partial \mathbf{z}}{\partial t})\ dt$$
$$= \int_0^T (\mathbf{r}_h, \mathbf{z})\ dt = \int_0^T (\mathbf{r}_h, \mathbf{z} - \mathbf{z}_h)\ dt$$

for any $\mathbf{z}_h \in Y_h$. Noting the approximation property (c), we deduce that

$$|(\mathbf{e}_h(\cdot,T), \psi)| \le C_6 \left(\|h\mathbf{e}_h(\cdot,0)\|_{L_2(\Omega)} \|\mathbf{z}(\cdot,0)\|_{H^1(\Omega)} \right.$$
$$\left. + \|h\mathbf{r}_h\|_{L_2(0,T;L_2(\Omega))} \|\mathbf{z}\|_{L_2(0,T;H^1(\Omega))} \right).$$

According to a hyperbolic regularity theorem due to Rauch [50],

$$\left(\|\mathbf{z}(\cdot,0)\|^2_{H^1(\Omega)} + \|\mathbf{z}\|^2_{L_2(0,T;H^1(\Omega))} \right)^{\frac{1}{2}} \le C_7 \|\psi\|_{H^1(\Omega)},$$

and hence the required result with $C_5 = C_6 C_7$. □

In the next section, we consider the *a posteriori* error analysis of a fully discrete method for an unsteady hyperbolic equation.

7.2 A posteriori analysis of evolution-Galerkin methods

Here, we develop the *a posteriori* error analysis of evolution-Galerkin finite element methods for unsteady scalar hyperbolic problems. The presentation closely follows the article of Süli and Houston [57], albeit with some abbreviations. For a detailed study of the (unconditional) stability and accuracy properties of evolution-Galerkin methods we refer to [46].

Given a final time $T > 0$, a function $f \in L_2(I; L_2(\Omega))$ with $I = (0,T]$, and $u_0 \in L_2(\Omega)$, we consider the hyperbolic initial-value problem

$$\frac{\partial u}{\partial t} + \mathbf{a} \cdot \nabla u = f, \qquad \mathbf{x} \in \Omega,\ t \in I, \tag{56}$$
$$u(\mathbf{x}, 0) = u_0(\mathbf{x}), \quad \mathbf{x} \in \Omega, \tag{57}$$

where $\Omega = \mathbb{R}^n$. For the sake of simplicity we assume that the velocity field $\mathbf{a}$ is in $C([0,T]; C_0^1(\Omega)^n)$, that it is incompressible, i.e. $\nabla\cdot\mathbf{a} = 0$ on $\Omega\times[0,T]$, and that the supports of u_0 and f are compact subsets in Ω and $\Omega\times[0,T]$, respectively. This problem has a unique weak solution u in $L_\infty(I; L_2(\Omega))$; moreover, if $u_0 \in H^1(\Omega)$ and $f \in L_2(I; H^1(\Omega))$ then u belongs to $L_2(I; H^1(\Omega))$.

We consider a subdivision (not necessary uniform) of the time interval $I = [0,T]$ given by $0 = t^0 < t^1 < \ldots < t^M < t^{M+1} = T$; we define time intervals $I^m = (t^{m-1}, t^m]$ and time steps $k_m = t^m - t^{m-1}$. For each m, let $\mathcal{T}^m = \{\kappa\}$ be a partition of Ω into closed simplices κ, with corresponding mesh function h_m, piecewise continuous on Ω, satisfying

$$C_\dagger h_\kappa^n \le \text{meas}(\kappa) \quad \forall\kappa \in \mathcal{T}^m, \tag{58}$$

$$C_* h_\kappa \le h_m(\mathbf{x}) \le h_\kappa \qquad \forall\mathbf{x}\in\kappa \quad \forall\kappa \in \mathcal{T}^m, \tag{59}$$

where h_κ is the diameter of κ, and $C_\dagger$ and C_* are positive constants. Further, h will denote the global mesh function given by $h(\mathbf{x},t) = h_m(\mathbf{x})$ for $(\mathbf{x},t)$ in $\Omega \times I^m$, and we define the corresponding time step function $k = k(t)$ by $k(t) = k_m$, $t \in I^m$. Let $\Lambda^m = \Omega \times I^m$; for $p, q \in \mathbb{N}$, let

$$S^{h_m} = \{v \in H_0^1(\Omega) : v|_\kappa \in \mathcal{P}_p(\kappa)\ \forall\kappa \in \mathcal{T}^m\},$$

$$V^{h_m} = \{v : v(\mathbf{x},t)|_{\Lambda^m} = \sum_{j=0}^{q} t^j v_j,\ v_j \in S^{h_m}\},$$

$$V^h = \{v : v(\mathbf{x},t)|_{\Lambda^m} \in V^{h_m},\ m = 1,\ldots,M+1\},$$

where $\mathcal{P}_p(\kappa)$ denotes the set of polynomials of degree at most p over κ. In the following, we shall assume that $p = 1$ and $q = 0$. We note that if $v \in V^{h_m}$ for $m = 1,\ldots,M+1$, then v is continuous in space at any time, but may be discontinuous in time at the discrete time levels t^m. To account for this, we introduce the notation $v_\pm^m := \lim_{s\to 0^+} v(t^m \pm s)$, and $[v^m] := v_+^m - v_-^m$.

The evolution-Galerkin approximation of (56), (57) makes use of the particle trajectories (or characteristics) associated with equation (56): the path $\mathbf{X}(\mathbf{x}, s; \cdot)$ of a particle located at position $\mathbf{x} \in \Omega$ at time $s \in [0,T]$ is defined as the solution of the initial value problem

$$\frac{\mathrm{d}}{\mathrm{d}t}\mathbf{X}(\mathbf{x},s;t) = \mathbf{a}(\mathbf{X}(\mathbf{x},s;t),t), \tag{60}$$

$$\mathbf{X}(\mathbf{x},s;s) = \mathbf{x}. \tag{61}$$

For u smooth enough, the *material derivative* $D_t u$ is then defined by

$$D_t u(\mathbf{x},s) := \frac{\mathrm{d}}{\mathrm{d}t} u(\mathbf{X}(\mathbf{x},s;t),t)\ |_{t=s}$$

$$= \frac{\partial}{\partial t} u(\mathbf{x},s) + \mathbf{a}(\mathbf{x},s)\cdot\nabla u(\mathbf{x},s) \quad \forall\mathbf{x}\in\Omega,\ s \in I. \tag{62}$$

The evolution-Galerkin time-discretisation is based on approximating the material derivative by a divided difference operator along particle trajectories. The simplest appropriate scheme arises from using Euler's method, giving, for $m = 0, \ldots, M$,

$$D_t u(\cdot, t^{m+1}) \approx \frac{u(\cdot, t^{m+1}) - u(\mathbf{X}(\cdot, t^{m+1}; t^m), t^m)}{k_{m+1}}.$$

Suppose that u_h^m denotes the approximation to $u(\cdot, t^m)$ at time t^m; then, applying the finite element method in space results in what is known as the evolution-Galerkin discretisation of the scalar linear hyperbolic equation (56): find $u_h^{m+1} \in S^{h_{m+1}}$, for $m = 0, \ldots, M$, such that

$$\left(\frac{u_h^{m+1} - u_h^m(\mathbf{X}(\cdot, t^{m+1}; t^m))}{k_{m+1}}, v \right) = (\bar{f}, v) \quad \forall v \in S^{h_{m+1}}, \tag{63}$$

$$(u_h^0, v) = (u_0, v) \quad \forall v \in S^{h_0}, \tag{64}$$

where $\bar{f}|_{\Lambda^{m+1}} := f(\cdot, t^{m+1})$. Alternatively, by integrating (63) with respect to t over I^{m+1}, we obtain the following equivalent formulation: find u_h such that, for $m = 0, 1, \ldots, M$, $u_h|_{\Lambda^{m+1}} \in V^{h_{m+1}}$ and satisfies

$$(D_t^h u_h, v)_{m+1} = (\bar{f}, v)_{m+1} \quad \forall v \in V^{h_{m+1}}, \tag{65}$$

$$(u_{h-}^0, v) = (u_0, v) \qquad \forall v \in V^{h_0}, \tag{66}$$

where

$$D_t^h u_h|_{\Lambda^{m+1}} = (u_{h-}(\mathbf{X}(\mathbf{x}, t^{m+1}; t^{m+1}), t^{m+1}) - u_{h-}(\mathbf{X}(\mathbf{x}, t^{m+1}; t^m), t^m))/k_{m+1};$$

here, for $v, w \in L_2(I^{m+1}; L_2(\Omega))$, we have used the notation

$$(v, w)_{m+1} = \int_{t^m}^{t^{m+1}} (v, w)\, dt.$$

Before stating the relevant *a posteriori* error bound for this method, we note that in (65), (66) the space discretisation may vary in both space and time, but the time steps are only variable in time and not in space, so the corresponding space-time mesh will not be fully optimal. The method obeys the following *a posteriori* error bound.

Theorem 24. *Let u and u_h be solutions of (56), (57) and (65), (66), respectively. Then*

$$\|u - u_h\|_{L_\infty(0,T;H^{-1}(\Omega))} \leq \mathring{\mathcal{E}}(u_h, h, k, f), \tag{67}$$

where

$$\begin{aligned}
\mathring{\mathcal{E}}(u_h, h, k, f) &= \mathcal{E}(u_h, h, k, f) + \mathcal{E}_0(u_0, u_{h-}^0, h), \\
\mathcal{E}(u_h, h, k, f) &= C_1\|hR_1\|_Q + C_2\|kR_1\|_Q \\
&\quad + C_3\|kR_2\|_Q + C_4\|kR_3\|_Q + C_5\|kR_4\|_Q, \qquad (68)\\
\mathcal{E}_0(u_0, u_{h-}^0, h) &= C_6\|u_0 - u_{h-}^0\|, \qquad (69)
\end{aligned}$$

and

$$R_1|_{\Lambda^{m+1}} = [u_h^m]/k_{m+1} + \mathbf{a}\cdot\nabla u_h - f,$$
$$R_2|_{\Lambda^{m+1}} = (D_t^h u_h - ([u_h^m]/k_{m+1} + \mathbf{a}\cdot\nabla u_h))/k_{m+1},$$
$$R_3|_{\Lambda^{m+1}} = [u_h^m]/k_{m+1},$$
$$R_4 = (f - \bar{f})/k,$$

and C_i, $i = 1,\ldots,6$, are (computable) positive constants.

Here and below $\|\cdot\| \equiv \|\cdot\|_{L_2(\Omega)}$ and $\|\cdot\|_Q \equiv \|\cdot\|_{L_2(Q)}$, where $Q = \Omega\times(0,T)$. For a proof of this result the reader is referred to Süli and Houston [57], where the precise values of the constants $C_1,\ldots,C_6$ appearing in the error bound are also specified.

In the remainder of this section we consider the computational implementation of the *a posteriori* error bound stated in the last theorem. Much of what will be said, however, applies in a more general setting.

For a given tolerance, TOL, we consider the problem of finding a discretisation in space and time $\mathcal{S}^h = \{(\mathcal{T}^m, t^m)\}_{n\geq 0}$ such that:

1. $\|u - u_h\|_{L_\infty(I;H^{-1}(\Omega))} \leq \mathrm{TOL}$;
2. $\mathcal{S}^h$ is optimal in the sense that the number of degrees of freedom is minimal.

In order to satisfy these criteria we shall use the *a posteriori* error estimate (67) to choose $\mathcal{S}^h$ such that:

1. $\overset{\circ}{\mathcal{E}}(u_h, h, k, f) \leq \mathrm{TOL}$;
2. The number of degrees of freedom in $\mathcal{S}^h$ is minimal.

The term $\mathcal{E}_0(u_0, u_{h-}^0, h)$ is easily controlled at the start of a computation; so here we shall only consider the problem of constructing $\mathcal{S}^h$ in an efficient way to ensure that

$$\mathcal{E}(u_h, h, k, f) \leq \mathrm{TOL}',$$

where $\mathrm{TOL} = \mathrm{TOL}' + \mathcal{E}_0(u_0, u_{h-}^0, h)$. To do so, we first write $\mathcal{E}$ symbolically in terms of two residual terms: one that controls the spatial mesh and one that controls the temporal mesh, i.e. we let

$$\mathcal{E}(u_h, h, k, f) \equiv C_1'\|hR_1'\|_Q + C_2'\|kR_2'\|_Q. \tag{70}$$

Simultaneously, we split the tolerance TOL' into a spatial part, TOL_h, and a temporal part, TOL_k. Thus, for reliability we require that the following conditions hold:

$$C_1'\|hR_1'\|_Q \leq \mathrm{TOL}_h, \tag{71}$$
$$C_2'\|kR_2'\|_Q \leq \mathrm{TOL}_k. \tag{72}$$

To design the space-time mesh $\mathcal{S}^h$, at each time level t^m we decompose the norm in (71) into norms over elements $\kappa \in \mathcal{T}^m$, and the norm in (72) into norms over time slabs as follows:

$$
\begin{aligned}
C_1'\|hR_1'\|_Q &\le C_1'\sqrt{T} \max_{1\le m\le M+1} \|h_m R_1'(u_h^m)\| \\
&\le C_1'\sqrt{NT} \max_{1\le m\le M+1} \left(\max_{\kappa\in\mathcal{T}^m} \|h_m R_1'(u_h^m)\|_{L_2(\kappa)} \right), \\
C_2'\|kR_2'\|_Q &\le C_2'\sqrt{T} \max_{1\le m\le M+1} \|k_m R_2'(u_h^m)\|,
\end{aligned}
$$

where N is the number of elements in the spatial mesh at time t^m. Thus, if

$$
\begin{aligned}
C_1'\sqrt{NT}\|h_m R_1'(u_h^m)\|_{L_2(\kappa)} &\le \mathrm{TOL}_h \quad \forall \kappa \in \mathcal{T}^m, \text{ for } m = 1,\dots,M+1, \\
C_2'\sqrt{T}\|k_m R_2'(u_h^m)\| &\le \mathrm{TOL}_k, \quad \text{for } m = 1,\dots,M+1,
\end{aligned}
$$

then (71) and (72) will automatically hold.

For the practical implementation of this method, we consider the following adaptive algorithm for constructing $\mathcal{S}^h$, under the assumption that the final time T is fixed: for each $m = 1,2,\dots,M+1$, with $\mathcal{T}_0^m$ a given initial mesh and $k_{m,0}$ an initial time step, determine meshes $\mathcal{T}_j^m$ with N_j elements of size $h_{m,j}(\mathbf{x})$ and time steps $k_{m,j}$ and the corresponding approximate solution $u_{h,j}$ defined on I_j^m such that, for $j = 0,1,\dots,\hat{m}-1$,

$$
C_1\|h_{m,j+1}R_1(u_{h,j}^m)\|_{L_2(\kappa)} = \frac{\mathrm{TOL}_h}{\sqrt{N_j T}} \quad \forall \kappa \in \mathcal{T}_j^m, \tag{73}
$$

$$
\begin{aligned}
&C_2\|k_{m,j+1}R_1(u_{h,j}^m)\| + C_3\|k_{m,j+1}R_2(u_{h,j}^m)\| \\
&+C_4\|k_{m,j+1}R_3(u_{h,j}^m)\| + C_5\|k_{m,j+1}R_4(u_{h,j}^m)\| = \frac{\mathrm{TOL}_k}{\sqrt{T}},
\end{aligned} \tag{74}
$$

where $I_j^m = (t^{m-1}, t^{m-1} + k_{m,j}]$ and $\mathrm{TOL}' = \mathrm{TOL}_h + \mathrm{TOL}_k$. We define $\mathcal{T}^m = \mathcal{T}_{\hat{m}}^m$, $k_m = k_{m,\hat{m}}$ and $h_m = h_{m,\hat{m}}$, where for each m, the number of trials $\hat{m}$ is the smallest integer such that for $j = \hat{m}$ the following stopping condition is satisfied:

$$
C_1\|h_{m,\hat{m}}R_1(u_{h,\hat{m}}^m)\|_{L_2(\kappa)} \le \frac{\mathrm{TOL}_h}{\sqrt{N_{\hat{m}} T}} \quad \forall \kappa \in \mathcal{T}_{\hat{m}}^m, \tag{75}
$$

$$
\begin{aligned}
&C_2\|k_{m,\hat{m}}R_1(u_{h,\hat{m}}^m)\| + C_3\|k_{m,\hat{m}}R_2(u_{h,\hat{m}}^m)\| \\
&+C_4\|k_{m,\hat{m}}R_3(u_{h,\hat{m}}^m)\| + C_5\|k_{m,\hat{m}}R_4(u_{h,\hat{m}}^m)\| \le \frac{\mathrm{TOL}_k}{\sqrt{T}}.
\end{aligned} \tag{76}
$$

By construction, this stopping condition will guarantee reliability of the adaptive algorithm; for efficiency, we try to ensure that (75) and (76) are satisfied with near equality. Since the final time T is fixed, the time step given by (74) may need to be limited to ensure that $t^M + k_{M+1,\hat{m}} = T$.

For the implementation of this adaptive algorithm, we shall assume that $\mathcal{T}_0^m = \mathcal{T}^{m-1}$ for $m = 2, 3, \ldots$.

Having described the construction of the space-time mesh $\mathcal{S}^h$ to achieve the required error control,

$$\|u - u_h\|_{L_\infty(I;H^{-1}(\Omega))} \leq \mathrm{TOL},$$

we note that, in order to generate the desired mesh we need a suitable mesh modification technique.

Temporal adaptation is quite straightforward, since the time step can just be set equal to $k_{n,j+1}$ given by (74). For constructing the spatial mesh in two space dimensions we use the red-green isotropic refinement strategy of Bank [6]. Here, the user must first specify a (coarse) *background* mesh upon which any future refinement will be based. Red refinement corresponds to dividing a certain triangle (father) into four similar triangles (sons) by connecting the midpoints of the sides. Green refinement is only temporary and is used to remove any hanging nodes caused by a red refinement. We note that green refinement is only used on elements which have one hanging node. For elements with two or more hanging nodes a red refinement is performed. The advantage of this refinement strategy is that the degradation of the 'quality' of the mesh is limited since red refinement is obviously harmless and green triangles can *never* be further refined. Within this mesh modification strategy it is also possible to de-refine the mesh by removing redundant elements, provided that these do not belong to the original background mesh. Thus, to prevent an overly refined mesh in regions where the solution is smooth the background mesh should be chosen suitably coarse. For the practical implementation of this mesh modification strategy we have used the FEMLAB package developed by Kenneth Eriksson (Chalmers University).

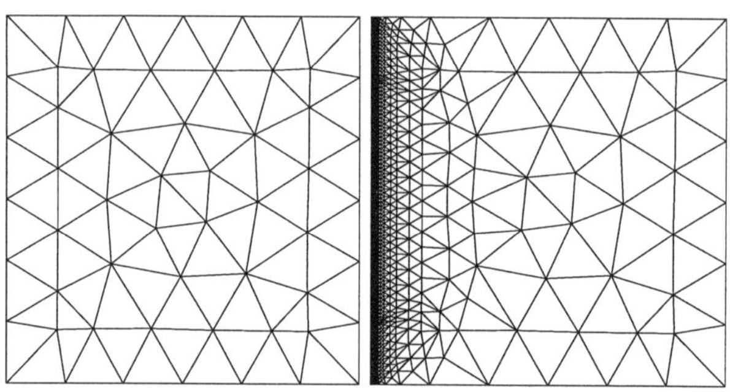

(a) (b)

Fig. 3. (a) Background mesh, with 56 nodes and 86 elements; (b) Background mesh adapted to resolve the initial condition, with 8335 nodes and 15860 elements.

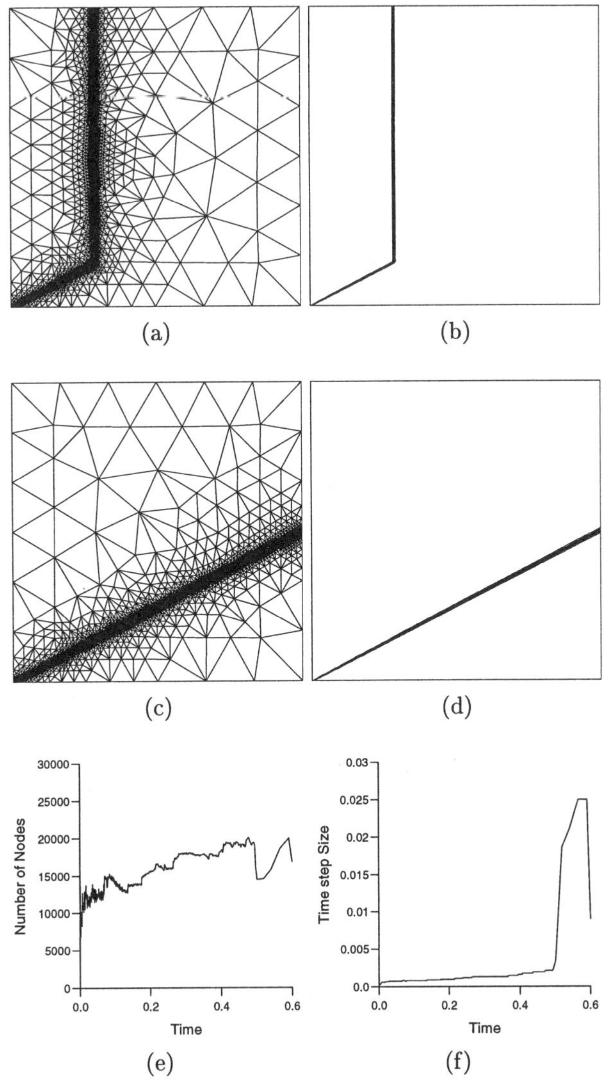

Fig. 4. Layer problem for $\mathrm{TOL}_h = 0.007$ and $\mathrm{TOL}_k = 0.25$ with $T = 0.6$: (a) & (b) Mesh and solution (resp.) at time, $t = 0.1428$, with 13596 nodes and 27109 elements; (c) & (d) Mesh and solution (resp.) at final time, $t = 0.6$, with 16829 nodes and 33566 elements; (e) History of nodes against time; (f) History of time step size against time.

7.3 Numerical experiments

In this section, we present some numerical experiments to illustrate the performance of the adaptive algorithm (73) on the model hyperbolic test problem:

$$\frac{\partial u}{\partial t} + \mathbf{a} \cdot \nabla u = f, \qquad \mathbf{x} \in \Omega,\ t \in I, \tag{77}$$

$$u(\mathbf{x}, 0) = u_0(\mathbf{x}), \quad \mathbf{x} \in \Omega, \tag{78}$$

where $\Omega = (0,1)^2$, $f = 0$, $\mathbf{a} = (2,1)$, subject to the boundary condition $u(0,y) = 1$ for $0 \le y \le 1$, $u(x,0) = (\delta - x)^+/\delta$ for $0 \le x \le 1$. The function u_0 appearing in the initial condition is defined as follows: $u_0(\mathbf{x}) = 0$ for $\mathbf{x} \in \Omega_\delta = (\delta, 1) \times (0,1)$; and for $\mathbf{x} \in \Omega \backslash \Omega_\delta$, $u_0(\mathbf{x})$ is chosen to be the linear function that satisfies the boundary conditions at inflow. We note that initially, for δ small, the solution to this problem has a boundary layer along $x = 0$; this layer then propagates into the domain Ω, and eventually exits through $x = 1$. In the following, we shall let $\delta = 7.8125 \times 10^{-3}$ and $T = 0.6$. First, we specify the background mesh as the one shown in Figure 3(a); this is initially refined in order to resolve the boundary layer along $x = 0$ at time $t = 0$, as shown in Figure 3(b). Numerical results are presented in Figure 4 for $\mathrm{TOL}_h = 0.007$ and $\mathrm{TOL}_k = 0.25$.

In Figures 4(a), 4(b) and 4(c), 4(d) we see that the adaptive algorithm has refined the spatial mesh in parts of the domain where the solution has a steep layer, and has kept the mesh coarse elsewhere. Figures 4(e), 4(f) show the history of the number of nodes in the spatial mesh against time, and the size of the time step against time, respectively.

We note that the artificial diffusion model introduced in [27] was employed in this experiment with $C_1^{\hat{\epsilon}} = C_2^{\hat{\epsilon}} = 0.2$ and $\hat{\epsilon}_{\max} = 7.0 \times 10^{-4}$.

8 Nonlinear conservation laws

In this concluding section we discuss briefly the extension of the approach to *a posteriori* error analysis described earlier in the case of linear problems to nonlinear hyperbolic equations; for a survey of the theory and numerical analysis of conservation laws, we refer to the recent monographs of Godlewski and Raviart [21], and Kröner [33]. For the sake of simplicity we shall restrict ourselves to scalar nonlinear conservation laws of the form

$$\frac{\partial}{\partial t} u(x,t) + \frac{\partial}{\partial x} f(u(x,t)) = 0, \qquad -\infty < x < \infty, \quad 0 < t \le T, \tag{79}$$

with strictly convex flux function f (i.e. $f'' \ge \alpha > 0$), subject to the initial condition

$$u(x,0) = u_0(x), \qquad -\infty < x < \infty, \tag{80}$$

where u_0 is a compactly supported Lip^+ bounded function. We say that a function $x \mapsto w(x)$ is Lip^+ bounded if

$$\|w\|_{Lip^+(\mathbb{R})} \equiv \text{ess.sup}_{x \neq y} \left(\frac{w(x) - w(y)}{x - y} \right)^+ < \infty,$$

where $(\cdot)^+ \equiv \max(\cdot, 0)$.

A weak solution to this initial-value problem obeys the identity

$$(u(\cdot,T), v(\cdot,T)) = (u_0, v(\cdot,0)) + \int_0^T (u(\cdot,t), v_t(\cdot,t)) + (f(u(\cdot,t)), v_x(\cdot,t))\, \mathrm{d}t$$
$$v \in C^1([0,T], C_0^1(\mathbb{R})), \qquad (81)$$

where $(\cdot,\cdot)$ denotes the inner product of $L_2(\mathbb{R})$.

We recall that the entropy solution of the nonlinear conservation law (79), (80) satisfies the estimate (see [10], [59])

$$\|u(\cdot,t)\|_{Lip^+(\mathbb{R})} \leq \frac{1}{\|u_0\|^{-1}_{Lip^+(\mathbb{R})} + \alpha t}, \qquad t \geq 0; \qquad (82)$$

the case of $\|u_0\|_{Lip^+(\mathbb{R})} = \infty$ is included in this estimate and it corresponds to the exact t^{-1} decay rate of an initial rarefaction. We also note that since u_0 has compact support the same is true of the function $x \mapsto u(x,t)$ for each $t \in (0,T]$.

Identity (81) is the starting point for the construction of a finite element approximation to the initial value problem (79), (80). We consider the (non-uniform) mesh $-\infty < \ldots < x_{-l} < \ldots < x_0 < \ldots < x_l < \ldots < \infty$ on the real line, and the (non-uniform) mesh $0 = t_0 < \ldots < t_m < \ldots < t_M = T$ on the interval $[0,T]$. We define the piecewise constant mesh function $x \mapsto h(x)$ whose value on $(x_{l-1}, x_l]$ is $x_l - x_{l-1}$; similarly, we define the piecewise constant mesh function $t \mapsto k(t)$ whose value on $(t_{m-1}, t_m]$ is $t_m - t_{m-1}$. On the associated partitions of $\mathbb{R}$ and $[0,T]$ we consider a pair of finite element spaces $U_h \subset L_\infty(\mathbb{R})$ and $V_k \subset L_\infty(0,T)$, respectively, and define the finite element trial space as

$$X_{hk} = U_h \otimes V_k.$$

We shall suppose that each element of U_h has compact support in $\mathbb{R}$, which is a reasonable assumption given that $u(\cdot,t)$ has compact support for each $t \in [0,T]$. We adopt the convention that elements of U_h and V_k are chosen to be continuous from the left. We then consider a (possibly different) pair of finite element spaces $S_h \subset W^1_\infty(\mathbb{R})$ and $W_k \subset W^1_\infty(0,T)$ on these two partitions, respectively, and define the finite element test space

$$Y_{hk} = S_h \otimes W_k;$$

we shall suppose that each element of S_h has compact support in $\mathbb{R}$.

The finite element approximation to (79), (80) is defined as follows: find $u_{hk} \in X_{hk}$ such that, for each $v_{hk} \in Y_{hk}$,

$$(u_{hk}(\cdot,T), v_{hk}(\cdot,T)) = (u_0, v_{hk}(\cdot,0)) + \int_0^T (u_{hk}(\cdot,t), v_{hk,t}(\cdot,t)) + (f(u_{hk}(\cdot,t)), v_{hk,x}(\cdot,t))\,\mathrm{d}t. \qquad (83)$$

In what follows, we shall suppose that a numerical solution u_{hk} exists and that there is a constant L_{hk} (possibly dependent on h and k) such that

$$\text{ess.sup}_{t\in[0,T]}\|u_{hk}(\cdot,t)\|_{Lip^+(\mathbb{R})} \le L_{hk}. \qquad (84)$$

We remark that if U_h is a finite element space consisting of continuous piecewise polynomials then this condition is trivially satisfied. The *a posteriori* error analysis of this method relies on considering a dual problem defined as follows:

$$\begin{aligned} -\frac{\partial z}{\partial t} - a(x,t)\frac{\partial z}{\partial x} &= 0, \qquad -\infty < x < \infty,\ 0 \le t < T, \\ z(x,T) &= \psi(x), \qquad -\infty < x < \infty, \end{aligned} \qquad (85)$$

where $\psi \in C_0^\infty(\mathbb{R})$ and

$$a(x,t) = \begin{cases} \frac{f(u(x,t)) - f(u_{hk}(x,t))}{u(x,t) - u_{hk}(x,t)} & \text{if } u(x,t) \ne u_{hk}(x,t), \\ f'(u(x,t)) & \text{otherwise.} \end{cases}$$

We note that, despite being linear, this is a non-standard problem because $a(\cdot,\cdot)$ may be discontinuous and then the classical theory of well-posedness of linear hyperbolic equations does not apply. Nevertheless, under certain hypotheses on a, it can be shown that this problem is meaningful; the next theorem, due to Tadmor (see [59], Theorem 2.2) will make this more precise.

Theorem 25. *Suppose that:*

i) $a \in L_\infty(Q)$ *where* $Q = \mathbb{R} \times (0,T)$*;*
ii) a *satisfies the following one-sided Lipschitz condition:*

$$\|a(\cdot,t)\|_{Lip^+(\mathbb{R})} \le m(t), \qquad m \in L_1(0,T). \qquad (86)$$

Then there exists a unique Lipschitz continuous function $(x,t) \mapsto z(x,t)$ *defined on* $\mathbb{R} \times [0,T]$ *which solves the backward transport equation (85); moreover,* z *obeys the estimate*

$$\|z(\cdot,t)\|_{W^1_\infty(\mathbb{R})} \le \|\psi\|_{W^1_\infty(\mathbb{R})} e^{\mu(t)}, \quad \mu(t) \equiv \int_t^T m(\tau)\,d\tau, \quad 0 \le t \le T. \qquad (87)$$

Next we verify that the hypotheses of this theorem are satisfied with our choice of a. First note that we can write

$$a(x,t) = \int_0^1 f'((1-\xi)u(x,t) + \xi u_{hk}(x,t))\,\mathrm{d}\xi. \qquad (88)$$

Recalling that $f''(w) \geq \alpha > 0$ for all real w, a simple calculation shows that

$$\|a(\cdot,t)\|_{Lip^+(\mathrm{I\!R})} \leq A_{hk} \max\left(\|u(\cdot,t)\|_{Lip^+(\mathrm{I\!R})}, \|u_{hk}(\cdot,t)\|_{Lip^+(\mathrm{I\!R})}\right) \equiv m_{hk}(t),$$

where

$$A_{hk} = \max_{|w|\leq K_{hk}} f''(w), \qquad K_{hk} \equiv \max(\|u\|_{L_\infty(Q)}, \|u_{hk}\|_{L_\infty(Q)}).$$

Now (82) and (84) imply that (86) is satisfied; further, since both u and u_{hk} are in $L_\infty(Q)$, hypothesis i) of Theorem 25 is a trivial consequence of (88).

Now we turn to the error analysis. First, we derive a representation formula for the error. Suppose that $\psi \in C_0^\infty(\mathrm{I\!R})$ and let z denote the solution of the backward transport problem (85) with final data ψ. Then,

$$\begin{aligned}(e_h(\cdot,T),\psi) &= (e_h(\cdot,T), z(\cdot,T)) - \int_0^T (e_h(\cdot,t), z_t + a z_x)\,\mathrm{d}t \\ &= (u(\cdot,T), z(\cdot,T)) - \int_0^T (u, z_t) + (f(u), z_x)\,\mathrm{d}t \\ &\quad -(u_{hk}(\cdot,T), z(\cdot,T)) + \int_0^T (u_{hk}, z_t) + (f(u_{hk}), z_x)\,\mathrm{d}t.\end{aligned}$$

Noting that u obeys (81), we deduce that

$$\begin{aligned}(e_h(\cdot,T),\psi) &= (u_0, z(\cdot,0)) - (u_{hk}(\cdot,T), z(\cdot,T)) \\ &\quad + \int_0^T (u_{hk}, z_t) + (f(u_{hk}), z_x)\,\mathrm{d}t \\ &= (u_0, z(\cdot,0) - z_{hk}(\cdot,0)) - (u_{hk}(\cdot,T), z(\cdot,T) - z_{hk}(\cdot,T)) \\ &\quad + \int_0^T (u_{hk}, z_t - z_{hk,t}) + (f(u_{hk}), z_x - z_{hk,x})\,\mathrm{d}t \\ &\quad + (u_0, z_{hk}(\cdot,0)) - (u_{hk}(\cdot,T), z_{hk}(\cdot,T)) \\ &\quad + \int_0^T (u_{hk}, z_{hk,t}) + (f(u_{hk}), z_{hk,x})\,\mathrm{d}t,\end{aligned}$$

for any z_{hk} in Y_{hk}. By virtue of (83), the expression on the right can be further reduced to give

$$\begin{aligned}(e_h(\cdot,T),\psi) &= (u_0, z(\cdot,0) - z_{hk}(\cdot,0)) - (u_{hk}(\cdot,T), z(\cdot,T) - z_{hk}(\cdot,T)) \\ &\quad + \int_0^T (u_{hk}, z_t - z_{hk,t}) + (f(u_{hk}), z_x - z_{hk,x})\,\mathrm{d}t.\end{aligned}$$

Next we integrate by parts in order to recover the residual on the right-hand side: thus, noting that z and z_{hk} are continuous functions of x and t, that $u_{hk}(x,\cdot)$ is continuous from the left for each $x \in \mathbb{R}$ and that $u_{hk}(\cdot,t)$ is continuous from the left at each $t \in (0,T]$, we have that

$$\begin{aligned}(e_h(\cdot,T),\psi) = (u_0 - u_{hk}(\cdot,0+), z(\cdot,0) - z_{hk}(\cdot,0))\\ - \sum_{m=1}^{M-1} ([u_{hk}(\cdot,t_m)], z(\cdot,t_m) - z_{hk}(\cdot,t_m))\\ - \sum_{l=-\infty}^{\infty} \int_0^T [u_{hk}(x_l,t)](z(x_l,t) - z_{hk}(x_l,t))\,\mathrm{d}t\\ - \sum_{l=-\infty}^{\infty} \sum_{m=1}^{M} \int_{x_{l-1}}^{x_l} \int_{t_{m-1}}^{t_m} (u_{hk,t} + f(u_{hk})_x)(z - z_{hk})\,\mathrm{d}x\,\mathrm{d}t,\end{aligned}$$

where

$$[u_{hk}(\cdot,t_m)] = u_{hk}(\cdot,t_m+) - u_{hk}(\cdot,t_m)$$

and

$$[u_{hk}(x_l,\cdot)] = u_{hk}(x_l+,\cdot) - u_{hk}(x_l,\cdot).$$

Thus, letting

$$r_{hk}(x,0) = u_0(x) - u_{hk}(x,0+),$$

$$\begin{aligned}r_{hk}(x,t_m) &= [u_{hk}(x,t_m)], \quad x \in \mathbb{R}, \quad m = 1,\ldots,M-1,\\ r_{hk}(x_l,t) &= [u_{hk}(x_l,t)], \quad l = \ldots,-1,0,1,\ldots, \quad t \in \bigcup_{m=1}^{M} (t_{m-1},t_m),\\ r_{hk}(x,t) &= u_{hk,t}(x,t) + f(u_{hk}(x,t))_x, \quad (x,t) \in \bigcup_{l,m} (x_{l-1},x_l) \times (t_{m-1},t_m),\end{aligned}$$

we have that

$$\begin{aligned}|(e_h(\cdot,T),\psi)| \le \sum_{m=0}^{M-1} \|h|r_{hk}(\cdot,t_m)\|_{L_1(\mathbb{R})} \|h^{-1}(z(\cdot,t_m) - z_{hk}(\cdot,t_m))\|_{L_\infty(\mathbb{R})}\\ + \sum_{l=-\infty}^{\infty} \|kr_{hk}(x_l,\cdot)\|_{L_1(0,T)} \|k^{-1}(z(x_l,\cdot) - z_{hk}(x_l,\cdot))\|_{L_\infty(0,T)}\\ + \sum_{l=-\infty}^{\infty} \sum_{m=1}^{M} \|r_{hk}\|_{L_1(Q_{lm})} \|z - z_{hk}\|_{L_\infty(Q_{lm})},\end{aligned}$$

where $Q_{lm} = (x_{l-1},x_l) \times (t_{m-1},t_m)$.

In order to proceed, we make some weak assumptions on the approximation properties of the test space: we suppose that there exists z_{hk} in Y_{hk} and positive constants C_8 and C_9, independent of h, k, z and z_{hk} such that

$$\|h^{-1}(z(\cdot,t_m) - z_{hk}(\cdot,t_m))\|_{L_\infty(\mathbb{R})} \le C_8\|z_x(\cdot,t_m)\|_{L_\infty(\mathbb{R})},$$
$$\|k^{-1}(z(x_l,\cdot) - z_{hk}(x_l,\cdot))\|_{L_\infty(0,T)} \le C_9\|z_t(x_l,\cdot)\|_{L_\infty(0,T)},$$
$$\|z - z_{hk}\|_{L_\infty(Q_{lm})} \le C_8\|hz_x\|_{L_\infty(Q_{lm})} + C_9\|kz_t\|_{L_\infty(Q_{lm})}.$$

Any standard finite element space consisting of continuous piecewise polynomials will satisfy these inequalities (see [12] or [11]). Hence,

$$\begin{aligned}|(e_h(\cdot,T),\psi)| &\le C_8 \sum_{m=0}^{M-1} \|hr_{hk}(\cdot,t_m)\|_{L_1(\mathbb{R})}\|z_x(\cdot,t_m)\|_{L_\infty(\mathbb{R})}\\ &+C_9 \sum_{l=-\infty}^{\infty} \|kr_{hk}(x_l,\cdot)\|_{L_1(0,T)}\|z_t(x_l,\cdot)\|_{L_\infty(0,T)}\\ &+ \sum_{l=-\infty}^{\infty}\sum_{m=1}^{M} \|r_{hk}\|_{L_1(Q_{lm})}\left(C_8\|hz_x\|_{L_\infty(Q_{lm})} + C_9\|kz_t\|_{L_\infty(Q_{lm})}\right).\end{aligned}$$

Recalling the strong stability result (87), we have that

$$\begin{aligned}|(e_h(\cdot,T),\psi)| &\le C_8 \mathrm{e}^{\mu_{hk}}\|\psi\|_{W^1_\infty(\mathbb{R})} \sum_{m=0}^{M-1} \|hr_{hk}(\cdot,t_m)\|_{L_1(\mathbb{R})}\\ &+C_9\mathrm{e}^{\mu_{hk}}\|\psi\|_{W^1_\infty(\mathbb{R})} \sum_{l=-\infty}^{\infty} \|a(x_l,\cdot)\|_{L_\infty(0,T)}\|kr_{hk}(x_l,\cdot)\|_{L_1(0,T)}\\ &+\mathrm{e}^{\mu_{hk}}\|\psi\|_{W^1_\infty(\mathbb{R})} \sum_{l=-\infty}^{\infty}\sum_{m=1}^{M} \left(C_8\|hr_{hk}\|_{L_1(Q_{lm})}\right.\\ &\qquad \left.+C_9\|a\|_{L_\infty(Q_{lk})}\|kr_{hk}\|_{L_1(Q_{lm})}\right),\end{aligned}$$

where

$$\mu_{hk} = \int_0^T m_{hk}(t)\,\mathrm{d}t.$$

Upon dividing by $\|\psi\|_{W^1_\infty(\mathbb{R})}$ and taking the supremum over all ψ, we deduce that

$$\begin{aligned}\|u(\cdot,T) - u_{hk}(\cdot,T)\|_{Lip'(\mathbb{R})} &\le C_8\mathrm{e}^{\mu_{hk}} \sum_{m=0}^{M-1} \|hr_{hk}(\cdot,t_m)\|_{L_1(\mathbb{R})}\\ &+C_9\mathrm{e}^{\mu_{hk}} \sum_{l=-\infty}^{\infty} \|a(x_l,\cdot)\|_{L_\infty(0,T)}\|kr_{hk}(x_l,\cdot)\|_{L_1(0,T)}\\ &+C_8\mathrm{e}^{\mu_{hk}} \sum_{l=-\infty}^{\infty}\sum_{m=1}^{M} \|hr_{hk}\|_{L_1(Q_{lm})}\\ &+C_9\mathrm{e}^{\mu_{hk}} \sum_{l=-\infty}^{\infty}\sum_{m=1}^{M} \|a\|_{L_\infty(Q_{lk})}\|kr_{hk}\|_{L_1(Q_{lm})},\end{aligned} \tag{89}$$

where we used the notation

$$\|w\|_{Lip'(\mathbb{R})} = \sup_{\psi \in C_0^\infty(\mathbb{R})} \frac{|(w,\psi)|}{\|\psi\|_{W^1_\infty(\mathbb{R})}}.$$

Thus we have proved an *a posteriori* bound on the error between u and its finite element approximation u_{hk} of the form

$$\|u(\cdot,T) - u_{hk}(\cdot,T)\|_{Lip'(\mathbb{R})} \le \text{Const.}\, \eta_Q(u_{hk}),$$

where $\eta_Q(u_{hk})$ is the *a posteriori* error estimator on $Q = \mathbb{R}\times(0,T)$, appearing on the right-hand side of inequality (89). We note that since $u_{hk}(\cdot,t)$ has compact support in $\mathbb{R}$ for each $t \in [0,T]$, the same is true of r_{hk}, so each of the infinite sums appearing in (89) collapses to a summation over a finite number of terms.

Instead of using, as we have, a tensor-product grid on Q, we could have considered an unstructured space-time triangulation $\mathcal{T}_h$ of Q with associated finite element trial space $X_h \subset L_\infty(Q)$ and test space $Y_h \subset W^1_\infty(Q)$ instead of X_{hk} and Y_{hk}, respectively. Thus, repeating the same argument as above, we would have arrived at the following *a posteriori* error bound for the numerical solution $u_h \in X_h$ (defined similarly as u_{hk} before):

$$\|u(\cdot,T) - u_h(\cdot,T)\|_{Lip'(\mathbb{R})} \le \text{Const.}\, \eta_Q(u_h),$$

where

$$\eta_Q(u_h) = \|hr_h^0\|_{L_1(\mathbb{R})} + \sum_{\kappa\in\mathcal{T}_h}\left(\|hr_h\|_{L_1(\kappa)} + \sum_{e\subset\partial\kappa\setminus\partial Q}\|h\hat{r}_h\|_{L_1(e)}\right),$$

with e denoting an edge of a triangle κ in $\mathcal{T}_h$,

$$\begin{aligned}
r_h^0(x) &= u_0(x) - u_h(x,0+), \quad x\in\mathbb{R},\\
r_h(x,t) &= u_{h,t}(x,t) + f(u_h(x,t))_x, \quad (x,t)\in\kappa,\\
\hat{r}_h(x,t) &= [u_h(x,t)\nu_t(x,t) + f(u_h(x,t))\nu_x(x,t)], \quad (x,t)\in e\subset\partial\kappa\setminus\partial Q,
\end{aligned}$$

where (ν_x,ν_t) is the unit outward normal to edge e (with respect to the triangle κ), and $[w]$ signifies the jump of w across e. In the case of $X_h = Y_h$, a second-order numerical dissipation term could have also been included into the finite element method; the effects of this on the *a posteriori* error bound can be analysed similarly to the streamline diffusion stabilisation discussed earlier on.

To conclude this section, we note that a direct approach, different from ours, to the *a posteriori* error analysis of numerical approximations to scalar nonlinear conservation laws was pursued in [13], [14], [34].

9 Conclusions

We have presented an overview of recent developments which concern the *a posteriori* error analysis of finite element approximations to linear and nonlinear hyperbolic partial differential equations of first order. We derived various global and local bounds on the discretisation error, and investigated the question of error localisation and error propagation.

While for elliptic equations there is already a well-established theoretical framework of *a posteriori* error estimation which has been successfully implemented into working adaptive algorithms, very much less is known about these issues in the context of hyperbolic and nearly-hyperbolic problems. However, this is now a field of active research, and the outcome of these investigations can make a considerable impact on the design of reliable numerical algorithms for large-scale computations in engineering applications.

References

1. Adams, R.A. (1975). *Sobolev Spaces.* Academic Press.
2. Ainsworth, M. and Oden, T. (1996). A Posteriori Error Estimation in Finite Element Analysis. Series in Computational and Applied Maths., Elsevier.
3. Babuška, I. and Aziz, A.K. (1972). *Survey lectures on the mathematical foundation of the finite element method.* In: The Mathematical Foundations of the Finite Element Method, A.K. Aziz and I. Babuška, (Eds.), Academic Press.
4. Baiocchi, C. and Capelo, A. (1984). *Variational and Quasi-Variational Inequalities: Applications to Free Boundary Problems.* John Wiley & Sons.
5. Balland, P. and Süli, E. (1997). Analysis of the cell vertex scheme for hyperbolic problems with variable coefficients. SIAM J. Numer. Anal., **34**, 1127–1151.
6. Bank, R. (1985). PLTMG user's guide. Technical Report Edition 4, University of California, San Diego.
7. Becker, R. and Rannacher, R. (1996). Weighted a posteriori error control in finite element methods. Technical Report, Universität Heidelberg, Preprint No. 96-01.
8. Bergh, I. and Löfström, J. (1976). *Interpolation Spaces.* Springer-Verlag, Grundlehren der Mathematischen Wissenschaften 223.
9. Bernardi, C. (1989). Optimal finite-element interpolation on curved domains. SIAM J. Numer. Anal., **26**, 1212-1240.
10. Brenier, Y. and Osher, S. (1988). The discrete one-sided Lipschitz condition for convex scalar conservation laws. SIAM J. Numer. Anal., **25**, 8–23.
11. Brenner, S.C. and Scott, L.R. (1997). *The Mathematical Theory of Finite Element Methods.* 2nd corr. ed. Springer-Verlag. Texts in Applied Mathematics 15.
12. Ciarlet, P.G. (1978). *The Finite Element Method for Elliptic Problems.* North Holland, Amsterdam.

13. Cockburn, B. and Gau, H. (1995). A posteriori error estimates for general numerical methods for scalar conservation laws. Mat. Aplic. Comp., **14**, No. 1, 37–47
14. Cockburn, B. and Gremaud, P.-A. (1996). Error estimates for finite element methods for scalar conservation laws. SIAM J. Numer. Anal., **33**, 522–554.
15. Eriksson, K., Estep, D., Hansbo, P., and Johnson, C. (1995). Introduction to Adaptive Methods for Differential Equations. Acta Numerica. Cambridge University Press. 105–158.
16. Friedrichs, K.O. (1958). Symmetric positive linear differential equations. Comm. Pure Appl. Math., **11**, 333–418.
17. Führer, C. (1997). A posteriori error control for nonlinear hyperbolic problems. Ph.D. Thesis, SFB 359, Universität Heidelberg.
18. Giles, M.B. (1997). On adjoint equations for error analysis and optimal grid adaptation in CFD. Oxford University Computing Laboratory Technical Report, *NA97/11.*
19. Giles, M.B., Larson, M.G., Levenstam, M., and Süli, E. (1997). Adaptive error control for finite element approximations of the lift and drag in a viscous flow. Oxford University Computing Laboratory Technical Report, *NA97/06.*
20. Girault, V. and Raviart, P.-A. (1979). *Finite Element Approximation of the Navier-Stokes Equations.* Lecture Notes in Mathematics 749. Springer-Verlag.
21. Godlewski, E. and Raviart, P.-A. (1996). *Numerical Approximation of Hyperbolic Systems of Conservation Laws.* Series in Applied Mathematical Sciences 118. Springer-Verlag.
22. Hairer, E., Norsett, S., and Wanner, G. (1993). *Solving ordinary differential equations.* 2nd rev. ed. Series in Computational Mathematics 8. Springer-Verlag.
23. Handscomb, D.C. (1995). Error of linear interpolation on a triangle. Oxford University Computing Laboratory Technical Report, *NA95/09.*
24. Hebeker, F.-K, Führer, C., and Rannacher, R. (1997). An adaptive finite element method for unsteady convection-dominated flows with stiff source terms. Preprint (SFB 359), Universität Heidelberg.
25. Houston, P., Mackenzie, J., Süli, E., and Warnecke, G. (1999). A posteriori error analysis of Petrov-Galerkin approximations of Friedrichs systems. Numerische Mathematik (to appear).
26. Houston, P. and Süli, E. (1995). Adaptive Lagrange-Galerkin methods for unsteady convection–dominated diffusion problems. Oxford University Computing Laboratory Technical Report, *NA95/24.*
27. Houston, P. and Süli, E. (1996). On the design of an artificial diffusion model for the Lagrange-Galerkin method on unstructured triangular grids. Oxford University Computing Laboratory Technical Report, *NA96/07.*
28. Houston, P. and Süli, E. (1997). Local *a posteriori* error analysis for hyperbolic problems. Oxford University Computing Laboratory Technical Report, *NA97/14.*
29. Johnson, C. (1990). Adaptive finite element methods for diffusion and convection problems. Computer Methods in Applied Mechanics and Engineering, **82**, 301–322.

30. Johnson, C. (1994). A new paradigm for adaptive finite element methods. In: Whiteman, J.R., ed., The Mathematics of Finite Elements and Applications. Highlights 1993. John Wiley & Sons, 105–120.
31. Johnson, C. and Hansbo, P. (1992). Adaptive finite element methods in computational mechanics. Computer Methods in Applied Mechanics and Engineering, **101**, 143–181.
32. Johnson, C. and Szepessy, A. (1995). Adaptive finite element methods for conservation laws based on a posteriori estimates. Comm. Pure Appl. Math., **48**, 199–243.
33. Kröner, D. (1997). *Numerical Schemes for Conservation Laws.* John Wiley & Sons and B.G. Taubner Publishers.
34. Kröner, D. and Ohlberger, M. (1998). A posteriori error estimates for upwind finite volume schemes for nonlinear conservation laws in multi dimensions. Freiburg. Preprint 02–1998
35. Kufner, A., John, O., and Fučik, S. (1977) *Function Spaces.* Noordhoff International Publishing.
36. Lax, P.D. (1955). *On the Cauchy problem for hyperbolic equations and the differentiability of solutions of elliptic equations.* Comm. Pure. Appl. Math., **8**, 615–633.
37. Lax, P.D. and Phillips, R.S. (1960). Local boundary conditions for dissipative symmetric linear differential operators. Comm. Pure Appl. Math., **13**, 427–455.
38. Lesaint, P. (1973). Finite element methods for symmetric hyperbolic equations. Numer. Math., **21**, 244–255.
39. Lesaint, P. and Raviart, P.-A. (1979). Finite element collocation methods for first order systems. Math. Comput., **33**, 891–918.
40. Mackenzie, J., Sonar, T., and Süli, E. (1994). Adaptive finite volume methods for hyperbolic problems. In: Whiteman, J.R., ed., The Mathematics of Finite Elements and Applications. Highlights 1993. John Wiley & Sons, 289–298.
41. Mackenzie, J., Süli, E., and Warnecke, G. (1994). A posteriori error estimates for the cell-vertex finite volume method. In: Hackbusch, W. and Wittum, G., eds., Adaptive Methods: Algorithms, Theory and Applications. Vieweg, Braunschweig, **44**, 221–235.
42. Mackenzie, J., Süli, E., and Warnecke, G. (1995). A posteriori error analysis of Petrov-Galerkin approximations of Friedrichs systems. Oxford University Computing Laboratory Technical Report. *NA95/01.*
43. Melenk, J.M., and Schwab, C. (1997). An *hp* finite element method for convection-diffusion problems. Reseach Report No 97-05, Seminar für Angewandte Mathematik, ETH, Zürich.
44. Morton, K.W. and Süli, E. (1991). Finite volume methods and their analysis. IMA Journal of Numerical Analysis, **11**, 241–60.
45. Morton, K.W. and Süli, E. (1994). A posteriori and a priori error analysis of finite volume methods. In: Whiteman, J.R., ed., The Mathematics of Finite Elements and Applications. Highlights 1993. John Wiley & Sons, 267–288.
46. Morton, K.W. and Süli, E. (1995). Evolution Galerkin methods and their supraconvergence. Numerische Mathematik, **71**, 331–355.

47. Nečas, J. (1967). *Les méthodes directes en théorie des équations elliptiques.* Masson, Paris.
48. Peraire, J., Paraschivoiu, M., and Patera, A. (1996). A posteriori finite element bounds for linear functional outputs of elliptic partial differential equations. Symposium on Advances in Computational Mechanics. Submitted to Comp. Meth. Appl. Engnrg.
49. Rannacher, R. and Suttmeier F.-T. (1996). A feed-back approach to error control in finite element methods: application to linear elasticity. Preprint 96-42 (SFB 359), University of Heidelberg.
50. Rauch, J. (1972) $\mathcal{L}_2$ is a continuable initial condition for Kreiss' mixed problems. Comm. Pure Appl. Math., **25**, 265–285.
51. Sandboge, R. (1996). *Adaptive Finite Element Methods for Reactive Flow Problems.* Ph.D. Thesis. Department of Mathematics. Chalmers University Göteborg.
52. Sonar, T. and Süli, E. (1998). A dual graph-norm refinement indicator for finite volume approximations of the Euler equations. Numerische Mathematik, **78**, No. 4, 619–658.
53. Süli, E. (1989). Finite volume methods on distorted meshes: stability, accuracy, adaptivity. Oxford University Computing Laboratory Technical Report, *NA89/06.*
54. Süli, E. (1992). The accuracy of cell vertex finite volume methods on quadrilateral meshes. Math. Comput., **59**, 359–382.
55. Süli, E. (1991). The accuracy of finite volume methods on distorted partitions. In: Whiteman, J.R., ed., The Mathematics of Finite Elements and Applications VII, Academic Press, London, 253–260.
56. Süli, E. (1996). A posteriori error analysis and global error control for adaptive finite element approximations of hyperbolic problems. In: D.F. Griffiths and G.A. Watson, eds. Numerical Analysis 1995, Pitman Lecture Notes in Mathematics Series 344, 169–190.
57. Süli, E. and Houston, P. (1997). Finite element methods for hyperbolic problems: a posteriori error analysis and adaptivity. In: I.S. Duff and G.A. Watson, eds. The State of the Art in Numerical Analysis, Clarendon Press, Oxford, 441–471.
58. Tartakoff, D. (1972). Regularity of solutions to boundary value problems for first order systems. Indiana University Mathematics Journal, **21**, No. 12, 1113–1129.
59. Tadmor, E. (1991). Local error estimates for discontinuous solutions of nonlinear hyperbolic equations. SIAM J. Numer. Anal., **28**, 891–906.
60. Szabó, B. and Babuška, I. (1991). *Finite Element Analysis.* J. Wiley & Sons, New York.
61. Verfürth, R. (1996). *A Review of a Posteriori Error Estimation and Adaptive Mesh-Refinement Techniques.* B.G. Teubner, Stuttgart.
62. Winther, R. (1981). A stable finite element method for initial boundary value problems for first-order hyperbolic systems. Math. Comput., **36**, 65–86.

Numerical Methods for Gasdynamic Systems on Unstructured Meshes

Timothy J. Barth

NASA Ames Research Center, Moffett Field, CA 94035, USA

Abstract. This article considers stabilized finite element and finite volume discretization techniques for systems of conservation laws. Using newly developed techniques in entropy symmetrization theory, simplified forms of the Galerkin least-squares (GLS) and the discontinuous Galerkin (DG) finite element method are developed and analyzed. The use of symmetrization variables yields numerical schemes which inherit global entropy stability properties of the PDE system. Detailed consideration is given to symmetrization of the Euler, Navier-Stokes, and magnetohydrodynamic (MHD) equations. Numerous calculations are presented to evaluate the spatial accuracy and feature resolution capability of the simplified DG and GLS discretizations. Next, upwind finite volume methods are reviewed. Specifically considered are generalizations of Godunov's method to high order accuracy and unstructured meshes. An important component of high order accurate Godunov methods is the spatial reconstruction operator. A number of reconstruction operators are reviewed based on Green-Gauss formulas as well as least-squares approximation. Several theoretical results using maximum principle analysis are presented for the upwind finite volume method. To assess the performance of the upwind finite volume technique, various numerical calculations in computational fluid dynamics are provided.

Contents

Overview

The efficient, accurate, and robust numerical solution of systems of hyperbolic conservation laws on arbitrary unstructured meshes remains a challenging problem. The intent of this article is to present several relatively new design techniques used in the construction of stabilized finite element and finite volume methods for compressible fluid flow problems.

Section 1 briefly reviews the standard symmetrization theory for entropy endowed hyperbolic systems. This provides a rigorous framework for analyzing stabilized finite element methods using an entropy norm measure of stability. Next, an eigenvector scaling theorem is introduced for symmetrizable systems. The scaling theorem motivates simplified variants of the Galerkin least-squares (GLS) and the discontinuous Galerkin (DG) finite element methods which are discussed in Sect 2. Specific examples of the eigenvalue scaling theorem are given for the compressible Euler, Navier-Stokes, and magnetohydrodynamic (MHD) equations.

In Sect. 2, the Galerkin least-squares method advocated by Hughes and Mallet [34] and Johnson [36] as well as the discontinuous Galerkin method developed by Johnson and Pitkäranta [37], Bey and Oden [11], Cockburn et al. [17,18], and Bassi and Rebay [10] is examined. Both DG and GLS methods are capable of achieving $O(h^{k+1/2})$ accuracy using P_k polynomial elements in the natural norm induced by the bilinear variational form [37,36]. By exploiting the eigenvalue scaling theorem, it is shown how to construct simplified variants of the standard GLS and DG methods for systems of hyperbolic equations equipped with entropy extensions. Both simplified schemes utilize symmetrization variables for both implementational and theoretical reasons discussed below. In the context of the discontinuous Galerkin method, this appears to be the first faithful implementation of the discontinuous Galerkin method using symmetrization variables. Using results from basic symmetrization theory, global nonlinear entropy stability is shown. The nonlinear stability analysis also reveals a subtle deficiency in the Roe numerical flux [46] when used in the discontinuous Galerkin method. Spatial convergence studies and numerous calculations are also presented to validate the simplified formulations.

Sections 3 and 4 consider maximum principles and their use in the construction of stabilized finite volume methods. It is well known that high order spatial accuracy of these methods hinges heavily on the reconstruction operator associated with the scheme. A number of reconstruction techniques are reviewed based on Green-Gauss and least-squares approximation. Non-oscillatory numerical solutions are obtained by enforcing monotonicity conditions on the reconstructed functions. Using maximum principle analysis, several theoretical results for the upwind finite volume method are presented. Numerous calculations are shown to assess the performance of the upwind finite volume techniques on practical problems in computational fluid dynamics.

1 Symmetrization of Systems of Conservation Laws

This section reviews several related topics associated with the symmetrization of systems of conservation laws:

1. Basic entropy symmetrization theory
2. Symmetrization and the eigenvector scaling theorem
3. Symmetrization of the compressible Euler and Navier-Stokes equations
4. Symmetrization of the magnetohydrodynamic (MHD) equations

Using the tools developed in these sections, simplified forms of the Galerkin least-squares (GLS) and discontinuous Galerkin (DG) finite element methods are presented.

There are many motivations, some theoretical, some practical, for recasting conservation law equations into symmetric form. Three motivations are listed below. The first motivation is widely recognized while the remaining two are less often appreciated.

1. **Energy Considerations.** Consider the compressible Navier-Stokes equations in quasi-linear form with $\boldsymbol{u}$ the vector of conserved variables, $\boldsymbol{f}^i$ the flux vectors, and M_{ij} the viscosity matrices

$$\boldsymbol{u}_{,t} + \boldsymbol{f}^i_{,\boldsymbol{u}}\, \boldsymbol{u}_{,x_i} = (M_{ij} \boldsymbol{u}_{,x_j})_{,x_i}, \quad x \in \mathbb{R}^d \ . \tag{1}$$

In this form, the inviscid flux Jacobian matrices $\boldsymbol{f}^i_{,\boldsymbol{u}}$ are not symmetric and the viscosity matrices M_{ij} are neither symmetric nor positive semi-definite. This makes energy analysis very difficult. When recast in symmetric form, the inviscid coefficient matrices are symmetric and the viscosity matrices are symmetric positive semi-definite. As will be shown later, the use of symmetrization variables in the finite element method yields an *ab initio* form of global energy stability for the Galerkin least-squares [33] and discontinuous Galerkin finite element methods.

2. **Dimensional consistency.** As a representative example, consider the time derivative term from (1). The weak variational statement associated with this equation requires the integration of terms such as $\int \boldsymbol{w}^T_{,t}\boldsymbol{u}\, dx\, dt$. When $\boldsymbol{w}$ and $\boldsymbol{u}$ reside in the same space of functions, the inner product quantity $\boldsymbol{w}^T\boldsymbol{u}$ is dimensionally inconsistent. Consequently, errors made in a computation depend fundamentally on how the equations have been dimensionalized. When recast in symmetric form, the inner product $\boldsymbol{w}^T\boldsymbol{u}$ is dimensionally consistent with units of specific entropy.
3. **Eigenvector scaling.** Apart from degenerate scalings, any scaling of eigenvectors satisfies the eigenvalue problem. Unfortunately, numerical discretization techniques sometimes place additional demands on the form of right eigenvectors. As noted by Balsara [4] in his study of high order Godunov methods, several of the schemes he studied that interpolate "characteristic" data (see for example Harten *et al.* [32]) showed accuracy degradation that depended on the specific scaling of the eigenvectors of the inviscid flux Jacobians. In the characteristic interpolation approach, the solution data is projected onto the local right eigenvectors of the flux Jacobian, interpolated between cells, and finally transformed back. The interpolant thus depends on the eigenvector form. Entropy symmetrization theory provides a systematic approach to scaling eigenvectors. This is especially useful in MHD flow where the right eigenvectors must be carefully scaled, especially near the triple umbilic point where fast, slow and Alfén wave speeds coincide [12]. The eigenvector scaling theorem is also used heavily in later sections to construct simplified variants of two stabilized finite element methods.

1.1 A Brief Review of Entropy Symmetrization Theory

Consider a system of m coupled first order differential equations in d space coordinates and time which represents a conservation law process. Let $\boldsymbol{u}(x,t) : \mathbb{R}^d \times \mathbb{R}^+ \mapsto \mathbb{R}^m$ denote the dependent solution variables and $\boldsymbol{f}(\boldsymbol{u}) : \mathbb{R}^m \mapsto \mathbb{R}^{m\times d}$ the flux vector

$$\boldsymbol{u}_{,t} + \boldsymbol{f}^i_{,x_i} = 0 \tag{2}$$

with implied summation on the index i. Additionally, this system is assumed to possess the following properties:

1. Hyperbolicity. The linear combination

$$\boldsymbol{f}_{,\boldsymbol{u}}(\boldsymbol{n}) = n_i\, \boldsymbol{f}^i_{,\boldsymbol{u}}$$

has m real eigenvalues and a complete set of eigenvalues for all $\boldsymbol{n} \in \mathbb{R}^d$.
2. Entropy Inequality. Existence of entropy pairs $\{U(\boldsymbol{u}), F^i(\boldsymbol{u})\} : \mathbb{R}^m \mapsto \mathbb{R}$ such that in addition to (2) the following inequality holds

$$U_{,t} + F^i_{,x_i} \leq 0. \tag{3}$$

In the standard symmetrization theory [26,42,44,30], one seeks a change of variables $\boldsymbol{u}(\boldsymbol{v}) : \mathbb{R}^m \mapsto \mathbb{R}^m$ applied to (2) so that when transformed

$$\boldsymbol{u}_{,\boldsymbol{v}}\boldsymbol{v}_{,t} + \boldsymbol{f}^i_{,\boldsymbol{v}}\boldsymbol{v}_{,x_i} = 0 \tag{4}$$

the matrix $\boldsymbol{u}_{,\boldsymbol{v}}$ is symmetric positive definite and the matrices $\boldsymbol{f}^i_{,\boldsymbol{v}}$ are symmetric. Clearly, if functions $\mathcal{U}(\boldsymbol{v}), \mathcal{F}^i(\boldsymbol{v}) : \mathbb{R}^m \mapsto \mathbb{R}$ can be found such that

$$\boldsymbol{u}^T = \mathcal{U}_{,\boldsymbol{v}}, \quad (\boldsymbol{f}^i)^T = \mathcal{F}^i_{,\boldsymbol{v}}$$

then the matrices

$$\boldsymbol{u}_{,\boldsymbol{v}} = \mathcal{U}_{,\boldsymbol{v},\boldsymbol{v}}, \quad \boldsymbol{f}^i_{\boldsymbol{v}} = \mathcal{F}^i_{,\boldsymbol{v},\boldsymbol{v}}$$

are symmetric. To insure positive-definiteness of $\boldsymbol{u}_{,\boldsymbol{v}}$ so that mappings are one-to-one, convexity of $\mathcal{U}(\boldsymbol{v})$ is imposed. Since $\boldsymbol{v}$ is not yet known, little progress has been made. Introducing the following dual relationships

$$U(\boldsymbol{u}) = \boldsymbol{v}^T(\boldsymbol{u})\, \boldsymbol{u} - \mathcal{U}(\boldsymbol{v}(\boldsymbol{u})) \tag{5}$$

$$F^i(\boldsymbol{u}) = \boldsymbol{v}^T(\boldsymbol{u})\, \boldsymbol{f}^i(\boldsymbol{u}) - \mathcal{F}^i(\boldsymbol{v}(\boldsymbol{u})) \tag{6}$$

followed by differentiation yields

$$U_{,\boldsymbol{u}} = \boldsymbol{v}^T + \boldsymbol{u}^T\boldsymbol{v}_{,\boldsymbol{u}} - \mathcal{U}_{,\boldsymbol{v}}\boldsymbol{v}_{,\boldsymbol{u}} = \boldsymbol{v}^T$$

$$F^i_{,\boldsymbol{u}} = \boldsymbol{v}^T \boldsymbol{f}^i_{,\boldsymbol{u}} + (\boldsymbol{f}^i)^T\boldsymbol{v}_{,\boldsymbol{u}} - \mathcal{F}_{,\boldsymbol{v}}\boldsymbol{v}_{,\boldsymbol{u}} = \boldsymbol{v}^T \boldsymbol{f}^i_{,\boldsymbol{u}} \ .$$

These formulas give explicit expressions for the entropy variables $\boldsymbol{v}$ in terms of derivatives of the entropy function $U(\boldsymbol{u})$

$$\boldsymbol{v}^T = U_{,\boldsymbol{u}}$$

and the corresponding flux Jacobians

$$\boldsymbol{v}^T \boldsymbol{f}^i_{,\boldsymbol{u}} = F^i_{,\boldsymbol{u}} \ .$$

Symmetrization and generalized entropy functions are intimately linked via the following two theorems:

Theorem 1. *Godunov [26] If a hyperbolic system is symmetrized via change of variables, then there exists a generalized entropy pair for the system.*

Theorem 2. *Mock [44] If a hyperbolic system is equipped with a generalized entropy pair U, F^i, then the system is symmetrized under the change of variables $\boldsymbol{v}^T = U_{,\boldsymbol{u}}$.*

An important property of entropy-symmetrized systems emerges when inner products of the conservation law system are taken with respect to the entropy variables, i.e.

$$\boldsymbol{v}^T \left(\boldsymbol{u}_{,t} + \boldsymbol{f}^i_{,x_i}\right) = U_{,t} + F^i_{,x_i} = 0 \tag{7}$$

for smooth solutions. This property is a crucial component in the proof of global energy stability of the stabilized Galerkin and Petrov-Galerkin methods discussed in later sections.

Remark 3. For many physical systems, entropy inequalities of the form (3) can be derived by appealing directly to the conservation law system and the second law of thermodynamics. Using this strategy, specific entropy functions for the Euler, Navier-Stokes, and MHD equations are considered in Secs. 1.4, 1.5 respectively.

1.2 Symmetrization and Eigenvector Scaling

In this section, an important property of right (or left) symmetrizable systems is given. Simplifying upon the previous notation, let $\tilde{A}_0 = \boldsymbol{u}_{,\boldsymbol{v}}$, $A_i = \boldsymbol{f}^i_{,\boldsymbol{v}}$ and rewrite (4)

$$\underbrace{\tilde{A}_0}_{\text{SPD}} \boldsymbol{v}_{,t} + \underbrace{A_i \tilde{A}_0}_{\text{Symm}} \boldsymbol{v}_{,x_i} = 0 \tag{8}$$

or more compactly as

$$\tilde{A}_0 \boldsymbol{v}_{,t} + \tilde{A}_i \boldsymbol{v}_{,x_i} = 0 \tag{9}$$

with $\tilde{A}_i = A_i \tilde{A}_0$. The following theorem states a property of the symmetric matrix products $A_i \tilde{A}_0$ symmetrized via the symmetric positive definite matrix $\tilde{A}_0$.

Theorem 4 (Eigenvector Scaling). *Let $A \in \mathbb{R}^{n \times n}$ be an arbitrary diagonalizable matrix and S the set of all right symmetrizers:*

$$S = \{B \in \mathbb{R}^{n \times n} \mid B \ \text{ SPD}, \ AB \ \text{ symmetric}\}.$$

Further, let $R \in \mathbb{R}^{n \times n}$ denote the right eigenvector matrix which diagonalizes A

$$A = R \Lambda R^{-1}$$

with r distinct eigenvalues, $\Lambda = \mathrm{Diag}(\lambda_1 I_{m_1\times m_1}, \lambda_2 I_{m_2\times m_2}, \ldots, \lambda_r I_{m_r\times m_r})$. *Then for each* $B \in S$ *there exists a symmetric block diagonal matrix* $T = \mathrm{Diag}(T_{m_1\times m_1}, T_{m_2\times m_2}, \ldots, T_{m_r\times m_r})$ *that block scales columns of* R, $\tilde{R} = RT$, *such that*

$$B = \tilde{R}\tilde{R}^T, \quad A = \tilde{R}\Lambda\tilde{R}^{-1}$$

which imply

$$AB = \tilde{R}\Lambda\tilde{R}^T.$$

Proof. The symmetry of B and AB implies that

$$AB - BA^T = R\Lambda R^{-1}B - BR^{-T}\Lambda R^T = 0$$

or equivalently for $Y \in \mathbb{R}^{n\times n}$

$$\Lambda Y - Y\Lambda = 0, \quad Y = R^{-1}BR^{-T}. \tag{10}$$

Partition Y into $r \times r$ blocks, $Y_{m_i\times m_j}$, with block dimensions corresponding to eigenvalue multiplicities. Equation (10) then reduces to the following set of decoupled systems

$$\lambda_i I_{m_i\times m_i} Y_{m_i\times m_j} - \lambda_j Y_{m_i\times m_j} I_{m_j\times m_j} = 0, \quad \forall i,j \le r$$

or simply as

$$(\lambda_i - \lambda_j) Y_{m_i\times m_j} = 0, \quad \forall i,j \le r. \tag{11}$$

This implies that Y is of block diagonal form since $Y_{m_i\times m_j} = 0, i \neq j$. From the definition $Y = R^{-1}BR^{-T}$, Y is congruent to B, hence symmetric positive definite (SPD). Given the block diagonal structure of Y, the square root (or Cholesky factorization) exists globally as well as for each diagonal block, e.g. $Y_{m_i\times m_i} = Y^{1/2}_{m_i\times m_i} Y^{1/2}_{m_i\times m_i}$. This yields the stated theorem with $T = Y^{1/2}$.
□

This theorem is a variant of the well known theory developed for the commuting matrix equation

$$AX - XA = 0, \quad A, X \in \mathbb{R}^{n\times n},$$

see for example Gantmacher [24]. Note that the theory addresses the more general situation for which the matrix A can only be reduced to Jordan canonical form.

Remark 5. Note that the Eigenvector Scaling Theorem can also be derived using the theory of simultaneous diagonalization by congruence. Given $\tilde{A}_0 \in$

$\mathbb{R}^{n\times n}$ SPD and $\tilde{A} \in \mathbb{R}^{n\times n}$ symmetric, the theory of simultaneous diagonalization by congruence states that one can generically find an $\tilde{A}_0^{-1}$-orthogonal matrix C and diagonal matrix Ω such that

$$\tilde{A}_0 = C\,C^T, \quad \text{and}\, \tilde{A} = C\Omega C^T \ .$$

With the added assumption $\tilde{A} = A\tilde{A}_0$ and the diagonalization $A = R\Lambda R^{-1}$, it then follows that

$$\begin{aligned} \tilde{A} &= A\tilde{A}_0 \\ C\Omega C^T &= R\Lambda R^{-1} C\,C^T \\ \Omega &= C^{-1} R\Lambda R^{-1} C. \end{aligned}$$

By similarity, it is concluded at $\Omega = \Lambda$ and C differs from R by at most a scaling of eigenvectors.

Remark 6. Considering the hyperbolic system in $\mathbb{R}^d$, it is clear that the right eigenvectors associated with each A_i can be scaled so that $A_i\tilde{A}_0 = \tilde{R}_i\Lambda_i\tilde{R}_i^T$ which produces a revealing form of the symmetric quasi-linear form

$$\tilde{A}_0 \boldsymbol{v}_{,t} + \tilde{R}_i\Lambda_i\tilde{R}_i^T \boldsymbol{v}_{,x_i} = 0$$

with $\tilde{A}_0 = \tilde{R}_1\tilde{R}_1^T = \cdots = \tilde{R}_d\tilde{R}_d^T$.

1.3 Generalized Matrix Functions with Respect to the Riemannian Metric Tensor

To better understand the consequences of the Eigenvector Scaling Theorem, consider the linear symmetric hyperbolic system in one space dimension and time

$$\tilde{A}_0 \boldsymbol{v}_{,t} + \tilde{A}\boldsymbol{v}_{,x} = 0 \ . \tag{12}$$

Let $\tilde{R}$ denote the scaled eigenvectors of A so that

$$A = \tilde{R}\Lambda\tilde{R}^{-1}, \quad \tilde{A}_0 = \tilde{R}\tilde{R}^T, \quad \tilde{A} = A\,\tilde{A}_0 = \tilde{R}\Lambda\tilde{R}^T \ .$$

This implies that the symmetric system (12) with constant coefficient matrices is diagonalized via characteristic-like variables $\mathbf{z} = \tilde{R}^T \boldsymbol{v}$

$$\mathbf{z}_{,t} + \Lambda \mathbf{z}_{,x} = 0 \ .$$

Next, separate the diagonalized spatial operator in terms of positive and negative characteristic components

$$\mathbf{z}_{,t} + \Lambda^+ \mathbf{z}_{,x} + \Lambda^- \mathbf{z}_{,x} = 0$$

and recompose in original form by multiplication from the left by $\tilde{R}$

$$\tilde{A}_0 \boldsymbol{v}_{,t} + \tilde{R}\Lambda^+\tilde{R}^T \boldsymbol{v}_{,x} + \tilde{R}\Lambda^-\tilde{R}^T \boldsymbol{v}_{,x} = 0 \ .$$

Observe that the decomposition of $\tilde{A}$ satisfies

$$\tilde{A} = \tilde{R}\Lambda^{+}\tilde{R}^{T} + \tilde{R}\Lambda^{-}\tilde{R}^{T}$$

but that

$$\tilde{A}^{\pm} \neq \tilde{R}\Lambda^{\pm}\tilde{R}^{T}$$

in the usual matrix sense. This inequality is a direct result of the presence of a Riemannian metric tensor $\tilde{A}_0$. To properly account for the appearance of a metric tensor, it is useful to define the notion of a matrix function $f(\tilde{A})$ with respect to a metric tensor $\tilde{A}_0$

$$f_{\tilde{A}_0}(\tilde{A}) \equiv \tilde{A}_0 f(\tilde{A}_0^{-1}\tilde{A}) \ . \tag{13}$$

This definition reflects the following steps: (1) multiplication of the system by $\tilde{A}_0^{-1}$ in order to restore a Euclidean metric tensor, (2) invocation of the matrix function on the matrix product $\tilde{A}_0^{-1}\tilde{A}$, (3) multiplication of the result by $\tilde{A}_0$ to restore the original metric tensor. Proposition 7 shows that this generalized matrix function is symmetric and has a rather simple construction for symmetrizable systems. This basic construction is used repeatedly in later discussions.

Proposition 7. *Let $\tilde{A}_0$ denote the SPD right symmetrizer of A such that $\tilde{A} = A\tilde{A}_0$ $\tilde{A}_0 = \tilde{R}\tilde{R}^T$, and $A = \tilde{R}\Lambda\tilde{R}^{-1}$. The generalized matrix function $f_{\tilde{A}_0}(\tilde{A})$ is symmetric and defined equivalently as*

$$f_{\tilde{A}_0}(\tilde{A}) = f(A)\,\tilde{A}_0 \tag{14}$$

or in terms of scaled eigenvectors $\tilde{R}$

$$f_{\tilde{A}_0}(\tilde{A}) = \tilde{R}f(\Lambda)\tilde{R}^T \ . \tag{15}$$

Proof. The desired results are obtained using similarity and the Eigenvector Scaling Theorem. Symmetry is observed after the following rearrangement

$$\begin{aligned}
f_{\tilde{A}_0}(\tilde{A}) &= \tilde{A}_0 f(\tilde{A}_0^{-1}\tilde{A}) \\
&= \tilde{A}_0 f(\tilde{A}_0^{-1}\tilde{A}\tilde{A}_0^{-1}\tilde{A}_0) \\
&= \tilde{A}_0^{1/2} f(\tilde{A}_0^{-1/2}\tilde{A}\tilde{A}_0^{-1}\tilde{A}_0^{1/2})\tilde{A}_0^{1/2} \\
&= \tilde{A}_0^{1/2} f(\tilde{A}_0^{-1/2}\tilde{A}\tilde{A}_0^{-1/2})\tilde{A}_0^{1/2} \\
&= \tilde{A}_0^{1/2} U f(\Omega) U^T \tilde{A}_0^{1/2}
\end{aligned}$$

where U is an orthonormal matrix diagonalizing the symmetric matrix combination $\tilde{A}_0^{-1/2}\tilde{A}\tilde{A}_0^{-1/2}$ with eigenvalues Ω_{ii}. The two proposed forms emerge after manipulation and use of the Eigenvector Scaling Theorem

$$f_{\tilde{A}_0}(\tilde{A}) = \tilde{A}_0 f(\tilde{A}_0^{-1}\tilde{A})$$

$$\begin{aligned}
&= \tilde{A}_0 f(\tilde{A}_0^{-1} A \tilde{A}_0) \\
&= \tilde{A}_0 \tilde{A}_0^{-1} f(A) \tilde{A}_0 \\
&= f(A) \tilde{A}_0 \\
&= \tilde{R} f(\Lambda) \tilde{R}^{-1} \tilde{R} \tilde{R}^T \\
&= \tilde{R} f(\Lambda) \tilde{R}^T \ .
\end{aligned}$$

□

In later sections, the generalized matrix absolute value function $|\tilde{A}|_{\tilde{A}_0}$ will be required. From proposition 7

$$|\tilde{A}|_{\tilde{A}_0} = |A|\,\tilde{A}_0 = \tilde{R}|\Lambda|\tilde{R}^T \ . \tag{16}$$

The matrix absolute value function has a natural generalization to $\mathbb{R}^d$ using an L_p-like norm definition

$$|\tilde{A}|_{p,\tilde{A}_0} = \left(\sum_{i=1}^{d} |A_i|^p\right)^{1/p} \tilde{A}_0 \tag{17}$$

which has a particularly simple form when $p = 1$

$$|\tilde{A}|_{1,\tilde{A}_0} = \sum_{i=1}^{d} \tilde{R}_i\,|\Lambda_i|\,\tilde{R}_i^T \ . \tag{18}$$

This formula suggests a simplified variant of the GLS finite element method when cast in symmetric form. Simplified forms of GLS and DG are discussed in Sect. 2.

1.4 Scaling Theorem Example: 3-D Compressible Euler and Navier-Stokes Equations

Consider the compressible Navier-Stokes equations, $x \in \mathbb{R}^d$,

$$\frac{\partial}{\partial t}\begin{pmatrix} \rho \\ \rho \boldsymbol{V} \\ E \end{pmatrix} + \nabla \cdot \begin{pmatrix} \rho \boldsymbol{V} \\ \rho \boldsymbol{V}\boldsymbol{V} + \mathbf{I}\,p \\ \boldsymbol{V}\,(E+p) \end{pmatrix} = \nabla \cdot \begin{pmatrix} 0 \\ \boldsymbol{\tau} \\ \boldsymbol{\tau}\boldsymbol{V} - \boldsymbol{q} \end{pmatrix} \tag{19}$$

where $\boldsymbol{V} \in \mathbb{R}^d$ is the velocity vector, ρ and p the density and pressure of the fluid, E the specific total energy defined as

$$E = \frac{p}{\gamma - 1} + \frac{1}{2}\rho \boldsymbol{V}^2 \tag{20}$$

and $\boldsymbol{\tau}$ is the viscous stress tensor

$$\boldsymbol{\tau} = \lambda\left(\frac{\partial \boldsymbol{V}_i}{\partial x_i}\right) + \mu\left(\frac{\partial \boldsymbol{V}_i}{\partial x_j} + \frac{\partial \boldsymbol{V}_j}{\partial x_i}\right) . \tag{21}$$

In addition, an ideal gas is assumed $p = \rho R T$ as well as Fourier heat conduction $\boldsymbol{q} = -\kappa \nabla T$. In these equations λ and μ are diffusion coefficients, κ the coefficient of thermal conductivity, γ the ratio of specific heats, and R the ideal gas law constant. The Euler equations are easily obtained by dropping the viscous stress tensor terms.

In 1983, Harten [30] proposed the generalized convex entropy function

$$U(\boldsymbol{u}) = -\rho g(s), \quad g' > 0, \; \frac{g''}{g'} < \gamma^{-1}$$

for the compressible Euler equations. In this equation, s is the thermodynamic entropy of the fluid. This choice was motivated from the well known entropy transport inequality for the inviscid Euler equations

$$s_{,t} + \boldsymbol{V}_i \, s_{,x_i} \geq 0$$

which generalizes to

$$g(s)_{,t} + \boldsymbol{V}_i \, g(s)_{,x_i} \geq 0, \quad g' > 0$$

or after combining with the continuity equations

$$(\rho g(s))_{,t} + (\rho \boldsymbol{V}_i \, g(s))_{,x_i} \geq 0 \; .$$

Comparing this with the required entropy inequality

$$U_{,t} + F^i_{,x_i} \leq 0,$$

it becomes clear that $\{U, F^i\} = \{-\rho g(s), -\rho \boldsymbol{V}_i \, g(s)\}$ is an acceptable entropy pair. Hughes, Franca, and Mallet [33] removed the arbitrariness of $g(s)$ by showing that symmetrization of the Navier-Stokes equations with heat conduction places the additional restriction that $g(s)$ be at most affine in s, i.e. $g(s) = c_0 + c_1 s$. A convenient choice is given by $U(s) = \frac{-\rho s}{\gamma - 1}$ which yields the following entropy variables:

$$\boldsymbol{v} = U^T_{,\boldsymbol{u}} = \begin{pmatrix} -\frac{s}{\gamma-1} + \frac{\gamma+1}{\gamma-1} - \frac{E}{p} \\ \frac{\rho \boldsymbol{V}}{p} \\ -\frac{\rho}{p} \end{pmatrix}$$

The change of variable matrix $\boldsymbol{u}_{,\boldsymbol{v}}$ takes a particularly simple form

$$\boldsymbol{u}_{,\boldsymbol{v}} = \begin{pmatrix} \rho & \rho \boldsymbol{V}^T & E \\ \rho \boldsymbol{V} & \rho \boldsymbol{V}\boldsymbol{V} + p\mathbf{I} & \rho H \boldsymbol{V} \\ E & \rho H \boldsymbol{V}^T & \rho H^2 - \frac{a^2 p}{\gamma - 1} \end{pmatrix}$$

with a the sound speed, $a^2 = \gamma p / \rho$, and H the specific total enthalpy, $H = a^2/(\gamma - 1) + \boldsymbol{V}^2/2$.

Consider the application of the Eigenvector Scaling Theorem 4 to the inviscid Euler equation terms appearing in the Navier-Stokes equations

$$\tilde{A}_0 \boldsymbol{v}_{,t} + A_i \tilde{A}_0 \boldsymbol{v}_{,x_i} = 0 \tag{22}$$

with $A_i = \boldsymbol{f}^i_{,\boldsymbol{u}}$ and $\tilde{A}_0 = \boldsymbol{u}_{,\boldsymbol{v}}$. It is sufficient to consider symmetrization of the arbitrary linear combinations of the form

$$A(\boldsymbol{n}) = n_i A_i, \quad \|\boldsymbol{n}\| = 1 \tag{23}$$

and scaled right eigenvectors $\tilde{R}(\boldsymbol{n})$ such that

$$A(\boldsymbol{n}) = \tilde{R}(\boldsymbol{n}) \Lambda(\boldsymbol{n}) \tilde{R}^{-1}(\boldsymbol{n}), \quad \tilde{A}_0 = \tilde{R}(\boldsymbol{n}) \tilde{R}^T(\boldsymbol{n}) \ .$$

From this result, the symmetric coefficient matrices are given by

$$\tilde{A}(\boldsymbol{n}) = A(\boldsymbol{n}) \tilde{A}_0 = \tilde{R}(\boldsymbol{n}) \Lambda(\boldsymbol{n}) \tilde{R}^T(\boldsymbol{n}) \ .$$

From a derivational point-of-view, it is advantageous to first compute the eigensystem associated with the system in primitive variables, $\boldsymbol{w} = (\rho, \boldsymbol{V}, p)^T$,

$$\boldsymbol{w}_{,t} + \boldsymbol{u}^{-1}_{,\boldsymbol{w}} A_i \, \boldsymbol{u}_{,\boldsymbol{w}} \, \boldsymbol{w}_{,x_i} = 0 \tag{24}$$

and then to compute the scaling of the right eigenvectors, $\tilde{r}(\boldsymbol{n})$, of the primitive variable system. Once the scaled right eigenvectors of the primitive variable system have been computed, scaled right eigenvectors of the conservative system are easily recovered from

$$\tilde{R}(\boldsymbol{n}) = \boldsymbol{u}_{,\boldsymbol{w}} \, \tilde{r}(\boldsymbol{n})$$

with

$$\boldsymbol{u}_{,\boldsymbol{w}} = \begin{pmatrix} 1 & \mathbf{0}^T & 0 \\ \boldsymbol{V} & \rho \mathbf{I} & \mathbf{0} \\ \frac{1}{2}\boldsymbol{V}^2 & \rho \boldsymbol{V}^T & \frac{1}{\gamma - 1} \end{pmatrix} . \tag{25}$$

Following the procedure described in Theorem 4, after some straightforward algebraic manipulation, the following scaled right eigenvectors of the primitive variable system have been obtained:

Entropy and Shear Waves: $\lambda_{1,2,3} = \boldsymbol{V} \cdot \boldsymbol{n}$

$$\tilde{r}_{1,2,3} = \sqrt{\frac{1}{\gamma \rho}} \begin{pmatrix} \sqrt{\gamma - 1}\, \rho\, \boldsymbol{n}^T \\ a\, [\mathcal{C}] \\ \mathbf{0}^T \end{pmatrix} \tag{26}$$

where $[\mathcal{C}(\boldsymbol{n})] = n_i \epsilon_{ijk}$ and ϵ_{ijk} is the usual alternation tensor.

Acoustic Waves: $\lambda_{\pm} = \boldsymbol{V} \cdot \boldsymbol{n} \pm a$

$$\tilde{r}_{\pm} = \sqrt{\frac{1}{2\gamma\rho}} \begin{pmatrix} \rho \\ \pm a\,\boldsymbol{n} \\ \rho\,a^2 \end{pmatrix} . \tag{27}$$

In Fig. 1, $\|\tilde{R}^T(\boldsymbol{n})\|_2$, is graphed for $\boldsymbol{n} = (\cos\theta, \sin\theta)^T, \theta \in [0, 2\pi]$ with constant $(\rho, \boldsymbol{V}, p)$ using entropy scaled eigenvectors and the "naturally" scaled eigenvectors given in Struijs [52]. The entropy scaled eigenvectors produce a constant result since $\tilde{R}(\boldsymbol{n})\tilde{R}^T(\boldsymbol{n}) = \boldsymbol{u}_{,\boldsymbol{v}}$ which is independent of $\boldsymbol{n}$. This provides a simple numerical check of the scaling factors. Although the naturally scaled eigenvectors seem to differ in minor ways from the entropy scaled counterparts, the MHD example given in the next section provides a more convincing argument for the proposed eigenvector scaling technique.

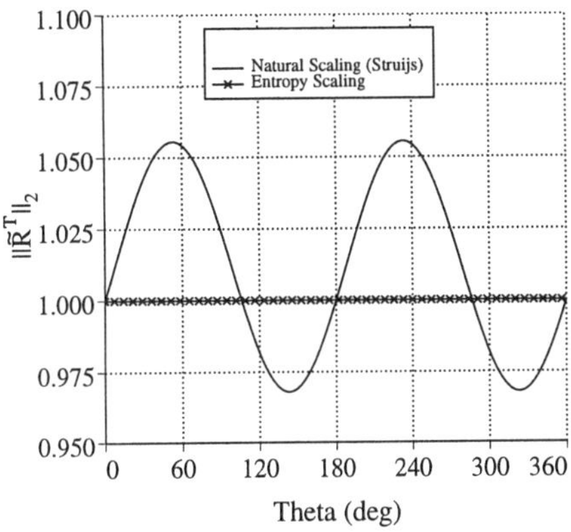

Fig. 1. The matrix norm $\|\tilde{R}^T(\boldsymbol{n})\|_2$ dependency on $\boldsymbol{n}$, $\left(\rho/\rho_\infty, \boldsymbol{V}^T/a_\infty, p/p_\infty\right) = (1, (.8, -.6), 2.4)$.

As mentioned in the introduction, symmetrization of the Navier-Stokes equations results in the pleasing form

$$\underbrace{\boldsymbol{u}_{,\boldsymbol{v}}}_{SPD} \boldsymbol{v}_{,t} + \underbrace{A_i \boldsymbol{u}_{,\boldsymbol{v}}}_{Symm} \boldsymbol{v}_{,x_i} = (\underbrace{M_{ij} \boldsymbol{u}_{,\boldsymbol{v}}}_{SPSD} \boldsymbol{v}_{,x_j})_{,x_i} \; .$$

Of significant importance in finite element methods, taking the inner product with entropy variables

$$\boldsymbol{v}^T(\boldsymbol{u}_{,\boldsymbol{v}}\boldsymbol{v}_{,t} + A_i \boldsymbol{u}_{,\boldsymbol{v}}\boldsymbol{v}_{,x_i} - (M_{ij}\boldsymbol{u}_{,\boldsymbol{v}}\boldsymbol{v}_{,x_j})_{,x_j}) = 0$$

yields the famous Clausius-Duhem statement [33] for constant specific heat C_v at constant volume

$$(\rho s)_{,t} + (\rho \boldsymbol{V}_i\, s)_{,x_i} + \left(\frac{q_i}{C_v T}\right)_{,x_i} = \boldsymbol{v}^T_{,x_i} (M \boldsymbol{u}_{,\boldsymbol{v}})_{ij} \boldsymbol{v}_{,x_j} \geq 0 \ .$$

Using this result, the nonlinear stability analysis of Sect. 2.3 generalizes naturally from the compressible Euler equations to the compressible Navier-Stokes equations.

1.5 Scaling Theorem Example: Magnetohydrodynamic Equations

As a second example, consider the ideal (nonrelativistic) MHD equations, $x \in \mathbb{R}^d$,

$$\frac{\partial}{\partial t}\begin{pmatrix} \rho \\ \rho \boldsymbol{V} \\ E \\ \boldsymbol{B} \end{pmatrix} + \nabla \cdot \begin{pmatrix} \rho \boldsymbol{V} \\ \rho \boldsymbol{V}\boldsymbol{V} + \mathbf{I}\ (p + \frac{1}{2}\boldsymbol{B}^2) - \boldsymbol{B}\boldsymbol{B} \\ \boldsymbol{V}\ (E + p + \frac{1}{2}\boldsymbol{B}^2) - \boldsymbol{B}\ (\boldsymbol{V} \cdot \boldsymbol{B}) \\ \boldsymbol{V}\boldsymbol{B} - \boldsymbol{B}\boldsymbol{V} \end{pmatrix} = 0. \tag{28}$$

In addition to the variables defined for the Navier-Stokes equations, $\boldsymbol{B} \in \mathbb{R}^d$ is the magnetic field, and E is the specific total energy redefined as

$$E = \frac{p}{\gamma - 1} + \rho \frac{1}{2} \boldsymbol{V}^2 + \frac{1}{2} \boldsymbol{B}^2 \ . \tag{29}$$

Unlike the Navier-Stokes equations, the MHD equations have the additional PDE constraint

$$\nabla \cdot \boldsymbol{B} = 0 \tag{30}$$

so that the magnetic field is divergence free for all times. This constraint is consistent with the conservation law system in the sense that if one takes the divergence of the $\boldsymbol{B}$-field equation in (28)

$$\nabla \cdot \left(\frac{\partial \boldsymbol{B}}{\partial t} + \nabla \cdot (\boldsymbol{V}\boldsymbol{B} - \boldsymbol{B}\boldsymbol{V})\right) = \frac{\partial}{\partial t}(\nabla \cdot \boldsymbol{B}) = 0 \ . \tag{31}$$

Consequently, if the prescribed initial data satisfies $\nabla \cdot \boldsymbol{B}(x, 0) = 0$, then it must remain so for all time.

Symmetrization of the MHD Equations The symmetrization and constraint embedding technique presented in the following paragraphs was originally developed by Godunov in 1972 [27]. Using Godunov's symmetrization technique and the Eigenvector Scaling Theorem of Sect. 1, scaled eigenvectors of the MHD system are derived. This is particularly useful in computer implementations of MHD since the eigenvectors require careful scaling, especially near the triple umbilic point where fast, slow and Alfén wave speeds coincide [12]. The paper by Roe and Balsara [47] addresses the issue of eigenvector scaling for MHD using somewhat *ad hoc* arguments concerning bi-orthogonality and singularity removal. The Eigenvector Scaling Theorem provides a systematic technique for scaling flux Jacobian eigenvectors which is uniquely determined for a given entropy function (whenever the corresponding eigenvalues are distinct). The resulting scaled eigenvectors are similar (but not the same) as those given by Roe and Balsara.

In ideal MHD, entropy is again given by $s = \log(p\rho^{-\gamma})$ so that

$$ds = -\frac{\gamma}{\rho} d\rho + \frac{1}{p} dp \ .$$

Inserting terms from (28), the following transport equation for smooth flow results:

$$s_{,t} + \boldsymbol{V} \cdot \nabla s + (\gamma - 1)\frac{\boldsymbol{V} \cdot \boldsymbol{B}}{p}(\nabla \cdot \boldsymbol{B}) = 0 \tag{32}$$

or after combining with the continuity equation

$$(\rho s)_{,t} + \nabla \cdot (\rho \boldsymbol{V} s) + (\gamma - 1)\frac{\rho \boldsymbol{V} \cdot \boldsymbol{B}}{p}(\nabla \cdot \boldsymbol{B}) = 0 \ . \tag{33}$$

Clearly, if $\nabla \cdot \boldsymbol{B} = 0$ then entropy is simply translated along streamlines

$$s_{,t} + \boldsymbol{V} \cdot \nabla s = 0 \ . \tag{34}$$

In the context of symmetrization theory, this equation suggests that a suitable entropy pair for MHD is given by

$$\{U, F^i\} = \{-\rho s, -\rho \boldsymbol{V}_i \, s\} \ . \tag{35}$$

Even so, starting from the MHD equations (28) in quasi-linear form

$$\boldsymbol{u}_{,t} + A_i \boldsymbol{u}_{,x_i} = 0, \quad A_i = \boldsymbol{f}^i_{,\boldsymbol{u}} \tag{36}$$

followed by the change of variables to

$$\tilde{A}_0 \boldsymbol{v}_{,t} + A_i \tilde{A}_0 \boldsymbol{v}_{,x_i} = 0, \quad \tilde{A}_0 = \boldsymbol{u}_{,\boldsymbol{v}} \tag{37}$$

does *not* symmetrize the system. For example, in 2-D the following result is obtained:

$$A_1 \tilde{A}_0 - (A_1 \tilde{A}_0)^T = (\gamma - 1)\begin{pmatrix} 0 & 0 & 0 & 0 & 0 & 0 \\ 0 & 0 & 0 & -\frac{p\boldsymbol{B}_1^2}{\rho} & -\frac{p\boldsymbol{B}_1}{\rho} & 0 \\ 0 & 0 & 0 & -\frac{p\boldsymbol{B}_1\boldsymbol{B}_2}{\rho} & -\frac{p\boldsymbol{B}_2}{\rho} & 0 \\ 0 & \frac{p\boldsymbol{B}_1^2}{\rho} & \frac{p\boldsymbol{B}_1\boldsymbol{B}_2}{\rho} & 0 & -\frac{p\boldsymbol{V}_2\boldsymbol{B}_2}{\rho} & -\frac{p\boldsymbol{V}_2\boldsymbol{B}_1}{\rho} \\ 0 & \frac{p\boldsymbol{B}_1}{\rho} & \frac{p\boldsymbol{B}_2}{\rho} & \frac{p\boldsymbol{V}_2\boldsymbol{B}_2}{\rho} & 0 & \frac{p\boldsymbol{V}_2}{\rho} \\ 0 & 0 & 0 & -\frac{p\boldsymbol{V}_2\boldsymbol{B}_1}{\rho} & -\frac{p\boldsymbol{V}_2}{\rho} & 0 \end{pmatrix} \tag{38}$$

This failure to symmetrize the MHD system can be explained by the simple observation that the constraint $\nabla \cdot \boldsymbol{B} = 0$ has not been used.

Godunov's Symmetrization of the MHD Equations To symmetrize the MHD equations, Godunov [27] considered the generic divergence constraint

$$(\boldsymbol{C}_i)_{,x_i} = 0$$

and recognized that the MHD equations are of the prototype form

$$(\mathcal{U}_{,\boldsymbol{v}})_{,t} + (\mathcal{F}^i_{,\boldsymbol{v}} + \boldsymbol{C}_i \Psi)_{,x_i} = 0, \quad \Psi \in \mathbb{R}^m \tag{39}$$

with $\boldsymbol{u} = \mathcal{U}^T_{,\boldsymbol{v}}$, $\boldsymbol{f}^i = (\mathcal{F}^i_{,\boldsymbol{v}} + \boldsymbol{C}_i\Psi)^T$, and $\{\mathcal{U}, \mathcal{F}^i\}$ the dual functions associated with the entropy pair $\{U, F^i\}$. Godunov then observed that whenever Ψ can be written as the $\boldsymbol{v}$-gradient primitive of another scalar function Φ,

$$\Psi = \Phi_{,\boldsymbol{v}} \ , \quad \Phi \in \mathbb{R}$$

then the system is symmetrizable in $\boldsymbol{v}$ variables by subtracting a $\Phi_{,\boldsymbol{v}}$ multiple of the constraint equation from the conservation law system, i.e.

$$(\mathcal{U}_{,\boldsymbol{v}})_{,t} + (\mathcal{F}^i_{,\boldsymbol{v}} + \boldsymbol{C}_i \Phi_{,\boldsymbol{v}})_{,x_i} - \Phi_{,\boldsymbol{v}}\,(\boldsymbol{C}_i)_{,x_i} = 0 \ . \tag{40}$$

To see this symmetry, rewrite the equation in quasi-linear form

$$\mathcal{U}_{,\boldsymbol{v},\boldsymbol{v}}\,\boldsymbol{v}_{,t} + \left(\mathcal{F}^i_{,\boldsymbol{v},\boldsymbol{v}} + \Phi_{,\boldsymbol{v},\boldsymbol{v}} \boldsymbol{C}_i\right) \boldsymbol{v}_{,x_i} = 0 \ . \tag{41}$$

By construction, the matrices $\mathcal{U}_{,\boldsymbol{v},\boldsymbol{v}}$, $\mathcal{F}^i_{,\boldsymbol{v},\boldsymbol{v}}$, and $\Phi_{,\boldsymbol{v},\boldsymbol{v}}$ are all symmetric. Finally, taking the inner product of (40) with respect to the $\boldsymbol{v}$ variables

$$\boldsymbol{v}^T \left((\mathcal{U}_{,\boldsymbol{v}})_{,t} + (\mathcal{F}^i_{,\boldsymbol{v}} + \Phi_{,\boldsymbol{v}} \boldsymbol{C}_i)^T_{,x_i} - \Phi_{,\boldsymbol{v}} (\boldsymbol{C}_i)_{,x_i} \right)^T = U_{,t} + F^i_{,x_i} + \boldsymbol{C}_i (\Phi_{,\boldsymbol{v}} \boldsymbol{v} - \Phi) = 0$$

shows that the entropy equation is obtained if $\Phi(\boldsymbol{v})$ is a homogeneous of degree one in $\boldsymbol{v}$. From Euler's theorem for homogeneous functions of degree one

$$\Phi = \Phi_{,\boldsymbol{v}} \boldsymbol{v}$$

so that

$$\boldsymbol{v}^T \left((\mathcal{U}_{,\boldsymbol{v}})_{,t} + (\mathcal{F}^i_{,\boldsymbol{v}} + \Phi_{,\boldsymbol{v}} \boldsymbol{C}_i)^T_{,x_i} - \Phi_{,\boldsymbol{v}} (\boldsymbol{C}_i)_{,x_i} \right) = U_{,t} + F^i_{,x_i} = 0 \ .$$

Turning now to the actual MHD equations, the unknown function $\Phi(\boldsymbol{v})$ can be determined with very little effort. Observe that the original and Godunov modified MHD system differ only by the term $\Phi_{,\boldsymbol{v}}\,(\boldsymbol{C}_i)_{,x_i}$. From (33) it is already known that

$$\boldsymbol{v}^T \left((\mathcal{U}_{,\boldsymbol{v}})_{,t} + (\mathcal{F}^i_{,\boldsymbol{v}} + \boldsymbol{C}_i \Psi)_{,x_i} \right) = U_{,t} + F^i_{,x_i} - (\gamma - 1) \frac{\rho \boldsymbol{V} \cdot \boldsymbol{B}}{p} (\nabla \cdot \boldsymbol{B}) = 0 \tag{42}$$

Therefore, it follows from homogeneity of $\Phi(\boldsymbol{v})$

$$\boldsymbol{v}^T \Phi_{,\boldsymbol{v}} (\nabla \cdot \boldsymbol{B}) = \Phi (\nabla \cdot \boldsymbol{B}) = (\gamma - 1) \frac{\rho \boldsymbol{V} \cdot \boldsymbol{B}}{p} (\nabla \cdot \boldsymbol{B})$$

so that

$$\Phi = (\gamma - 1) \frac{\rho \boldsymbol{V} \cdot \boldsymbol{B}}{p} \ .$$

Next, calculate the entropy variables from $U(\boldsymbol{u}) = -\rho s$,

$$\boldsymbol{v} = U_{\boldsymbol{u}}^T = (\gamma - 1) \begin{pmatrix} -\frac{s}{\gamma-1} + \frac{\gamma+1}{\gamma-1} - \frac{E}{p} + \frac{\boldsymbol{B}^2}{2p} \\ \frac{\rho \boldsymbol{V}}{p} \\ -\frac{\rho}{p} \\ \frac{\rho \boldsymbol{B}}{p} \end{pmatrix} .$$

A simple calculation verifies that $\Phi(\boldsymbol{v})$ is homogeneous of degree one as required

$$(\gamma - 1) \frac{\rho \boldsymbol{V} \cdot \boldsymbol{B}}{p} = -\frac{\boldsymbol{v}_2 \cdot \boldsymbol{v}_4}{v_3} \ .$$

This permits a direct calculation of the added term needed for symmetrization

$$\Phi_{,\boldsymbol{v}}(\nabla \cdot \boldsymbol{B}) = -\begin{pmatrix} 0 \\ \boldsymbol{B} \\ \boldsymbol{V} \cdot \boldsymbol{B} \\ \boldsymbol{V} \end{pmatrix} (\nabla \cdot \boldsymbol{B})$$

with the resulting modified form of the MHD equations

$$\frac{\partial}{\partial t}\begin{pmatrix} \rho \\ \rho \boldsymbol{V} \\ E \\ \boldsymbol{B} \end{pmatrix} + \nabla \cdot \begin{pmatrix} \rho \boldsymbol{V} \\ \rho \boldsymbol{V}\boldsymbol{V} + \mathbf{I}\ (p + \frac{1}{2}\boldsymbol{B}^2) - \boldsymbol{B}\boldsymbol{B} \\ \boldsymbol{V}\ (E + p + \frac{1}{2}\boldsymbol{B}^2) - \boldsymbol{B}\ (\boldsymbol{V} \cdot \boldsymbol{B}) \\ \boldsymbol{V}\boldsymbol{B} - \boldsymbol{B}\boldsymbol{V} \end{pmatrix} + \begin{pmatrix} 0 \\ \boldsymbol{B} \\ \boldsymbol{V} \cdot \boldsymbol{B} \\ \boldsymbol{V} \end{pmatrix} (\nabla \cdot \boldsymbol{B}) = 0. \quad (43)$$

Remark 8. This modified form of MHD is identical to that proposed by Powell [45]. Powell's derivation was based on the idea of removing the rank deficiency found in the flux Jacobians associated with (28) by adding a "divergence wave" to the flux Jacobian eigensystem.

Both Godunov and Powell note that the modified system satisfies

$$\nabla \cdot \left(\frac{\partial \boldsymbol{B}}{\partial t} + \boldsymbol{V} \cdot \nabla \boldsymbol{B} + \boldsymbol{B} \nabla \cdot \boldsymbol{V} - \boldsymbol{B} \cdot \nabla \boldsymbol{V} \right) = \frac{\partial}{\partial t}(\nabla \cdot \boldsymbol{B}) + \nabla \cdot (\boldsymbol{V}\, \nabla \cdot \boldsymbol{B}) = 0 \quad (44)$$

or after combining with the continuity equation

$$\frac{\partial}{\partial t}\zeta + \boldsymbol{V} \cdot \nabla \zeta = 0, \quad \zeta = \frac{\nabla \cdot \boldsymbol{B}}{\rho} \ . \quad (45)$$

Equation (45) states that $(\nabla \cdot \boldsymbol{B})/\rho$ is a passive scalar for the system. Any local $\nabla \cdot \boldsymbol{B}$ is simply advected away. Powell conjectures that this is numerically a more stable process.

Eigensystem of the Modified MHD Equations Let A_i^G denote the coefficient matrices of the Godunov MHD system (43) written in terms of conservation variables

$$\boldsymbol{u}_{,t} + A_i^G \boldsymbol{u}_{,x_i} = 0$$

or in symmetric form

$$\tilde{A}_0 \boldsymbol{v}_{,t} + \tilde{A}_i^G \boldsymbol{v}_{,x_i} = 0$$

with $\tilde{A}_i^G = A_i^G \tilde{A}_0$. Consider arbitrary combinations $A^G(\boldsymbol{n}) = \boldsymbol{n}_i A_i^G$. Recalling the Eigenvector Scaling Theorem, it follows that

$$\tilde{A}_0(\boldsymbol{n}) = \tilde{R}(\boldsymbol{n})\tilde{R}^T(\boldsymbol{n}), \quad \tilde{A}^G(\boldsymbol{n}) = \tilde{R}(\boldsymbol{n})\Lambda(\boldsymbol{n})\tilde{R}^T(\boldsymbol{n}) \ .$$

To determine the eigenstructure of the coefficient matrices A_i^G and $\tilde{A}_i^G$, it is again most convenient to write the equations in the primitive variable form $\boldsymbol{w} = (\rho, \boldsymbol{V}, p, \boldsymbol{B})^T$

$$\boldsymbol{w}_{,t} + \boldsymbol{u}_{,\boldsymbol{w}}^{-1} A_i^G \boldsymbol{u}_{,\boldsymbol{w}} \boldsymbol{w}_{,x_i} = 0 \ . \tag{46}$$

The Godunov coefficient matrices can then be written in the following compact form:

$$\boldsymbol{u}_{,\boldsymbol{w}}^{-1} A^G(\boldsymbol{n}) \boldsymbol{u}_{,\boldsymbol{w}} = \begin{pmatrix} \boldsymbol{V}\cdot\boldsymbol{n} & \rho\boldsymbol{n}^T & 0 & \boldsymbol{0}^T \\ \boldsymbol{0} & (\boldsymbol{V}\cdot\boldsymbol{n})I & \frac{1}{\rho}\boldsymbol{n} & \frac{1}{\rho}(\boldsymbol{n}\boldsymbol{B}^T - \boldsymbol{B}\cdot\boldsymbol{n}) \\ 0 & \gamma p \boldsymbol{n}^T & \boldsymbol{V}\cdot\boldsymbol{n} & 0 \\ \boldsymbol{0} & \boldsymbol{B}\boldsymbol{n}^T - \boldsymbol{B}\cdot\boldsymbol{n} & \boldsymbol{0} & (\boldsymbol{V}\cdot\boldsymbol{n})I \end{pmatrix} \tag{47}$$

with

$$\boldsymbol{u}_{,\boldsymbol{w}} = \begin{pmatrix} 1 & \boldsymbol{0}^T & 0 & \boldsymbol{0}^T \\ \boldsymbol{V} & \rho\mathbf{I} & \boldsymbol{0} & \boldsymbol{0}\boldsymbol{0}^T \\ \frac{1}{2}\boldsymbol{V}^2 & \rho\boldsymbol{V}^T & \frac{1}{\gamma-1} & \boldsymbol{B}^T \\ \boldsymbol{0} & \boldsymbol{0}\boldsymbol{0}^T & 0 & I \end{pmatrix} . \tag{48}$$

The Riemannian metric matrix $\boldsymbol{u}_{,\boldsymbol{v}}$ is also written simply as

$$\boldsymbol{u}_{,\boldsymbol{v}} = (\gamma - 1) \begin{pmatrix} \rho & \rho\boldsymbol{V}^T & E - \frac{1}{2}\boldsymbol{B}^2 & \boldsymbol{0}^T \\ \rho\boldsymbol{V} & \rho\boldsymbol{V}\boldsymbol{V} + p\mathbf{I} & \rho H\boldsymbol{V} & \boldsymbol{0}\boldsymbol{0}^T \\ E - \frac{1}{2}\boldsymbol{B}^2 & \rho H \boldsymbol{V}^T & \rho H^2 - \frac{a^2 p}{\gamma-1} + \frac{a^2\boldsymbol{B}^2}{\gamma} & \frac{p}{\rho}\boldsymbol{B}^T \\ \boldsymbol{0} & \boldsymbol{0}\boldsymbol{0}^T & \frac{p}{\rho}\boldsymbol{B} & \frac{p}{\rho} I \end{pmatrix}$$

with a the sound speed, $a^2 = \gamma p/\rho$, and H the specific total enthalpy, $H = a^2/(\gamma-1) + \boldsymbol{V}^2/2$.

In order to give the eigensystem for the MHD equations, it is useful to define $\boldsymbol{b} \equiv \boldsymbol{B}/\sqrt{\rho}$ and the fast and slow speeds

$$c_{f,s}^2 = \frac{1}{2}\left(a^2 + \boldsymbol{b}^2\right) \pm \frac{1}{2}\sqrt{\left(a^2 + \boldsymbol{b}^2\right)^2 - 4a^2(\boldsymbol{b}\cdot\boldsymbol{n})^2} \ . \tag{49}$$

The eigenvalues and naively scaled eigenvectors of the matrix (47) are written compactly as

Entropy and Divergence Waves: $\lambda_{1,2} = \boldsymbol{V} \cdot \boldsymbol{n}$

$$r_1 = \begin{pmatrix} 1 \\ \mathbf{0} \\ 0 \\ \mathbf{0} \end{pmatrix}, \quad r_2 = \begin{pmatrix} 0 \\ \mathbf{0} \\ 0 \\ \boldsymbol{n} \end{pmatrix} \tag{50}$$

Alfvén Waves: $\lambda_{\pm a} = \boldsymbol{V} \cdot \boldsymbol{n} \pm (\boldsymbol{b} \cdot \boldsymbol{n})$

$$r_{\pm a} = \begin{pmatrix} 0 \\ \pm(\boldsymbol{n} \times \boldsymbol{B}) \\ 0 \\ \sqrt{\rho}(\boldsymbol{n} \times \boldsymbol{B}) \end{pmatrix} \tag{51}$$

Magneto-acoustic Waves: $\lambda_{\pm f, \pm s} = \boldsymbol{V} \cdot \boldsymbol{n} \pm c_{f,s}$

$$r_{\pm f, \pm s} = \begin{pmatrix} \rho \\ \pm c_{f,s} \frac{c_{f,s}^2 \boldsymbol{n} - (\boldsymbol{b} \cdot \boldsymbol{n}) \boldsymbol{b}}{c_{f,s}^2 - (\boldsymbol{b} \cdot \boldsymbol{n})^2} \\ \rho a^2 \\ c_{f,s}^2 \frac{(\boldsymbol{B}\boldsymbol{n} - \boldsymbol{n}\boldsymbol{B}) \boldsymbol{n}}{c_{f,s}^2 - (\boldsymbol{b} \cdot \boldsymbol{n})^2} \end{pmatrix} \tag{52}$$

In this form, the magneto-acoustic eigenvectors exhibit several forms of degeneracy as carefully described in Roe and Balsara [47]. In the following paragraphs, the entropy scaled form of the MHD eigenvectors associated with the primitive variable system is given.

Entropy Scaled Eigenvectors of the Modified MHD Equations After consider algebraic manipulation, entropy scaled eigenvectors corresponding to the Godunov modified MHD equations have been obtained. Using the notation of Roe and Balsara define

$$\alpha_f^2 = \frac{a^2 - c_s^2}{c_f^2 - c_s^2} \quad \alpha_s^2 = \frac{c_f^2 - a^2}{c_f^2 - c_s^2} \tag{53}$$

and $\boldsymbol{n}^\perp$, a unit vector orthogonal to $\boldsymbol{n}$ lying in the plane spanned by $\boldsymbol{n}$ and $\boldsymbol{b}$, i.e. $\boldsymbol{n}^\perp \cdot \boldsymbol{n} = 0, \ \|\boldsymbol{n}^\perp\| = 1, \ \boldsymbol{n}^\perp \in \text{span}\{\boldsymbol{n}, \boldsymbol{b}\}$.

Entropy and Divergence Waves: $\lambda_{1,2} = \boldsymbol{V} \cdot \boldsymbol{n}$

$$\tilde{r}_1 = \sqrt{\frac{\gamma - 1}{\gamma}} \begin{pmatrix} \sqrt{\rho} \\ \mathbf{0} \\ 0 \\ \mathbf{0} \end{pmatrix}, \quad \tilde{r}_2 = \sqrt{\frac{1}{\gamma}} \begin{pmatrix} 0 \\ \mathbf{0} \\ 0 \\ a\,\boldsymbol{n} \end{pmatrix} \tag{54}$$

Alfvén Waves: $\lambda_{\pm a} = \boldsymbol{V} \cdot \boldsymbol{n} \pm \boldsymbol{b} \cdot \boldsymbol{n}$

$$\tilde{r}_{\pm a} = \sqrt{\frac{1}{2}} \begin{pmatrix} 0 \\ \mp \frac{\sqrt{p}}{\rho} (\boldsymbol{n}^{\perp} \times \boldsymbol{n}) \\ 0 \\ \sqrt{\frac{p}{\rho}} (\boldsymbol{n}^{\perp} \times \boldsymbol{n}) \end{pmatrix} \tag{55}$$

Fast Magneto-acoustic Waves: $\lambda_{\pm f} = \boldsymbol{V} \cdot \boldsymbol{n} \pm c_f$

$$\tilde{r}_{\pm f} = \sqrt{\frac{1}{2\gamma}} \begin{pmatrix} \alpha_f \sqrt{\rho} \\ \pm \frac{\alpha_f\, a^2\, \boldsymbol{n} + \alpha_s\, a\, ((\boldsymbol{b}\cdot\boldsymbol{n}^{\perp})\boldsymbol{n} - (\boldsymbol{b}\cdot\boldsymbol{n})\boldsymbol{n}^{\perp})}{\sqrt{\rho} c_f} \\ \alpha_f \sqrt{\rho} a^2 \\ \alpha_s a\, \boldsymbol{n}^{\perp} \end{pmatrix} \tag{56}$$

Slow Magneto-acoustic Waves: $\lambda_{\pm s} = \boldsymbol{V} \cdot \boldsymbol{n} \pm c_s$

$$\tilde{r}_{\pm s} = \sqrt{\frac{1}{2\gamma}} \begin{pmatrix} \alpha_s \sqrt{\rho} \\ \pm \mathrm{sgn}(\boldsymbol{b} \cdot \boldsymbol{n}) \frac{\alpha_s\, a\, (\boldsymbol{b}\cdot\boldsymbol{n})\, \boldsymbol{n} + \alpha_f\, c_f^2\, \boldsymbol{n}^{\perp}}{\sqrt{\rho} c_f} \\ \alpha_s \sqrt{\rho} a^2 \\ -\alpha_f a\, \boldsymbol{n}^{\perp} \end{pmatrix} \tag{57}$$

Next, the numerical experiment performed in Sec. 1.4 is repeated for the MHD equations. In Fig. 2, $\|\tilde{R}^T(\boldsymbol{n})\|_2$, is graphed for $\boldsymbol{n} = (\cos\theta, \sin\theta)^T, \theta \in [0, 2\pi]$ with constant $(\rho, \boldsymbol{V}, p, \boldsymbol{B})$ using entropy scaled eigenvectors, the naively scaled eigenvectors given earlier, and a slightly generalized form of the eigenvectors given in Roe and Balsara [47]. The singularities in the naively scaled eigenvectors are clearly seen. Once again note that the entropy scaled eigenvectors produce a constant result since $\tilde{R}(\boldsymbol{n})\tilde{R}^T(\boldsymbol{n}) = \boldsymbol{u}_{,\boldsymbol{v}}$ which is independent of $\boldsymbol{n}$.

2 Simplified Finite Element Methods for the Gasdynamic Equations

2.1 A Unified GLS and DG Formulation

Let $I^n =]t^n, t^{n+1}[$ denote the nth time interval and Ω the spatial domain composed of nonoverlapping elements T_i, $\Omega = \cup T_i$, $T_i \cap T_j = \emptyset$, $i \neq j$. To simplify the exposition, consider a single variational formulation with weakly enforced boundary conditions which includes both the GLS and DG schemes. Thus by choosing the correct space of functions (continuous or discontinuous) and/or omitting the least-squares variational term, one can switch from the GLS formulation to the DG formulation. In the GLS formulation, functions are continuous in space and discontinuous in time

$$\mathcal{V}^h = \left\{ \boldsymbol{v}^h \,|\, \boldsymbol{v}^h \in \left(C^0(\Omega \times I^n)\right)^m, \boldsymbol{v}^h_{|T \times I^n} \in \left(\mathcal{P}_k(T \times I^n)\right)^m \right\}$$

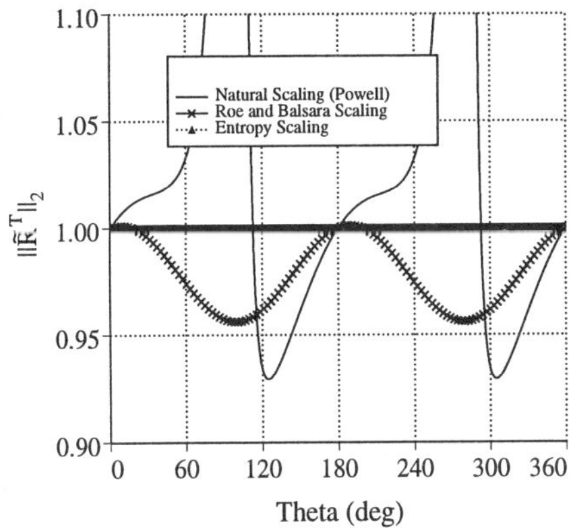

Fig. 2. The matrix norm $\|\tilde{R}^T(\boldsymbol{n})\|_2$ dependency on $\boldsymbol{n}$ for the Godunov modified MHD equations, $\left(\rho/\rho_\infty, \boldsymbol{V}^T/a_\infty, p/p_\infty, \boldsymbol{B}^T/\rho^{1/2}a_\infty\right) = (1.3, (.8, -.6), 2.0, (.2, 1.2))$.

where $\boldsymbol{v}$ denotes the entropy variables for the system. In the DG formulation, functions are discontinuous in space and time, i.e.

$$\mathcal{V}^h = \left\{\boldsymbol{v}^h \,|\, \boldsymbol{v}^h_{|T\times I^n} \in \left(\mathcal{P}_k(T\times I^n)\right)^m\right\} .$$

Consider the prototype hyperbolic system for the space-time domain $\Omega\times[0,T]$ with boundary data $\boldsymbol{g}$ imposed on Γ via admissibility condition

$$\begin{aligned} &\boldsymbol{u}_{,t} + \boldsymbol{f}^i_{,x_i} = 0 \quad \text{in } \Omega \\ &A^-(\boldsymbol{n})\boldsymbol{u} = A^-(\boldsymbol{n})\boldsymbol{g} \quad \text{on } \Gamma \end{aligned} \tag{58}$$

or in symmetric quasi-linear form for smooth solutions

$$\begin{aligned} &\tilde{A}_0\boldsymbol{v}_{,t} + \tilde{A}_i\boldsymbol{v}_{,x_i} = 0 \quad \text{in } \Omega \\ &\tilde{A}^-(\boldsymbol{n})\boldsymbol{v} = \tilde{A}^-(\boldsymbol{n})\tilde{\boldsymbol{g}} \quad \text{on } \Gamma \end{aligned} \tag{59}$$

with $A(\boldsymbol{n}) = \boldsymbol{n}_i\,A^i$ and $\tilde{A}(\boldsymbol{n}) = \boldsymbol{n}_i\,\tilde{A}^i$. The combined GLS and DG schemes are defined by the following stabilized variational formulation:

Find $\boldsymbol{v}^h \in \mathcal{V}^h$ such that for all $\boldsymbol{w}^h \in \mathcal{V}^h$

$$B(\boldsymbol{v}^h, \boldsymbol{w}^h)_{gal} + B(\boldsymbol{v}^h, \boldsymbol{w}^h)_{ls} + B(\boldsymbol{v}^h, \boldsymbol{w}^h)_{bc} = 0 \tag{60}$$

$$\begin{aligned} B(\boldsymbol{v},\boldsymbol{w})_{gal} = &\int_{I^n}\int_\Omega (-\boldsymbol{u}(\boldsymbol{v})\cdot\boldsymbol{w}_{,t} - \boldsymbol{f}^i(\boldsymbol{v})\cdot\boldsymbol{w}_{,x_i})\,dx\,dt \\ &+ \int_\Omega \left(\boldsymbol{w}(t^{n+1}_-)\cdot\boldsymbol{u}(\boldsymbol{v}(t^{n+1}_-)) - \boldsymbol{w}(t^n_+)\cdot\boldsymbol{u}(\boldsymbol{v}(t^n_-))\right)\,dx \end{aligned}$$

$$+ \int_{I^n} \sum_{e\in\mathcal{E}} \int_e (\boldsymbol{w}(x_-) - \boldsymbol{w}(x_+)) \cdot \boldsymbol{h}(\boldsymbol{v}(x_-), \boldsymbol{v}(x_+); \boldsymbol{n})\, dx\, dt$$

$$B(\boldsymbol{v}, \boldsymbol{w})_{ls} = \int_{I^n} \sum_{T\in\Omega} \int_T \left(\tilde{A}_0 \boldsymbol{w}_{,t} + \tilde{A}_i \boldsymbol{w}_{,x_i}\right) \cdot \boldsymbol{\tau} \left(\tilde{A}_0 \boldsymbol{v}_{,t} + \tilde{A}_i \boldsymbol{v}_{,x_i}\right) dx\, dt$$

$$B(\boldsymbol{v}, \boldsymbol{w})_{bc} = \int_{I^n} \int_\Gamma \boldsymbol{w} \cdot \boldsymbol{h}(\boldsymbol{v}, \tilde{\boldsymbol{g}}; \boldsymbol{n})\, dx\, dt$$

where $\boldsymbol{h}$ denotes a numerical flux function and $\boldsymbol{\tau}$ a small $m \times m$ SPD matrix for the least-squares term. For theoretical and practical analysis, three numerical flux functions are considered. All are of the form

$$\boldsymbol{h}(\boldsymbol{v}_-, \boldsymbol{v}_+; \boldsymbol{n}) = \frac{1}{2}\left(\boldsymbol{f}(\boldsymbol{v}_-; \boldsymbol{n}) + \boldsymbol{f}(\boldsymbol{v}_+; \boldsymbol{n})\right) - \frac{1}{2}\boldsymbol{h}^d(\boldsymbol{v}_-, \boldsymbol{v}_+; \boldsymbol{n}) \tag{61}$$

and consistent with the true flux is the sense that $\boldsymbol{h}(\boldsymbol{v}, \boldsymbol{v}; \boldsymbol{n}) = \boldsymbol{f}(\boldsymbol{v}; \boldsymbol{n})$. Specifically considered are the following forms:

1. **Symmetric Mean-Value Flux**. This flux is motivated from the nonlinear stability theory of Sect. 2.3. Define the parameterizations $\overline{\boldsymbol{v}}(\theta) = \boldsymbol{v}(x_+) - \theta\, [\boldsymbol{v}]_{x_-}^{x+}$ and $\overline{\overline{\boldsymbol{v}}}(\theta) = \boldsymbol{v}(x_-) + \theta\, [\boldsymbol{v}]_{x_-}^{x+}$. The symmetric mean-value flux is then given by

$$\boldsymbol{h}^d_{\mathrm{MV}}(\boldsymbol{v}_-, \boldsymbol{v}_+; \boldsymbol{n}) = |\tilde{A}(\boldsymbol{v}_-, \boldsymbol{v}_+; \boldsymbol{n})|_{\mathrm{MV}}\, [\boldsymbol{v}]_-^+$$

with

$$|\tilde{A}(\boldsymbol{v}_-, \boldsymbol{v}_+; \boldsymbol{n})|_{\mathrm{MV}} \equiv \int_0^1 (1-\theta) \left(|\tilde{A}(\overline{\overline{\boldsymbol{v}}}(\theta); \boldsymbol{n})|_{\tilde{A}_0} + |\tilde{A}(\overline{\boldsymbol{v}}(\theta); \boldsymbol{n})|_{\tilde{A}_0} \right) d\theta . \tag{62}$$

Recall that $|\tilde{A}|_{\tilde{A}_0}$ denotes the matrix absolute value with respect to the Riemannian matrix $\tilde{A}_0$. By construction, the matrix $|\tilde{A}(\boldsymbol{v}_-, \boldsymbol{v}_+; \boldsymbol{n})|_{\mathrm{MV}}$ is necessarily symmetric positive semi-definite. Using this form of flux dissipation (62), nonlinear entropy stability of the GLS and DG formulations is shown in Sect. 2.3. In addition, let

$$\tilde{A}(\boldsymbol{n})_{\mathrm{MV}} \equiv \int_0^1 (1-\theta) \left(\tilde{A}(\overline{\overline{\boldsymbol{v}}}(\theta); \boldsymbol{n}) + \tilde{A}(\overline{\boldsymbol{v}}(\theta); \boldsymbol{n}) \right) d\theta \tag{63}$$

from which the following useful property exists

$$[\boldsymbol{f}(\boldsymbol{n})]_{x_-}^{x+} = \tilde{A}(\boldsymbol{n})_{\mathrm{MV}}\, [\boldsymbol{v}]_{x_-}^{x+} \tag{64}$$

which is a necessary ingredient for optimal discontinuity resolution, see for example Barth [5] or Serre [48]. Note that the mean-value matrix described here should not be confused with the Volpert mean-value matrix

$$\tilde{A}(\boldsymbol{n})_{\mathrm{Volpert}} = \int_0^1 \tilde{A}(\overline{\overline{\boldsymbol{v}}}(\theta); \boldsymbol{n})\, d\theta \tag{65}$$

which also satisfies

$$[\boldsymbol{f}(\boldsymbol{n})]_{x_-}^{x+} = \tilde{A}(\boldsymbol{n})_{\text{Volpert}}\,[\boldsymbol{v}]_{x_-}^{x+} \tag{66}$$

but is incompatible with the energy analysis given in later sections. Also note that whenever $\tilde{A}(\boldsymbol{n})$ is constant with respect to $\boldsymbol{v}$ then

$$|\tilde{A}(\boldsymbol{v}_-,\boldsymbol{v}_+;\boldsymbol{n})|_{\text{MV}} = \int_0^1 2\,(1-\theta)\,|\tilde{A}(\boldsymbol{n})|_{\tilde{A}_0}\,d\theta = |\tilde{A}(\boldsymbol{n})|_{\tilde{A}_0}$$

as expected. To show stability of other (more practical) forms of flux dissipation, one needs only show that the new form is more entropy dissipative than the symmetric mean-value form, i.e.

$$[\boldsymbol{v}]_{x_-}^{x+}\cdot \boldsymbol{h}^d_{\text{MV}} \le [\boldsymbol{v}]_{x_-}^{x+}\cdot \boldsymbol{h}^d\ .$$

2. **Symmetric Variable Flux**. The symmetric variable flux function

$$\boldsymbol{h}^d_{\text{Symm}}(\boldsymbol{v}_-,\boldsymbol{v}_+;\boldsymbol{n}) = |\tilde{A}(\overline{\boldsymbol{v}}(\boldsymbol{v}_-,\boldsymbol{v}_+);\boldsymbol{n})|_{\tilde{A}_0}\ [\boldsymbol{v}]_-^+$$

is motivated from the symmetrization theory of Sect. 1.2. This flux function is of practical interest since it is easily formed and has a relatively straightforward Jacobian linearization as will be shown later. From the theory of Sect. 1.2

$$|\tilde{A}|_{\tilde{A}_0} = \tilde{R}\,|\Lambda|\,\tilde{R}^T\ .$$

In computations, any zero eigenvalues appearing in Λ are perturbed from zero so that $|\tilde{A}|_{\tilde{A}_0}$ is strictly SPD.

3. **Roe Flux** [46]. Lastly, we consider the conventional Roe flux function written with explicit dependence on the entropy variables $\boldsymbol{v}$

$$\boldsymbol{h}^d_{\text{Roe}}(\boldsymbol{v}_-,\boldsymbol{v}_+;\boldsymbol{n}) = |A(\overline{\boldsymbol{v}}_{\text{Roe}}(\boldsymbol{v}_-,\boldsymbol{v}_+);\boldsymbol{n})|\ [\boldsymbol{u}(\boldsymbol{v})]_-^+$$

where $\overline{\boldsymbol{v}}_{\text{Roe}}$ denotes the Roe-averaged state such that

$$[\boldsymbol{f}(\boldsymbol{n})]_-^+ = A(\overline{\boldsymbol{v}}_{\text{Roe}}(\boldsymbol{v}_-,\boldsymbol{v}_+);\boldsymbol{n})\ [\boldsymbol{u}(\boldsymbol{v})]_-^+\ .$$

In a later section, it is proven by combined theory and counterexample that the Roe flux fails to be entropy norm dissipative.

2.2 Energy Analysis (Linear Hyperbolic System)

Consider the prototype *linear* symmetric hyperbolic system for the space-time domain $\Omega\times[0,T]$ with Friedrichs admissible boundary data $\tilde{\boldsymbol{g}}$ imposed on Γ:

$$\begin{aligned} \tilde{A}_0\boldsymbol{v}_{,t} + \tilde{A}_i\boldsymbol{v}_{,x_i} &= 0 \quad \text{in } \Omega \\ \tilde{A}^-(\boldsymbol{n})\,\boldsymbol{v} &= \tilde{A}^-(\boldsymbol{n})\,\tilde{\boldsymbol{g}} \quad \text{on } \Gamma \end{aligned} \tag{67}$$

where $\boldsymbol{v}, \tilde{\boldsymbol{g}} \in \mathbb{R}^m$, $\tilde{A}_i \in \mathbb{R}^{m\times m}$ and $\tilde{A}(\boldsymbol{n}) = \boldsymbol{n}_i \tilde{A}_i$. As usual, $\tilde{A}_i$ are symmetric and $\tilde{A}_0$ is symmetric positive definite. To understand the energy consequences of the Galerkin Least-Squares and Discontinuous Galerkin methods, let $\mathcal{V}^h$ denote the finite-dimensional subspace of functions which vary discontinuously from element to element and time slab to time slab. Next, pose (67) in stabilized variational form:

Find $\boldsymbol{v}^h \in \mathcal{V}^h$ such that

$$B_{gal}(\boldsymbol{w}^h, \boldsymbol{v}^h) + B_{ls}(\boldsymbol{w}^h, \boldsymbol{v}^h) + B_{bc}(\boldsymbol{w}^h, \boldsymbol{v}^h) = 0, \quad \forall\, \boldsymbol{w}^h \in \mathcal{V}^h \tag{68}$$

with

$$\begin{aligned}
B_{gal}(\boldsymbol{w}, \boldsymbol{v}) =& \int_{I^n}\int_{\Omega} \boldsymbol{w}^T(\tilde{A}_0 \boldsymbol{v}_{,t} + \tilde{A}_i \boldsymbol{v}_{,x_i})\; dx\; dt + \int_{\Omega} \boldsymbol{w}^T(t_+^n)\tilde{A}_0[\boldsymbol{v}(t_+^n) - \boldsymbol{v}(t_-^n)]\; dx \\
&+ \int_{I^n} \sum_{e\in\mathcal{E}} \left(\boldsymbol{w}^T(x_-)\tilde{A}^-(\boldsymbol{n}) + \boldsymbol{w}^T(x_+)\tilde{A}^+(\boldsymbol{n})\right)[\boldsymbol{v}(x_+) - \boldsymbol{v}(x_-)]\; dx\; dt \\
B_{ls}(\boldsymbol{w}, \boldsymbol{v}) =& \int_{I^n}\int_{\Omega} \left(\tilde{A}_0 \boldsymbol{w}_{,t} + \tilde{A}_i \boldsymbol{w}_{,x_i}\right)^T \boldsymbol{\tau} \left(\tilde{A}_0 \boldsymbol{v}_{,t} + \tilde{A}_i \boldsymbol{v}_{,x_i}\right)\; dx\; dt \\
B_{bc}(\boldsymbol{w}, \boldsymbol{v}) =& \int_{I^n}\int_{\Gamma} \boldsymbol{w}^T \tilde{A}^-(\boldsymbol{n})(\tilde{\boldsymbol{g}} - \boldsymbol{v})\; dx\; dt
\end{aligned}$$

given $\boldsymbol{\tau}$ SPD and $I^n \in [t_+^n, t_-^{n+1}]$. It is relatively straightforward to derive the energy balance of this scheme in terms of the following norms: $\|\boldsymbol{v}\|^2_{\tilde{A}_0,\Omega} = \int_\Omega \boldsymbol{v}^T \tilde{A}_0 \boldsymbol{v}\; dx$, $\langle \boldsymbol{u}, \boldsymbol{v}\rangle_{\tilde{A}^\pm,\Gamma} = \int_\Gamma \boldsymbol{u}^T \tilde{A}^\pm(\boldsymbol{n})\boldsymbol{v}\; d\Gamma$, and $\langle \boldsymbol{v}\rangle^2_{|\tilde{A}|,\Gamma} = \langle \boldsymbol{v}, \boldsymbol{v}\rangle_{|\tilde{A}|,\Gamma}$.

Theorem 9. Global Energy Stability (Linear Hyperbolic System). *The variational formulation (68) is energy stable (modulo inflow data $\tilde{\boldsymbol{g}}$) with the following global energy balance:*

$$\begin{aligned}
&\sum_{n=0}^{N-1}\left(\|[\boldsymbol{v}]_{t_n^-}^{t_n^+}\|^2_{\tilde{A}_0,\Omega} + 2\|\tilde{A}_0 \boldsymbol{v}_{,t} + \tilde{A}_i \boldsymbol{v}_{,x_i}\|^2_{\boldsymbol{\tau},\Omega\times I^n} + \sum_{e\in\mathcal{E}} \left\langle [\boldsymbol{v}]_{x-}^{x+}\right\rangle^2_{|\tilde{A}|,e\times I^n} + \langle \boldsymbol{v}\rangle^2_{|\tilde{A}|,\Gamma\times I^n}\right) \\
&\quad + \|\boldsymbol{v}(t_-^N)\|^2_{\tilde{A}_0,\Omega} = \|\boldsymbol{v}(t_-^0)\|^2_{\tilde{A}_0,\Omega} + \sum_{n=0}^{N-1} 2\, \langle \boldsymbol{v}, \tilde{\boldsymbol{g}}\rangle_{(-\tilde{A}^-),\Gamma\times I^n}
\end{aligned} \tag{69}$$

where $[\boldsymbol{v}]_a^b = \boldsymbol{v}(b) - \boldsymbol{v}(a)$.

Proof. Construct the energy balance for the interval $[t_-^N, t_-^0] = \cup_{n=0}^{N-1} I^n$ by setting $\boldsymbol{w} = \boldsymbol{v}$ and evaluating the various integrals. The time derivative term

$$\int_\Omega \int_{I^n} \boldsymbol{v}^T \tilde{A}_0 \boldsymbol{v}_{,t}\; dt\; dx = \frac{1}{2}\left(\|\boldsymbol{v}(t_-^{n+1})\|^2_{\tilde{A}_0,\Omega} - \|\boldsymbol{v}(t_+^n)\|^2_{\tilde{A}_0,\Omega}\right)$$

and the jump integral across time slabs

$$\int_\Omega \boldsymbol{v}^T(t_+^n)\tilde{A}_0[\boldsymbol{v}]_{t_-^n}^{t_+^n}\; dx = \frac{1}{2}\left(\|[\boldsymbol{v}]_{t_-^n}^{t_+^n}\|^2_{\tilde{A}_0,\Omega} + \|\boldsymbol{v}(t_+^n)\|^2_{\tilde{A}_0,\Omega} - \|\boldsymbol{v}(t_-^n)\|^2_{\tilde{A}_0,\Omega}\right)$$

combine thus yielding

$$\int_\Omega \int_{I^n} \boldsymbol{v}^T \tilde{A}_0 \boldsymbol{v}_{,t}\, dt\, dx + \int_\Omega \boldsymbol{v}(t_+^n) \tilde{A}_0 [\boldsymbol{v}]_{t_-^n}^{t_+^n}\, dx$$
$$= \frac{1}{2} \| [\boldsymbol{v}]_{t_-^n}^{t_+^n} \|^2_{\tilde{A}_0, \Omega \times I^n} + \frac{1}{2} \left(\| \boldsymbol{v}(t_-^{n+1}) \|^2_{\tilde{A}_0, \Omega \times I^n} - \| \boldsymbol{v}(t_-^n) \|^2_{\tilde{A}_0, \Omega \times I^n} \right)$$

for a single time slab. Next, rewrite the spatial operator term

$$\int_{I^n} \int_\Omega \boldsymbol{v}^T \tilde{A}_i \boldsymbol{v}_{,x_i}\, dx\, dt$$
$$= \frac{1}{2} \left(\langle \boldsymbol{v}, \boldsymbol{v} \rangle_{\tilde{A}, \Gamma \times I^n} + \sum_{e \in \mathcal{E}} \left(\langle \boldsymbol{v}(x_-), \boldsymbol{v}(x_-) \rangle_{\tilde{A}, e \times I^n} - \langle \boldsymbol{v}(x_+), \boldsymbol{v}(x_+) \rangle_{\tilde{A}, e \times I^n} \right) \right)$$

so that

$$\int_{I^n} \int_\Omega \boldsymbol{v}^T \tilde{A}_i \boldsymbol{v}_{,x_i}\, dx\, dt - \int_{I^n} \int_\Gamma \boldsymbol{v}^T \tilde{A}^- \boldsymbol{v}\, dx\, dt$$
$$= \frac{1}{2} \left(\langle \boldsymbol{v}, \boldsymbol{v} \rangle_{|\tilde{A}|, \Gamma \times I^n} + \sum_{e \in \mathcal{E}} \left(\langle \boldsymbol{v}(x_-), \boldsymbol{v}(x_-) \rangle_{\tilde{A}, e \times I^n} - \langle \boldsymbol{v}(x_+), \boldsymbol{v}(x_+) \rangle_{\tilde{A}, e \times I^n} \right) \right).$$

Finally, consider the jump integral across interior edges

$$\int_{I^n} \sum_{e \in \mathcal{E}} \int_e \left(\boldsymbol{v}^T(x_-) \tilde{A}^- + \boldsymbol{v}^T(x_+) \tilde{A}^+ \right) [\boldsymbol{v}]_{x_-}^{x_+}\, dx\, dt$$
$$= \frac{1}{2} \sum_{e \in \mathcal{E}} \left(\left\langle [\boldsymbol{v}]_{x_-}^{x_+} \right\rangle^2_{|\tilde{A}|, e \times I^n} + \left(\langle \boldsymbol{v}(x_+), \boldsymbol{v}(x_+) \rangle_{\tilde{A}, e \times I^n} - \langle \boldsymbol{v}(x_-), \boldsymbol{v}(x_-) \rangle_{\tilde{A}, e \times I^n} \right) \right).$$

The least-squares integral produces a pure quadratic form without modification. Combining the above results, summing over time slabs, and multiplication by two yields the energy equation (69) which bounds the energy at the final time T in terms of initial energy and energy produced by inflow boundary data. □

From this result, one can restrict our attention to the GLS and DG scheme by simply dropping terms from the energy balance.

<u>Galerkin Least-Squares Energy Balance</u>

$$\| \boldsymbol{v}(t_-^N) \|^2_{\tilde{A}_0, \Omega} + \sum_{n=0}^{N-1} \left(\| [\boldsymbol{v}]_{t_-^n}^{t_+^n} \|^2_{\tilde{A}_0, \Omega} + 2 \| \tilde{A}_0 \boldsymbol{v}_{,t} + \tilde{A}_i \boldsymbol{v}_{,x_i} \|^2_{\boldsymbol{\tau}, \Omega \times I^n} \right)$$
$$+ \sum_{n=0}^{N-1} \langle \boldsymbol{v} \rangle^2_{|\tilde{A}|, \Gamma \times I^n} = \| \boldsymbol{v}(t_-^0) \|^2_{\tilde{A}_0, \Omega} + \sum_{n=0}^{N-1} 2 \langle \boldsymbol{v}, \tilde{\boldsymbol{g}} \rangle_{(-\tilde{A}^-), \Gamma \times I^n}$$

Discontinuous Galerkin Energy Balance

$$\|\boldsymbol{v}(t_-^N)\|^2_{\tilde{A}_0,\Omega} + \sum_{n=0}^{N-1}\left(\|[\boldsymbol{v}]_{t_-^n}^{t_+^n}\|^2_{\tilde{A}_0,\Omega} + \sum_{e\in\mathcal{E}}\left\langle[\boldsymbol{v}]_{x-}^{x+}\right\rangle^2_{|\tilde{A}|,e\times I^n} + \langle\boldsymbol{v}\rangle^2_{|\tilde{A}|,\Gamma\times I^n}\right)$$
$$= \|\boldsymbol{v}(t_-^0)\|^2_{\tilde{A}_0,\Omega} + \sum_{n=0}^{N-1} 2\,\langle\boldsymbol{v},\tilde{\boldsymbol{g}}\rangle_{(-\tilde{A}^-),\Gamma\times I^n}$$

The energy balance equation reveals how the discontinuous function space and the least-squares term all strengthen the energy boundedness of the formulation. In the GLS scheme, energy is removed by the least-squares term, outflow boundary conditions, and by a term proportional to the square in jump in the solution across time slabs. In the DG scheme, energy is removed by outflow boundary conditions and terms proportional to the square in jumps in the solution between elements in both space in time. The energy balance equation formally bounds the energy at time t_-^N in terms of initial data and inflow boundary data.

2.3 Energy Analysis (Nonlinear Hyperbolic System)

The combined GLS-DG formulation (60) permits a straightforward (nonlinear) global entropy stability result to be obtained. This analysis reveals the entropy properties needed in the numerical flux function for global entropy decay. Note that in actual numerical calculations, it is desirable to use the variational form given by (60) since integration by parts has been used to insure exact discrete conservation even with inexact numerical quadrature of the various integrals. For analysis purposes, however, it is desirable to use the following non-integrated-by-parts formulation:

Find $\boldsymbol{v}^h \in \mathcal{V}^h$ such that for all $\boldsymbol{w}^h \in \mathcal{V}^h$

$$B(\boldsymbol{v}^h,\boldsymbol{w}^h)_{gal} + B(\boldsymbol{v}^h,\boldsymbol{w}^h)_{ls} + B(\boldsymbol{v}^h,\boldsymbol{w}^h)_{bc} = 0 \tag{70}$$

$$\begin{aligned}
B(\boldsymbol{v},\boldsymbol{w})_{gal} &= \int_{I^n}\int_{\Omega} \boldsymbol{w}\cdot\left(\boldsymbol{u}_{,t} + \boldsymbol{f}^i_{,x_i}(\boldsymbol{v})\right) dx\,dt \\
&+ \int_{\Omega} \boldsymbol{w}(t_+^n)\cdot\left(\boldsymbol{u}(\boldsymbol{v}(t_+^n)) - \boldsymbol{u}(\boldsymbol{v}(t_-^n))\right)\, dx \\
&+ \int_{I^n}\sum_{e\in\mathcal{E}}\int_e \frac{1}{2}(\boldsymbol{w}(x_+) - \boldsymbol{w}(x_-))\cdot\boldsymbol{h}^d(\boldsymbol{v}(x_-),\boldsymbol{v}(x_+);\boldsymbol{n})\, dx\,dt \\
&+ \int_{I^n}\sum_{e\in\mathcal{E}}\int_e \frac{1}{2}(\boldsymbol{w}(x_+) + \boldsymbol{w}(x_-))\cdot[\boldsymbol{f}(\boldsymbol{v};\boldsymbol{n})]_{x-}^{x+}\, dx\,dt \\
B(\boldsymbol{v},\boldsymbol{w})_{ls} &= \int_{I^n}\sum_{T\in\Omega}\int_T \left(\tilde{A}_0\boldsymbol{w}_{,t} + \tilde{A}_i\boldsymbol{w}_{,x_i}\right)\cdot\boldsymbol{\tau}\left(\tilde{A}_0\boldsymbol{v}_{,t} + \tilde{A}_i\boldsymbol{v}_{,x_i}\right) dx\,dt
\end{aligned}$$

$$B(v,w)_{bc} = \int_{I^n}\int_{\Gamma} w \cdot \frac{1}{2}\Big(f(\tilde{g};n) - f(v;n) - h^d(v,\tilde{g};n)\Big)\ dx\ dt$$

Before proving the main result of this section, the following two lemmas are needed. These lemmas address entropy dissipation by discontinuous element approximations in time and space. Both lemmas are applications of Taylor's series with integral remainder.

Lemma 10. (Time Discontinuous Entropy Production). *Let $t_\pm$ denote a temporal slab interface. The following entropy jump identity holds across time slab boundaries*

$$\int_{\Omega}\left([U]_{t_-}^{t_+} - v^T(t_+)\,[u]_{t_-}^{t_+}\right)\ dx + |||\,[u]_{t_-}^{t_+}\,|||^2_{\tilde{A}_0^{-1},\Omega} = 0$$

where

$$|||\,[u]_{t_-}^{t_+}\,|||^2_{\tilde{A}_0^{-1},\Omega} \equiv \int_{\Omega}\int_0^1 (1-\theta)\,[u]_{t_-}^{t_+}\cdot \tilde{A}_0^{-1}(\overline{u}(\theta))\,[u]_{t_-}^{t_+}\,d\theta\ dx \geq 0 \quad (71)$$

and $\overline{u}(\theta) = u(t_+) - \theta\,[u]_{t_-}^{t_+}$.

Proof. Recall the following form of Taylor series with integral remainder

$$[U]_{t_-}^{t_+} - \frac{\partial U}{\partial u}(t_+)\ [u]_{t_-}^{t_+} + \int_0^1 (1-\theta)\ [u]_{t_-}^{t_+}\cdot\frac{\partial^2 U}{\partial u^2}\,(\overline{u}(\theta))\ [u]_{t_-}^{t_+}\ d\theta = 0\ .$$

Inserting the identities $\frac{\partial U}{\partial u} = v^T$, $\frac{\partial^2 U}{\partial u^2} = \tilde{A}_0^{-1}$, and integration over Ω yields the stated lemma, see also Shakib [49]. □

A related result follows for discontinuities across element boundaries in space.

Lemma 11. (Space Discontinuous Entropy Production). *Let $x_\pm$ denote a spatial element interface. The following entropy jump identity holds across spatial element boundaries*

$$\begin{aligned} -\left[F^i\right]_{x_-}^{x_+} &+ \frac{1}{2}\ \left(v^T(x_-) + v^T(x_+)\right)\ \left[f^i\right]_{x_-}^{x_+} \\ &= \int_0^1 \frac{1}{2}(1-\theta)\ [v]_{x_-}^{x_+}\cdot\left(\tilde{A}_i(\overline{\overline{v}}(\theta)) - \tilde{A}_i(\overline{v}(\theta))\right)[v]_{x_-}^{x_+}\ d\theta \end{aligned} \quad (72)$$

with $\overline{v}(\theta) = v(x_+) - \theta\,[v]_{x_-}^{x_+}$ and $\overline{\overline{v}}(\theta) = v(x_-) + \theta\,[v]_{x_-}^{x_+}$.

Proof. Recall the dual relationship $F^i(u) + \mathcal{F}^i(v) = v^T f^i$ and construct the jump identity

$$\left[F^i\right]_{x_-}^{x_+} + \left[\mathcal{F}^i\right]_{x_-}^{x_+} = \frac{1}{2}\left(v^T(x_-) + v^T(x_+)\right)\left[f^i\right]_{x_-}^{x_+}$$

$$+\frac{1}{2}\left((\boldsymbol{f}^i)^T(x_-)+(\boldsymbol{f}^i)^T(x_+)\right)[\boldsymbol{v}]_{x-}^{x+} \ .$$

Thus, the following equivalent form of the left-hand-side of (72) is considered:

$$\left[\mathcal{F}^i\right]_{x-}^{x+}-\frac{1}{2}\left((\boldsymbol{f}^i)^T(x_-)+(\boldsymbol{f}^i)^T(x_+)\right)[\boldsymbol{v}]_{x-}^{x+} \ .$$

Next, recall the Taylor series with integral remainder formulas

$$\left[\mathcal{F}^i\right]_{x-}^{x+}-\frac{\partial\mathcal{F}^i}{\partial \boldsymbol{v}}(x_+)\ [\boldsymbol{v}]_{x-}^{x+}+\int_0^1(1-\theta)\ [\boldsymbol{v}]_{x-}^{x+}\cdot\frac{\partial^2 F^i}{\partial \boldsymbol{v}^2}\left(\overline{\boldsymbol{v}}(\theta)\right)[\boldsymbol{v}]_{x-}^{x+}\ d\theta=0$$

and

$$\left[\mathcal{F}^i\right]_{x-}^{x+}-\frac{\partial\mathcal{F}^i}{\partial \boldsymbol{v}}(x_-)\ [\boldsymbol{v}]_{x-}^{x+}-\int_0^1(1-\theta)\ [\boldsymbol{v}]_{x-}^{x+}\cdot\frac{\partial^2 F^i}{\partial \boldsymbol{v}^2}\left(\overline{\overline{\boldsymbol{v}}}(\theta)\right)[\boldsymbol{v}]_{x-}^{x+}\ d\theta=0$$

with $\overline{\boldsymbol{v}}(\theta)$ and $\overline{\overline{\boldsymbol{v}}}(\theta)$ defined above. Inserting these formulas into (2.3) together with the identities $\frac{\partial\mathcal{F}^i}{\partial \boldsymbol{v}}=(\boldsymbol{f}^i)^T$ and $\frac{\partial^2\mathcal{F}^i}{\partial \boldsymbol{v}^2}=\tilde{A}_i$ produces the desired result. □

Remark 12. Since linear problems are modeled by $\boldsymbol{f}^i=\tilde{A}_i\boldsymbol{v}$ with constant A^i, the right-hand-side of (72) vanishes identically

$$-\left[F^i\right]_{x-}^{x+}+\frac{1}{2}\left(\boldsymbol{v}^T(x_-)+\boldsymbol{v}^T(x_+)\right)\left[\boldsymbol{f}^i\right]_{x-}^{x+}=0 \ .$$

This is consistent with the linear analysis given previously. In the nonlinear analysis, the dissipative part of the numerical flux $\boldsymbol{h}^d$ is essential to the entropy stability of the formulation as it must dominate the right-hand-side term in (72). The strategy used here is to define a minimum $\boldsymbol{h}^d_{min}$ needed from stability analysis so that the following combination of terms is positive

$$-\left[F^i\right]_{x-}^{x+}+\frac{1}{2}\left(\boldsymbol{v}^T(x_-)+\boldsymbol{v}^T(x_+)\right)\left[\boldsymbol{f}^i\right]_{x-}^{x+}+\frac{1}{2}[\boldsymbol{v}]_{x-}^{x+}\cdot\boldsymbol{h}^d_{min}\geq 0 \ . \quad (73)$$

This minimum flux dissipation $\boldsymbol{h}^d_{min}$ is the mean-value flux dissipation (62) defined earlier. Stability follows whenever the flux dissipation exceeds the mean-value minimum, i.e.

$$[\boldsymbol{v}]_{x-}^{x+}\cdot\boldsymbol{h}^d_{\mathrm{MV}}\leq[\boldsymbol{v}]_{x-}^{x+}\cdot\boldsymbol{h}^d \ . \quad (74)$$

Theorem 13. Global Entropy Stability (Nonlinear Hyperbolic System). *The variational formulation (60) with symmetric mean-value flux dissipation*

$$\boldsymbol{h}^d_{\mathrm{MV}}(\boldsymbol{v}_-,\boldsymbol{v}_+;\boldsymbol{n})=\int_0^1(1-\theta)\ \left(|\tilde{A}(\overline{\overline{\boldsymbol{v}}}(\theta);\boldsymbol{n})|_{\tilde{A}_0}+|\tilde{A}(\overline{\boldsymbol{v}}(\theta);\boldsymbol{n})|_{\tilde{A}_0}\right)[\boldsymbol{v}]_{x-}^{x+}\ d\theta$$

is entropy stable (modulo inflow data $\tilde{g}$) with the following global balance:

$$\sum_{n=0}^{N-1}\left(|||\,[u]_{t_-^n}^{t_+^n}\,|||^2_{\tilde{A}_0^{-1},\Omega}+2\|\tilde{A}_0 v_{,t}+\tilde{A}_i v_{,x_i}\|^2_{\tau,\Omega\times I^n}+\sum_{e\in\mathcal{E}}\left\langle [v]_{x-}^{x+}\right\rangle^2_{|\underline{\tilde{A}}|,e\times I^n}\right)$$
$$+\sum_{n=0}^{N-1}\langle v\rangle^2_{|\underline{\tilde{A}}|,\Gamma\times I^n}+\int_\Omega U(t_-^N)\,dx=\int_\Omega U(t_-^0)\,dx+\sum_{n=0}^{N-1}2\,G_\Gamma^n(\tilde{g},v;n)\tag{75}$$

with $|\underline{\tilde{A}}(n)|=\int_0^1 2\,(1-\theta)\left(\tilde{A}^+_{\tilde{A}_0}(\overline{\overline{v}}(\theta);n)-\tilde{A}^-_{\tilde{A}_0}(\overline{v}(\theta);n)\right)\,d\theta,\ \overline{v}(\theta)=v(x_+)-\theta\,[v]_{x-}^{x+},\ \overline{\overline{v}}(\theta)=v(x_-)+\theta\,[v]_{x-}^{x+}$, *and* $G_\Gamma^n(\tilde{g},v;n)=\int_{I^n}\int_\Gamma\Big(F(\tilde{g};n)+\frac{1}{2}\tilde{g}\cdot(|\underline{A}|-2\,A^+_{\mathrm{MV}})\,\tilde{g}-\tilde{g}\cdot(|\underline{A}|-A^+_{\mathrm{MV}})\,v\Big)\,dx\,dt.$

Proof. Construct the energy balance for the interval $[t_-^N,t_-^0]=\cup_{n=0}^{N-1}I^n$ by setting $w=v$ and evaluating the various integrals. Consider the time derivative integral first

$$\int_\Omega\int_{I^n} v^T u_{,t}\,dt\,dx=\int_\Omega\int_{I^n} U_{,t}\,dt\,dx=\int_\Omega\left([U]_{t_-^n}^{t_-^{n+1}}-[U]_{t_-^n}^{t_+^n}\right)\,dx$$

and combine with the jump integral across time slabs. By use of lemma 10

$$\int_\Omega\int_{I^n} U_{,t}\,dt\,dx+\int_\Omega v^T(t_+^n)[u]_{t_-^n}^{t_+^n}\,dx=\int_\Omega [U]_{t_-^n}^{t_-^{n+1}}\,dx+|||\,[u]_{t_-^n}^{t_+^n}\,|||^2_{\tilde{A}_0^{-1},\Omega}\,.$$

When summed over all time slabs, the first term on the right-hand-side of this equation vanishes except for initial and final time slab contributions. Next, rewrite the spatial operator term

$$\int_{I^n}\int_\Omega v^T f^i_{,x_i}\,dx\,dt=\int_{I^n}\int_\Omega F^i_{,x_i}\,dx\,dt$$

and apply the divergence theorem

$$\int_{I^n}\int_\Omega v^T f^i_{,x_i}\,dx\,dt=\int_{I^n}\sum_{e\in\mathcal{E}}\int_e-[F(n)]_{x-}^{x+}\,dx\,dt+\int_{I^n}\int_\Gamma F(n)\,dx\,dt$$

where $F(n)=n_i\,F^i$. From lemma 11 and the definition of $|\underline{\tilde{A}}|$, it follows that

$$\int_{I^n}\int_\Omega v^T f^i_{,x_i}\,dx\,dt+\int_{I^n}\sum_{e\in\mathcal{E}}\int_e\frac{1}{2}\,(v(x_+)+v(x_-))\,[f(n)]_{x-}^{x+}\,dx\,dt$$
$$+\int_{I^n}\sum_{e\in\mathcal{E}}\int_e\frac{1}{2}\,[v]_{x-}^{x+}\cdot h^d_{\mathrm{MV}}(v_-,v_+;n)\,dx\,dt$$
$$=\int_{I^n}\sum_{e\in\mathcal{E}}\int_e\Big(-[F(n)]_{x-}^{x+}+\frac{1}{2}\,(v(x_+)+v(x_-))\,[f(n)]_{x-}^{x+}$$

$$+\frac{1}{2}[v]_{x_-}^{x_+}\cdot|\tilde{A}(n)|_{\mathrm{MV}}\,[v]_{x_-}^{x_+}\Big)\;dx\;dt+\int_{I^n}\int_{\Gamma}F(n)\;dx\;dt$$
$$=\sum_{e\in\mathcal{E}}\frac{1}{2}\left\langle[v]_{x_-}^{x_+}\right\rangle^2_{|\underline{\tilde{A}}|,e\times I^n}+\int_{I^n}\int_{\Gamma}F(n)\;dx\;dt\;.$$

Finally, consider the boundary condition terms. After some rearrangement, the following form is obtained

$$\begin{aligned}B(v,v)_{bc}=&\int_{I^n}\int_{\Gamma}\Big(-[F(n)]_v^{\tilde{g}}+\frac{1}{2}(\tilde{g}+v)\cdot[f(n)]_v^{\tilde{g}}\\&+\frac{1}{2}(\tilde{g}-v)\cdot|A(n)|_{\mathrm{MV}}(\tilde{g}-v)\Big)\;dx\;dt\\&+\int_{I^n}\int_{\Gamma}\Big([F(n)]_v^{\tilde{g}}-\frac{1}{2}\tilde{g}\cdot[f(n)]_v^{\tilde{g}}-\frac{1}{2}|A(n)|_{\mathrm{MV}}(\tilde{g}-v)\Big)\;dx\;dt\;.\end{aligned}$$

The first group of terms appearing on the right-hand-side of this equation are identical to the terms arising for interior edges. Consequently,

$$\begin{aligned}B(v,v)_{bc}=&\frac{1}{2}\langle\tilde{g}-v\rangle^2_{|\underline{\tilde{A}}|,\Gamma\times I^n}\\&+\int_{I^n}\int_{\Gamma}\Big([F(n)]_v^{\tilde{g}}-\frac{1}{2}\tilde{g}\cdot[f(n)]_v^{\tilde{g}}-\frac{1}{2}\tilde{g}\cdot|A(n)|_{\mathrm{MV}}(\tilde{g}-v)\Big)\;dx\;dt\\=&\frac{1}{2}\langle v\rangle^2_{|\underline{\tilde{A}}|,\Gamma\times I^n}-G_\Gamma^n(\tilde{g},v;n)-\int_{I^n}\int_{\Omega}F(n)\;dx,\;dt\end{aligned}$$

where $G_\Gamma^n(\tilde{g},v;n)$ is defined above. The least-squares integral produces a pure quadratic form without modification. Combining the above result, summing over time slabs, and multiplication by two yields an entropy balance equation (75) which bounds the global entropy of the system at the final time T in terms of the initial entropy state and boundary entropy produced via $G_\Gamma^n(\tilde{g},v;n)$. □

Remark 14. The somewhat complicated form of $G_\Gamma^n(\tilde{g},v;n)$ is motivated by the observation that when the $\tilde{A}^i$ matrices are constant

$$G_\Gamma^n(\tilde{g},v;n)=\langle\tilde{g},v\rangle_{(-\tilde{A}^-),\Gamma\times I^n}$$

which agrees with the linear analysis.

2.4 Failure of the Roe Absolute Value Matrix to be Entropy Dissipative

From the nonlinear analysis of the previous section, the importance of the flux dissipation term h^d in the entropy stability of the discontinuous Galerkin method is apparent. In this section, the basic entropy dissipation inequality

$$[v]_-^+\cdot h^d\geq 0$$

is examined for the Roe flux dissipation [46]

$$h^d_{\rm Roe}(\boldsymbol{v}_-,\boldsymbol{v}_+;\boldsymbol{n}) = |A(\overline{\boldsymbol{v}}_{\rm Roe}(\boldsymbol{v}_-,\boldsymbol{v}_+);\boldsymbol{n})|\ [\boldsymbol{u}(\boldsymbol{v})]_-^+$$

where $\overline{\boldsymbol{v}}_{\rm Roe}$ denotes the Roe-averaged state such that

$$[\boldsymbol{f}(\boldsymbol{n})]_-^+ = A(\overline{\boldsymbol{v}}_{\rm Roe}(\boldsymbol{v}_-,\boldsymbol{v}_+);\boldsymbol{n})\ [\boldsymbol{u}(\boldsymbol{v})]_-^+ \ .$$

In this case, strict entropy dissipation requires that

$$[\boldsymbol{v}]_-^+ \cdot |A|_{\rm Roe}\ [\boldsymbol{u}(\boldsymbol{v})]_-^+ \geq 0 \ .$$

By Volpert path integration, a SPD matrix $\tilde{G}$ exists which is a mean value integrated form of $\tilde{A}_0$ such that

$$[\boldsymbol{u}]_-^+ = \tilde{G}(\boldsymbol{v}_-,\boldsymbol{v}_+)\ [\boldsymbol{v}]_-^+$$

so that

$$[\boldsymbol{v}]_-^+ \cdot |A|_{\rm Roe}\ [\boldsymbol{u}(\boldsymbol{v})]_-^+ = [\boldsymbol{v}]_-^+ \cdot |A|_{\rm Roe}\,\tilde{G}\ [\boldsymbol{v}]_-^+ \ .$$

Even so, there should be no expectation that $\tilde{G}$ is a general Lyapunov solution, i.e.

$$|A|_{\rm Roe}\,\tilde{G} + \tilde{G}\,|A|^T_{\rm Roe} \geq 0 \ .$$

To address this point, a counterexample is given for which the Roe absolute value matrix produces entropy dissipation of the incorrect sign. The counterexample is motivated by previous discrete shock profile analysis, see Barth [5] and Serre [48]. According to the established theory for the Roe flux

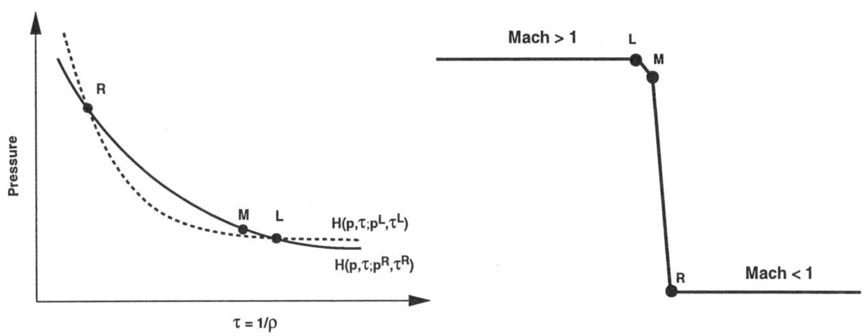

(a) Stationary shock Hugoniot curves. (b) Shock profile with transition point M.

Fig. 3. Stationary shock profile for the Roe matrix counterexample showing left (L), right (R), and transition point (M) states.

with piecewise constant data representation, stationary discontinuities are resolved with at most one intermediate shock transition point, see Fig. 3(b). As a stationary problem, the location of this transition point is not uniquely determined. Rather, a single degree of freedom exists in the system which is characterized by all transition point states that can connect to the post-shock state by a single (moving) compression shock. In other words, transition point states must lie on the shock Hugoniot $H(p,\tau;p^R;\tau^R)$. In reference [5], we showed using fixed-point iteration analysis that both Godunov and Roe fluxes admit locally unstable shock profiles when the pre-shock Mach number exceeds 6 for $\gamma = 7/5$ gases. In the following counterexample, a Mach 8 stationary shock is considered as depicted in Fig. 3. A transition point $\boldsymbol{u}_M$ was chosen close to the pre-shock left state, but perturbed from the shock Hugoniot by changing the transition point fluid density by .001%. The entropy dissipation associated with the jump from $\boldsymbol{u}_M$ to $\boldsymbol{u}_R$ was then computed. Specifically, the following data states were used in this counterexample

$$\begin{pmatrix} \rho \\ u \\ p \end{pmatrix}_- = \begin{pmatrix} 1.00182 \\ 7.99621 \\ .72240 \end{pmatrix}, \quad \begin{pmatrix} \rho \\ u \\ p \end{pmatrix}_+ = \begin{pmatrix} 5.56522 \\ 1.43750 \\ 53.21429 \end{pmatrix} .$$

A numerical calculation reveals that

$$[\boldsymbol{v}]_-^+ \cdot |A|_{\mathrm{Roe}} \, [\boldsymbol{u}(\boldsymbol{v})]_-^+ = -0.1355392$$

so that the Roe absolute value matrix fails to be strictly entropy dissipative. The implications of this have yet to be fully understood and/or appreciated.

2.5 A Simplified Galerkin Least-Squares Method in Symmetric Form

Consider a general isoparametrically-mapped element as shown in Fig. 4. For sake of generality, let x_0 denote the time coordinate. The analysis given below requires the unit direction vectors associated with the mapping $\boldsymbol{n}^i = \nabla\xi^i/|\nabla\xi^i|$ and the matrices

$$B(\boldsymbol{n}^i) = \sum_{j=0}^{d} \boldsymbol{n}_j^i A_j$$

and in scaled form

$$B^i = |\nabla\xi^i| B(\boldsymbol{n}^i) \; .$$

In the implementation of the Galerkin least-squares method, the most difficult computational aspect of the scheme is the calculation of the least-squares $\boldsymbol{\tau}$ matrix. In the papers by Hughes and Mallet [34] and Shakib [49], they proposed the following form for $\boldsymbol{\tau}$ on a mapped element

$$\boldsymbol{\tau}_p = |\tilde{B}|^{-1}_{p,\tilde{A}_0}, \quad |\tilde{B}|_{p,\tilde{A}_0} = \left(\sum_{i=0}^{d} |B^i|^p \right)^{1/p} \tilde{A}_0 \; . \tag{76}$$

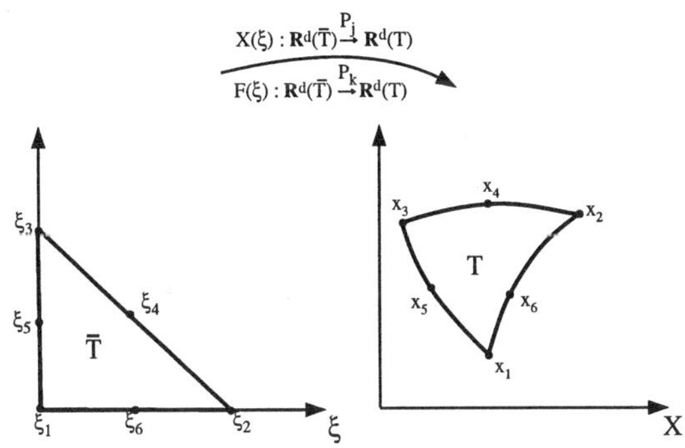

Fig. 4. Isoparametric element mapping $\boldsymbol{\xi} \mapsto \mathbf{x}$ from a unit element space $\boldsymbol{\xi}$ to a physical space $\mathbf{x}$. The simplex shape coordinates are mapped as P_j polynomials and approximation functions are mapped as P_k polynomials.

Equation (76) is of the same form given earlier in (17). Using the analysis techniques from Proposition 7, symmetry of $\boldsymbol{\tau}_p$ can be shown. This formula represents generalizations of the matrix absolute value with respect to the metric tensor matrix $\tilde{A}_0$ and multiple space dimension d. Historically, the value $p = 2$ has been used computing $\boldsymbol{\tau}$ for implementation in the Galerkin least-squares method. The $p = 2$ form necessitates the calculation of a matrix square root. Hughes advocates the use of the Cayley-Hamilton theorem for this computation. Unfortunately, the Cayley-Hamilton technique becomes unwieldy for matrices of dimension larger than 4 or 5. In light of the Eigenvector Scaling Theorem 4, it is useful to revisit the derivation of $\boldsymbol{\tau}$ with $p = 1$. Let $\tilde{B}^i = B^i \tilde{A}_0$, from (18) it follows that

$$\begin{aligned}\boldsymbol{\tau}_1 &= |\tilde{B}|^{-1}_{1,\tilde{A}_0} \\ &= \left[\sum_{i=0}^{d} |\tilde{B}^i|_{\tilde{A}_0}\right]^{-1} \\ &= \left[|\nabla\xi^0|\tilde{A}_0 + \sum_{i=1}^{d} |\nabla\xi^i|\, \tilde{R}(\boldsymbol{n}^i)\, |\Lambda(\boldsymbol{n}^i)|\, \tilde{R}^T(\boldsymbol{n}^i)\right]^{-1}\end{aligned}$$

using the entropy scaled eigenvectors $\tilde{R}(\boldsymbol{n}^i)$ of $\tilde{B}^i$. This represents a substantial simplification of the $\boldsymbol{\tau}$ matrix calculation.

In the remaining discussion, simplicial meshes in space are assumed. Time is included via space-time prisms such that the time coordinate x_0 is orthogonal to the space coordinates. In this case, it is natural to consider the $d+1$ barycentric (triangular) coordinates which satisfy $\sum_{i=1}^{d+1} \nabla\xi^i = 0$ and $\sum_{i=1}^{d+1} \tilde{B}^i = 0$. The motivation for constructing modified formulas using $d+1$ barycentric coordinates rather than the d mapping coordinates is that the resulting discretization becomes invariant to permutations of the simplex indices. For this reason, the following final form is used in computations with $\alpha > 0$

$$\begin{aligned}\boldsymbol{\tau}_1 &= \left[|\tilde{B}|^i_{\tilde{A}_0} + \alpha \sum_{i=1}^{d+1} |\tilde{B}^i|_{\tilde{A}_0} \right]^{-1} \\ &= \left[|\nabla \xi^0| \tilde{A}_0 + \alpha \sum_{i=1}^{d+1} |\nabla \xi^i| \, \tilde{R}(\boldsymbol{n}^i) \, |\Lambda(\boldsymbol{n}^i)| \, \tilde{R}^T(\boldsymbol{n}^i) \right]^{-1} .\end{aligned}$$

Remark 15. Since $|\nabla \xi^0| \propto (\Delta t)^{-1}$, for steady-state calculations $|\nabla \xi^0|$ is set to a small number $O(10^{-6})$. Given that $\tilde{A}_0$ is SPD, invertibility of $\boldsymbol{\tau}$ is then assured. In all 2-D calculations, $\alpha = 1$ has been used so that the scheme becomes fully upwind in 1-D.

Computing Jacobian derivatives of the simplified GLS scheme. One motivation for simplifying the $\boldsymbol{\tau}$ matrix in GLS concerns the computation of Jacobian derivatives of the GLS scheme. There are several computational scenarios which benefit greatly from having the ability to easily compute exact Jacobian derivatives: Newton's method, discrete flow/shape optimization, homotopy methods, *a posteriori* error estimation, etc. The most difficult aspect of computing Jacobian derivatives of the GLS discretization is the symbolic calculation of Jacobian derivatives of the $\boldsymbol{\tau}$ matrix itself. The matrix square root appearing in the standard definition of $\boldsymbol{\tau}$ makes the calculation of Jacobian derivatives impractical. This remains true even when the Cayley-Hamilton technique is used in the computation of the matrix square root. The simplified GLS $\boldsymbol{\tau}$ given in the previous section permits a straightforward calculation of Jacobian derivatives. For purposes of explanation, assume that the Jacobian derivative of $\boldsymbol{\tau}$ with respect to some unspecified variable $\boldsymbol{w}$ is required. Without stating the precise form of $\boldsymbol{w}$, a well-defined procedure exists for computing Jacobian derivatives of $\boldsymbol{\tau}$. Begin by recalling the following identity relating Jacobian derivatives of $\boldsymbol{\tau}$ and its inverse

$$\frac{\partial \boldsymbol{\tau}}{\partial \boldsymbol{w}} = -\boldsymbol{\tau} \frac{\partial \boldsymbol{\tau}^{-1}}{\partial \boldsymbol{w}} \boldsymbol{\tau} .$$

Next apply chainrule differentiation

$$\frac{\partial \boldsymbol{\tau}^{-1}}{\partial \boldsymbol{w}} = \frac{\partial |\tilde{B}^1|_{\tilde{A}_0}}{\partial \boldsymbol{w}} + \ldots + \frac{\partial |\tilde{B}^d|_{\tilde{A}_0}}{\partial \boldsymbol{w}}$$

with

$$\begin{aligned}\frac{\partial |\tilde{B}^i|_{\tilde{A}_0}}{\partial \boldsymbol{w}} = |\nabla \xi^i| \Bigg(& \frac{\partial \tilde{R}(\boldsymbol{n}^i)}{\partial \boldsymbol{w}} |\Lambda(\boldsymbol{n}^i)| \tilde{R}^T(\boldsymbol{n}^i) + \tilde{R}(\boldsymbol{n}^i) \frac{\partial |\Lambda(\boldsymbol{n}^i)|}{\partial \boldsymbol{w}} \tilde{R}^T(\boldsymbol{n}^i) \\ & + \tilde{R}(\boldsymbol{n}^i) |\Lambda(\boldsymbol{n}^i)| \frac{\partial \tilde{R}^T(\boldsymbol{n}^i)}{\partial \boldsymbol{w}} \Bigg) .\end{aligned}$$

Note that a high degree of computational efficiency can be achieved in the calculation of these Jacobian terms by exploiting the transpose symmetry of intermediate products.

2.6 A Simplified Discontinuous Galerkin Method in Symmetric Form

Simplification of the discontinuous Galerkin method follows by choosing the symmetric numerical flux function proposed in Sect. 2.1, i.e.

$$\boldsymbol{h}_{\text{Symm}}(\boldsymbol{v}_-, \boldsymbol{v}_+; \boldsymbol{n}) = \frac{1}{2}\left(\boldsymbol{f}(\boldsymbol{v}_-; \boldsymbol{n}) + \boldsymbol{f}(\boldsymbol{v}_+; \boldsymbol{n})\right) - \frac{1}{2}|\tilde{A}(\boldsymbol{v}_*; \boldsymbol{n})|_{\tilde{A}_0} \, [\boldsymbol{v}]_{x_-}^{x+} \quad (77)$$

with $|\tilde{A}|_{\tilde{A}_0} = \tilde{R}\,|\Lambda|\,\tilde{R}^T$, $\overline{\overline{\boldsymbol{v}}}(\theta) = \boldsymbol{v}(x_-) + \theta\,[\boldsymbol{v}]_{x_-}^{x+}$, and $\boldsymbol{v}_*$ such that

$$[\boldsymbol{v}]_{x_-}^{x+} \cdot |\tilde{A}(\boldsymbol{v}_*; \boldsymbol{n})| \; [\boldsymbol{v}]_{x_-}^{x+} = \sup_{0\le\theta\le1} [\boldsymbol{v}]_{x_-}^{x+} \cdot |\tilde{A}(\overline{\overline{\boldsymbol{v}}}(\theta); \boldsymbol{n})|_{\tilde{A}_0} \; [\boldsymbol{v}]_{x_-}^{x+} \; .$$

Nonlinear stability in the DG method follows from the analysis of Sect. 2.3.

Theorem 16. Symmetric Flux Dissipation. *The variational formulation (60) with numerical flux function (77) is entropy stable in the sense of Theorem 13.*

Proof. It is sufficient to show that the given flux dissipation

$$\boldsymbol{h}^d_{\text{Symm}} = |\tilde{A}(\boldsymbol{v}_*; \boldsymbol{n})|_{\tilde{A}_0} \; [\boldsymbol{v}]_{x_-}^{x+}$$

exceeds the mean-value value flux dissipation. This is reflected by the algebraic condition (74)

$$[\boldsymbol{v}]_{x_-}^{x+} \cdot \boldsymbol{h}^d_{\text{MV}} \le [\boldsymbol{v}]_{x_-}^{x+} \cdot \boldsymbol{h}^d_{\text{Symm}} \; .$$

From the mean-value flux definition

$$\begin{aligned}
[\boldsymbol{v}]_{x_-}^{x+} \cdot \boldsymbol{h}^d_{\text{MV}} &= [\boldsymbol{v}]_{x_-}^{x+} \cdot \int_0^1 (1-\theta)\left(|\tilde{A}(\overline{\overline{\boldsymbol{v}}}(\theta); \boldsymbol{n})|_{\tilde{A}_0} + |\tilde{A}(\overline{\boldsymbol{v}}(\theta); \boldsymbol{n})|_{\tilde{A}_0}\right) d\theta \; [\boldsymbol{v}]_{x_-}^{x+} \\
&\le \int_0^1 (1-\xi)\, d\xi \sup_{0\le\theta\le1} [\boldsymbol{v}]_{x_-}^{x+} \cdot \left(|\tilde{A}(\overline{\overline{\boldsymbol{v}}}(\theta); \boldsymbol{n})|_{\tilde{A}_0} + |\tilde{A}(\overline{\boldsymbol{v}}(\theta); \boldsymbol{n})|_{\tilde{A}_0}\right) [\boldsymbol{v}]_{x_-}^{x+} \\
&= [\boldsymbol{v}]_{x_-}^{x+} \cdot |\tilde{A}(\boldsymbol{v}_*; \boldsymbol{n})|_{\tilde{A}_0} \; [\boldsymbol{v}]_{x_-}^{x+} \\
&= [\boldsymbol{v}]_{x_-}^{x+} \cdot \boldsymbol{h}^d_{\text{Symm}}
\end{aligned}$$

where $\overline{\boldsymbol{v}}(\theta) = \boldsymbol{v}(x_+) - \theta\,[\boldsymbol{v}]_{x_-}^{x+}$, $\overline{\overline{\boldsymbol{v}}}(\theta) = \boldsymbol{v}(x_-) + \theta\,[\boldsymbol{v}]_{x_-}^{x+}$, and $\boldsymbol{v}_*$ as defined above. This establishes nonlinear stability of the DG method using the simplified flux function. □

Remark 17. Unfortunately, the state $\boldsymbol{v}_*$ is not generally known in closed form. Consequently, $\boldsymbol{v}_*$ has been approximated by a simple arithmetic average in all computations presented herein. Improved techniques for estimating $\boldsymbol{v}_*$ are a subject of future research.

Remark 18. Cockburn and Shu [18] have shown impressive results using the simpler Lax-Friedrichs flux. It is straightforward to derive a corresponding "symmetric Lax-Friedrichs" numerical flux function

$$\boldsymbol{h}_{SLF}(\boldsymbol{v}_-,\boldsymbol{v}_+;\boldsymbol{n}) = \frac{1}{2}\left(\boldsymbol{f}(\boldsymbol{v}_-;\boldsymbol{n}) + \boldsymbol{f}(\boldsymbol{v}_+;\boldsymbol{n})\right) - \frac{1}{2}\lambda_{max}(\boldsymbol{v}_*;\boldsymbol{n})\tilde{A}_0\left(\boldsymbol{v}_*\right)\,[\boldsymbol{v}]_{x-}^{x+} \tag{78}$$

with $\lambda_{max}(\boldsymbol{v}_*;\boldsymbol{n})$ the largest eigenvalue of $|A(\boldsymbol{v}_*;\boldsymbol{n})|$. Nonlinear entropy stability follows from

$$\begin{aligned}
[\boldsymbol{v}]_{x-}^{x+}\cdot\boldsymbol{h}^d_{\mathrm{MV}} &\le [\boldsymbol{v}]_{x-}^{x+}\cdot|\tilde{A}(\boldsymbol{v}_*;\boldsymbol{n})|_{\tilde{A}_0}\ [\boldsymbol{v}]_{x-}^{x+} \\
&= [\boldsymbol{v}]_{x-}^{x+}\cdot\tilde{R}(\boldsymbol{v}_*;\boldsymbol{n})\,|\Lambda(\boldsymbol{v}_*;\boldsymbol{n})|\,\tilde{R}^T(\boldsymbol{v}_*;\boldsymbol{n})\ [\boldsymbol{v}]_{x-}^{x+} \\
&\le [\boldsymbol{v}]_{x-}^{x+}\cdot\max_{1\le i\le m}(\Lambda_{ii}(\boldsymbol{v}_*;\boldsymbol{n}))\tilde{R}(\boldsymbol{v}_*;\boldsymbol{n})\,\tilde{R}^T(\boldsymbol{v}_*;\boldsymbol{n})\ [\boldsymbol{v}]_{x-}^{x+} \\
&= \lambda_{max}(\boldsymbol{v}_*;\boldsymbol{n})\,[\boldsymbol{v}]_{x-}^{x+}\cdot\tilde{A}_0\left(\boldsymbol{v}_*\right)\ [\boldsymbol{v}]_{x-}^{x+}
\end{aligned}$$

All calculations given below have been computed using the symmetric flux function (77).

Computing Jacobian derivatives of the simplified DG scheme. Using either flux function form, it becomes straightforward to construct Jacobian derivatives of the simplified DG scheme. The most difficult aspect of computing Jacobian derivatives of the DG discretization is the symbolic calculation of Jacobian derivatives of $|\tilde{A}|_{\tilde{A}_0}$ or $\tilde{A}_0$ appearing in the numerical flux functions. To calculate these derivatives, we employ essentially the same technique used in the simplified GLS method. For example, to compute derivatives of $|\tilde{A}|_{\tilde{A}_0}$ with respect to a vector $\boldsymbol{w}$, chainrule differentiation is used

$$\frac{\partial|\tilde{A}(\boldsymbol{n})|_{\tilde{A}_0}}{\partial \boldsymbol{w}} = \frac{\partial\tilde{R}(\boldsymbol{n})}{\partial \boldsymbol{w}}|\Lambda(\boldsymbol{n})|\tilde{R}^T(\boldsymbol{n}) + \tilde{R}(\boldsymbol{n})\frac{\partial|\Lambda(\boldsymbol{n})|}{\partial \boldsymbol{w}}\tilde{R}^T(\boldsymbol{n}) + \tilde{R}(\boldsymbol{n})|\Lambda(\boldsymbol{n})|\frac{\partial\tilde{R}^T(\boldsymbol{n})}{\partial \boldsymbol{w}}.$$

A similar procedure is used to obtain derivatives of the symmetric Lax-Friedrichs flux function.

The numerical results presented in Sects. 2.8 and 2.9 have been computed using a damped-Newton iteration scheme and the exact Jacobian derivatives described above. As a representative example, Fig. 5 shows the performance of Newton's method in solving typical simplified DG and GLS discretized flow problems. The test problem consists of subsonic flow past a single airfoil. This problem is described in detail in Sect. 2.9. The calculation begins using piecewise constant elements and the DG formulation. After achieving convergence with Newton's method, the calculation is halted and resumed using linear elements with the simplified GLS scheme. Again convergence of Newton's method is attained and the process repeated with quadratic elements and the simplified GLS discretization. For overall efficiency, the individual matrix problems produced by Newton's methods are solved to a tolerance of 10^{-3}. This prevents true quadratic convergence from being attained.

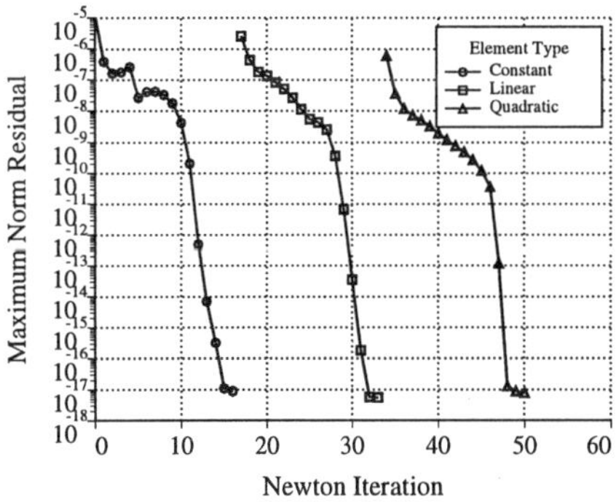

Fig. 5. Typical Newton's method convergence histories for a discretized subsonic airfoil flow problem using DG with constant elements and GLS with linear and quadratic elements.

2.7 Additional Stabilization Operators for the GLS and DG Schemes

To further enhance the stability of the GLS and DG schemes, additional (high order) stabilization terms are often added to the basic formulations [23,49,38]. In the paragraphs below, we consider an additional stabilization operator for the MHD equations which strengthens the $\nabla \cdot \boldsymbol{B}$ constraint and a general discontinuity capturing operator suitable for computing solutions with discontinuities or steep gradients.

$\nabla \cdot \boldsymbol{B}$ Stabilization in MHD Flow. Since the entropy function $U(\boldsymbol{u})$ is a convex combination of the conservation law variables, boundedness of all solution variables is controlled by the nonlinear entropy stability inherent in the stabilized schemes discussed earlier. In the context of MHD, this does not in itself guarantee satisfaction of the constraint equation $\nabla \cdot \boldsymbol{B} = 0$. In Sect. 2.9, it is found by numerical experiment that for simple smooth flows with well-posed boundary conditions that this constraint equation does in fact converge to zero at the optimal $O(h^k)$ rate using P_k polynomial approximation. For more general flows, the $\nabla \cdot \boldsymbol{B} = 0$ constraint satisfaction can be improved by adding additional stabilization terms to the symmetric GLS or DG formulation. The present choice of MHD stabilization strategy comes from looking at the basic entropy equation derived earlier for the MHD system before the constraint $\nabla \cdot \boldsymbol{B} = 0$ has been enforced (33)

$$(\rho s)_{,t} + \nabla \cdot (\rho \boldsymbol{V} s) + (\gamma - 1)\frac{\rho \boldsymbol{V} \cdot \boldsymbol{B}}{p}(\nabla \cdot \boldsymbol{B}) = 0$$

or in terms of the entropy pair $\{U, F^i\}$ for MHD

$$U_{,t} + F^i_{,x_i} - (\gamma - 1)\ \frac{\rho \boldsymbol{V} \cdot \boldsymbol{B}}{p}\ (\nabla \cdot \boldsymbol{B}) = 0\ .$$

When combined with Galerkin discretization using symmetrization variables, the two MHD stabilization operators described below both yield time and space integrated forms of the modified entropy equation (modulo boundary conditions)

$$\int_{I^n}\int_{\Omega}\left(U_{,t} + F^i_{,x_i} + \epsilon_{\rm mhd}\,(\gamma-1)\,\frac{\rho|\boldsymbol{V}\cdot\boldsymbol{B}|}{p}\,|\nabla\cdot\boldsymbol{B}|\right)\,dx\,dt = 0$$

so that for $\gamma > 1$ the entropy inequality is globally improved

$$\int_{I^n}\int_{\Omega}(U_{,t} + F^i_{,x_i})\;dx\;dt \leq 0$$

using a dimensionally consistent form of stabilization. The first MHD stabilization form is given by

$$B_{\rm mhd1}(\boldsymbol{w},\boldsymbol{v}) = \int_{I^n}\int_{\Omega}\epsilon_{\rm mhd}\;(\boldsymbol{w}\cdot\boldsymbol{S})\;{\rm sgn}(\boldsymbol{V}\cdot\boldsymbol{B})\;|\nabla\cdot\boldsymbol{B}|\;dx\;dt \tag{79}$$

for $\epsilon_{\rm mhd} \geq 0$ and $\boldsymbol{S} = (0, \boldsymbol{B}, \boldsymbol{B}\cdot\boldsymbol{V}, \boldsymbol{V})^T$. Looking at the energy associated with the operator,

$$B_{\rm mhd1}(\boldsymbol{v},\boldsymbol{v}) = \int_{I^n}\int_{\Omega}\epsilon_{\rm mhd1}\;(\gamma-1)\;\frac{|\boldsymbol{V}\cdot\boldsymbol{B}|}{p}\;|\nabla\cdot\boldsymbol{B}|\;dx\;dt \geq 0$$

it is clear that this operator strictly removes energy from the system. This form of stabilization is straightforward to implement in the GLS and DG schemes since the major components of (79) are already needed in the basic MHD Galerkin discretization. The nondifferentiability and nonlinearity of (79) can sometimes hinder convergence to steady-state in numerical computations. For this reason, a related stabilization term with improved numerical behavior is considered

$$B_{\rm mhd2}(\boldsymbol{w},\boldsymbol{v}) = \int_{I^n}\int_{\Omega}\epsilon_{\rm mhd2}\;(\boldsymbol{v}\cdot\boldsymbol{S})\;{\rm sgn}(\nabla\cdot\boldsymbol{B})\;(\boldsymbol{c}_i(\boldsymbol{v})\,\boldsymbol{w}_{,x_i})\;dx\;dt \tag{80}$$

where $\nabla\cdot\boldsymbol{B} = \boldsymbol{c}_i(\boldsymbol{v})\,\boldsymbol{v}_{,x_i}$. By construction, this term produces energy similar to the previous form

$$B_{\rm mhd2}(\boldsymbol{v},\boldsymbol{v}) = \int_{I^n}\int_{\Omega}\epsilon_{\rm mhd2}\;(\gamma-1)\;\frac{|\boldsymbol{V}\cdot\boldsymbol{B}|}{p}\;|\nabla\cdot\boldsymbol{B}|\;dx\;dt \geq 0\ .$$

This second form requires additional computational effort but performs better in actual numerical calculations in the sense of compability with Newton's method for solving the resulting discretized equations. Even so, further work in this area seems warranted.

Discontinuity Capturing Operators The Galerkin least-squares method produces excellent results in regions of smooth flow. In non-smooth regions, oscillations may appear in the direction of the solution gradient that are not controlled by least-squares stabilization. In the present calculations, the discontinuity capturing term due to Galeão and Dutra do Carmo [23] has been used

$$B_{dc}(\boldsymbol{w},\boldsymbol{v}) = \int_{I^n} \sum_{T\in\Omega} \int_T \frac{(\mathcal{L}\boldsymbol{v})^T\, \boldsymbol{\tau}\, \mathcal{L}\boldsymbol{v}}{(\nabla \boldsymbol{v})^T[\tilde{A}_0]\,\nabla \boldsymbol{v}}\; (\nabla \boldsymbol{w})^T[\tilde{A}_0]\,\nabla \boldsymbol{v}\; d\,x\; d\,t \qquad (81)$$

where $\mathcal{L}$ denotes the symmetric PDE operator. From an energy analysis perspective, the operator clearly removes energy from the system since

$$B_{dc}(\boldsymbol{v},\boldsymbol{v}) = \int_{I^n} \sum_{T\in\Omega} \int_T (\mathcal{L}\boldsymbol{v})^T \boldsymbol{\tau}\, \mathcal{L}\boldsymbol{v}\; d\,x\; d\,t \geq 0 \ .$$

This high order operator formally acts only in the direction of the solution gradient, see for example [23], thus providing high order *cross-wind* stabilization to the scheme. Cross-wind stabilization is extremely important in the accurate computation of shockwave phenomena. From a computational perspective, the stabilization term (81) has a straightforward Jacobian derivative calculation using the techniques described earlier. The numerical calculations presented in Sect. 2.9 demonstrate the effectiveness of this term in controlling oscillations at or near flow discontinuities.

2.8 Spatial Convergence Studies for Simplified GLS and DG

To evaluate accuracy and performance of the simplified GLS and DG schemes, Ringleb flow (an exact solution of the 2D Euler equations [13]) is computed in the channel geometry shown in Fig. 6(a). Although the exact solution contains a small supersonic region at inflow, the flow smoothly (isentropically) recompresses to a subsonic Mach number before exiting the channel. This flow provides a challenging test problem for numerical methods since small solution errors in the recompression region can become shockwave discontinuities with relatively large solution error. All integrations required in the present implementation of the GLS and DG methods are approximated using numerical quadrature. For this reason, it is extremely important that the integrated by parts formulation (60) be used since this form insures exact discrete conservation using inexact numerical integration. In performing these numerical integrations, quadrature rules have been chosen which meet the guidelines given by Cockburn et al. [16] for the DG method:

- Element Quadratures: Integrate $2k$ order polynomials exactly
  - $k = 1$, 3 mid-point quadrature $O(h^3)$
  - $k = 2$, 7 point quadrature $O(h^5)$
  - $k = 3$, 16 point quadrature $O(h^7)$

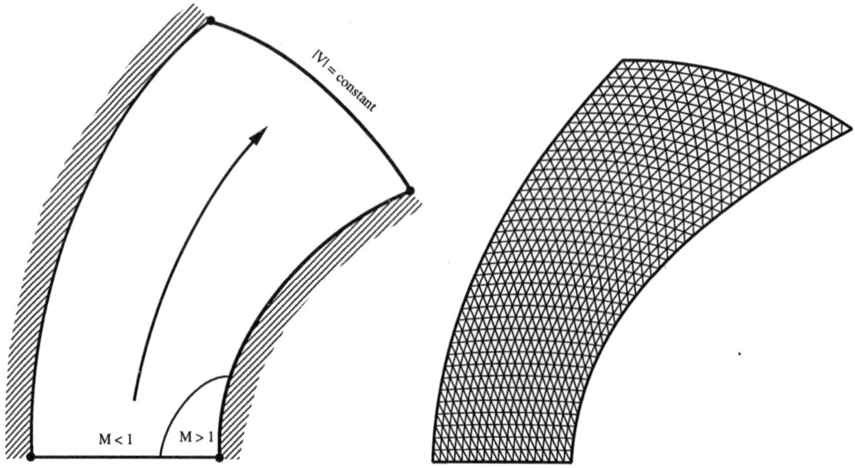

(a) Schematic of the Ringleb flow problem.

(b) Coarse triangulation of the problem domain (1600 elements).

Fig. 6. Ringleb flow test problem.

- Edge Quadratures: Integrate $2k+1$ order polynomials exactly
  - $k = 1$, 2 point Gaussian quadrature $O(h^4)$
  - $k = 2$, 3 point Gaussian quadrature $O(h^6)$
  - $k = 3$, 4 point Gaussian quadrature $O(h^8)$

In curved boundary domains, isoparametric (curved) element approximations are used. In this case, it is well known that the basic interpolation estimates hold for isoparametric elements if the element distortion is of order $O(h^2)$ or smaller, see Ciarlet and Raviart [14].

Remark 19. Note that super-isoparametric element approximations are actually used in conjunction with low order element approximation. In super-isoparametric element approximation the shape of the simplex is represented by a higher degree polynomial than is used for function approximation. The use of super-isoparametric elements is motivated by the discontinuous Galerkin results of Bassi and Rebay [10]. Numerical results shown later reveal that GLS computations using low order elements benefit similarly from super-isoparametric approximation.

Let $\boldsymbol{v}$ denote the exact solution and $\boldsymbol{v}^h$ the numerical approximation on a mesh with characteristic element size h, absolute solution errors have been computed for the Ringleb flow problem in the dimensionally consistent entropy norm

$$\|\boldsymbol{v} - \boldsymbol{v}^h\|_{\tilde{A}_0}^2 \equiv \int_{\Omega} (\boldsymbol{v} - \boldsymbol{v}^h)^T \tilde{A}_0 (\boldsymbol{v} - \boldsymbol{v}^h)\, dx$$

using $O(h^7)$ numerical quadrature on successively refined meshes with spacing $\{h, h/2, h/4\}$ containing approximately 1600, 6400, and 25600 elements. In Fig. 7 the effect of super-isoparametric approximation is examined using P_1 function representation and P_1, P_2, and P_3 shape representations together with simplified DG and GLS discretization. As this figure shows, the use of standard linear Lagrange elements for this curved boundary geometry degrades the overall accuracy to approximately $O(h^{3/2})$. Approximating the boundary element shape by quadratic and cubic polynomials increases the accuracy to $O(h^2)$ with a sizeable reduction in the absolute level of solution error when compared to the linear Lagrange element. The degradation in

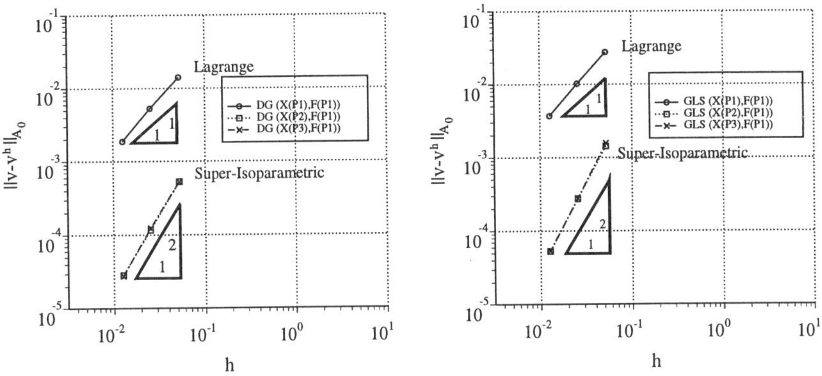

(a) Simplified discontinuous Galerkin. (b) Simplified Galerkin least-squares.

Fig. 7. The effect of super-isoparametric elements. Absolute solution error for Ringleb flow problem using P_1 function representation and P_1, P_2, and P_3 shape representation.

accuracy near the curved boundary using linear Lagrange elements is clearly observed in the DG and GLS isodensity contours shown in Figs. 8(a) and 9(a) when compared to the super-isoparametric results of Figs. 8(b) and 9(b). Next, higher order accurate computations of Ringleb flow have been computed using quadratic and cubic isoparametric element representations. Although not shown here, the effect of super-isoparametric approximation was found to be negligible for these higher order elements. Figure 10 shows the improved spatial convergence rates attained using P_2 and P_3 isoparametric elements. For both DG and GLS, the measured convergence rates fall slightly short of $O(h^3)$ and $O(h^4)$ using quadratic and cubic approximations respectively. Observe that the discontinuous Galerkin method produces lower levels of solution error using linear elements when compared to GLS. This is understandable given that the DG scheme uses more degrees of freedom than

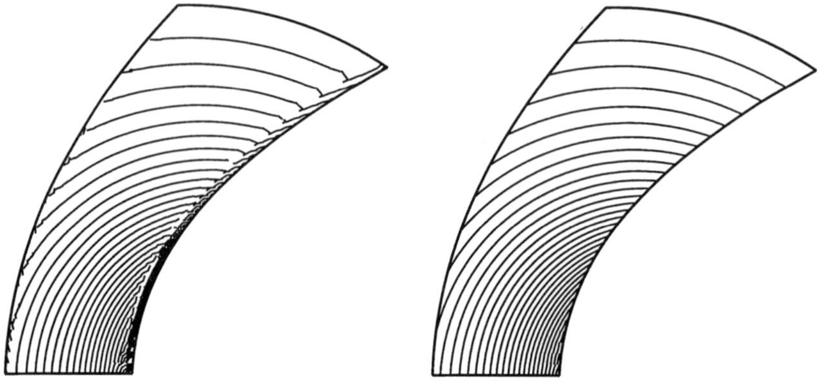

(a) Linear Lagrange element approximation, $(X(P_1), F(P_1))$. (b) Linear super-isoparametric element approximation, $(X(P_2), F(P_1))$.

Fig. 8. The effect of super-isoparametric elements (1600 element mesh). Isodensity contours for the DG discretization of Ringleb flow using Lagrange and super-isoparametric elements.

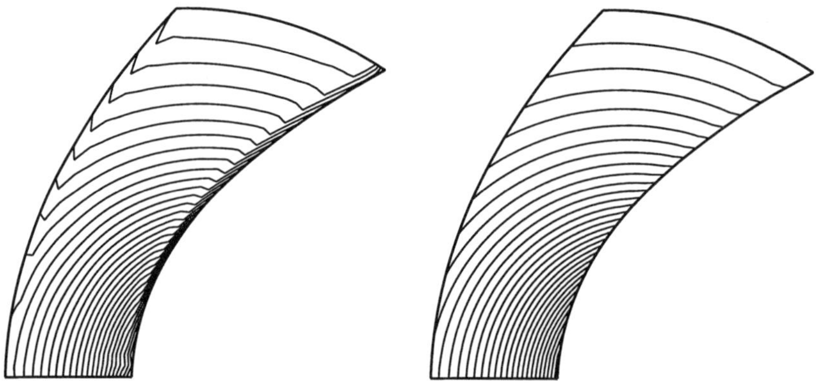

(a) Linear Lagrange element approximation, $(X(P_1), F(P_1))$. (b) Linear super-isoparametric element approximation, $(X(P_2), F(P_1))$.

Fig. 9. The effect of super-isoparametric elements (1600 element mesh). Isodensity contours for the GLS discretization of Ringleb flow using Lagrange and super-isoparametric elements.

GLS for the same 2D triangulation in the approximate ratios of $6, 3, 20/9$ for linear, quadratic, and cubic elements respectively. It is somewhat surprising that for quadratic and cubic elements, the GLS scheme attains similar or lower levels of error when compared with DG using the same mesh. It is conjectured that this could possibly be improved by adding least-squares stabilization of element interiors into the DG method.

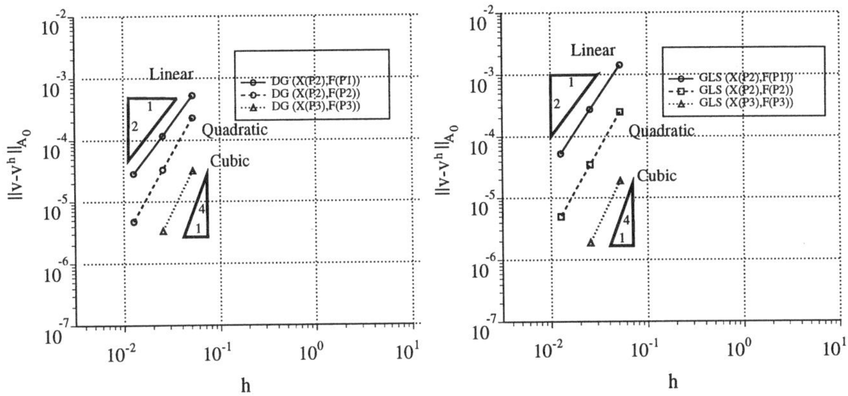

(a) Simplified discontinuous Galerkin. (b) Simplified Galerkin least-squares.

Fig. 10. Effect of higher order element approximation. Absolute solution error for the Ringleb flow problem using P_1, P_2, and P_3 elements.

2.9 Numerical Calculations Using Simplified GLS and DG

A number of example calculations will now be presented using the simplified DG and GLS discretization of Euler and MHD flow. Qualitative convergence of the simplified DG and GLS schemes is shown via mesh adaptation and/or higher order polynomial approximation.

Example 1: Subsonic Airfoil Flow As a first test example, subsonic Mach 0.6 flow over a NACA 0012 airfoil at 2° freestream angle of attack is computed on a relatively coarse triangulation containing 1500 elements. Figure 11(a) shows the coarse triangulation in the near airfoil region. Flow calculation were performed using DG and GLS with linear super-isoparametric $(X(P_2), F(P_1))$ elements, quadratic isoparametric $(X(P_2), F(P_2))$ elements, and cubic isoparametric $(X(P_3), F(P_3))$ elements. Figure 11(b) graphs the surface pressure coefficient distribution on the airfoil surface for DG and GLS solutions computed using linear super-isoparametric elements. The corresponding solution obtained using cubic isoparametric elements is included

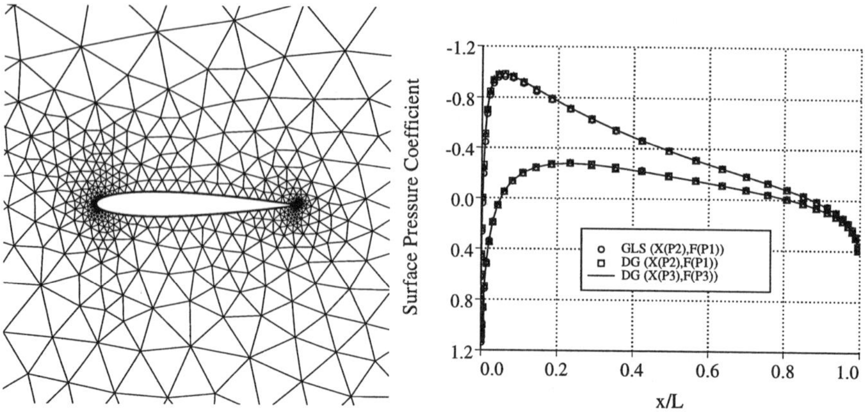

(a) Airfoil triangulation (1500 elements). (b) Surface pressure coefficient comparison.

Fig. 11. NACA 0012 airfoil test problem showing (a) coarse triangulation and (b) surface pressure coefficient comparison obtained from solutions computed using DG and GLS with linear $(X(P_2), F(P_1))$ super-isoparametric elements and DG with cubic $(X(P_3), F(P_3))$ isoparametric elements.

for comparison. Although this figure shows little or no differences between the different computations, the isoMach contour plots shown in Figs. 12 and 13 do show a noticeable improvement in both DG and GLS computed solution contours when the polynomial order is increased from linear to quadratic. The entropy distribution on the airfoil surface provides another discerning measure of accuracy. Assume that a constant freestream value of entropy is imposed. In this case, entropy remains constant everywhere in the domain. Since the freestream level of entropy can arbitrary be defined as zero by subtracting a suitable constant, surface entropy magnitude can be used as a measure of error. In Fig. 14, surface entropy distributions are plotted for simplified DG and GLS calculations using linear, quadratic, and cubic element approximations. Both DG and GLS show measurable improvement in surface entropy levels using higher order elements. While the improvement when using quadratic elements over linear elements is quite significant, the difference between quadratic and cubic element solutions is less impressive. To understand this, observe that the level of entropy on the airfoil surface is largely dictated by the initial entropy loss at the leading edge of the airfoil. The quadratic and cubic element solutions both produce roughly similar entropy losses at the airfoil leading edge stagnation point. Consequently, the overall improvement between quadratic and cubic elements is small. Close inspection of the geometry (not shown here) reveals that the cubic element approximation actually introduces spurious oscillations in the geometry near

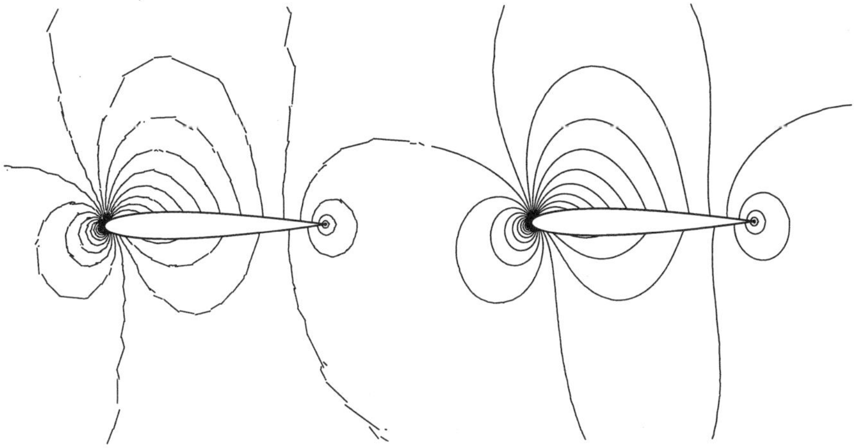

(a) DG isoMach contours using linear $(X(P_2), F(P_1))$ elements.

(b) DG isoMach contours using quadratic $(X(P_2), F(P_2))$ elements.

Fig. 12. IsoMach number solution contours obtained from the DG computation of subsonic flow over the NACA 0012 geometry.

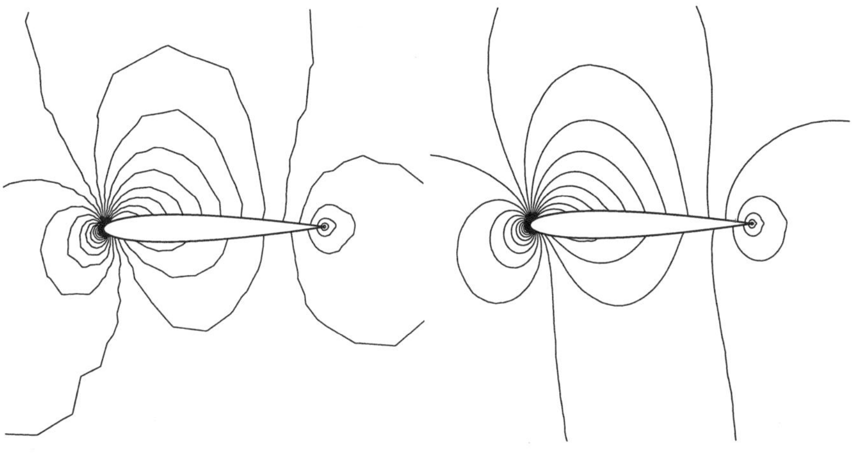

(a) GLS isoMach contours using linear $(X(P_2), F(P_1))$ elements.

(b) GLS isoMach contours using quadratic $(X(P_2), F(P_2))$ elements.

Fig. 13. IsoMach number contours computed obtained from the GLS computation of subsonic flow over the NACA 0012 geometry.

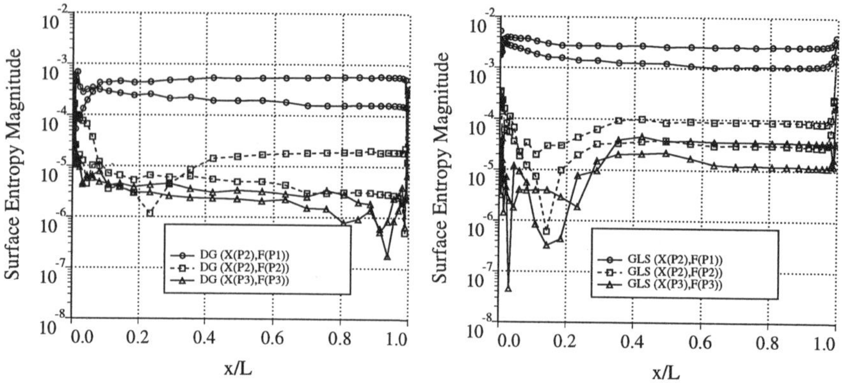

(a) DG surface entropy distribution. (b) GLS surface entropy distribution.

Fig. 14. Surface entropy distribution comparison using simplified DG and GLS with linear, quadratic, and cubic elements.

the leading edge which in turn degrades the solution accuracy. A remedy to problem has not yet been pursued.

Example 2: Simplified GLS Computation of Supersonic Flow Over a Step Geometry As a second example, Mach 3 supersonic Euler flow has been computed over a step geometry using the simplified GLS scheme with added discontinuity capturing operator (81). In this problem, the Mach 3 flow travels from left to right over a step geometry, see Fig. 15. The problem

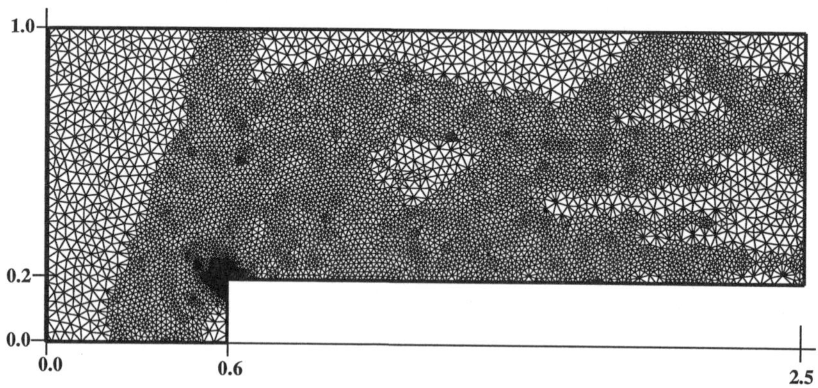

Fig. 15. Triangulation of step geometry with 1 level of conformal adaptive refinement (7592 vertices).

geometry and test conditions are similar to that given in Woodward and Collela [56] and Cockburn and Shu [18] but differs fundamentally in that time asymptotic solutions are sought rather than a time-accurate integration to a prescribed final time. This necessitates a truncation of the domain at 2.5 units, as shown in Fig. 15, so that the flow remains supersonic at outflow. The calculation of time asymptotic solutions was chosen so that large time step implicit integration techniques could be used. The computational mesh shown in Fig. 15 contains 7592 vertices and contains 1 level of adaptive refinement based on a heuristic density gradient criteria (not discussed here). The step corner creates a singularity in the solution. For this reason, the mesh is locally refined in the corner region. Other than this, no special modifications of the scheme or boundary conditions have been used in the corner region. Consequently, an entropy layer emanates from the corner geometry. Figure 16 shows isodensity contours of the solution computed using the simplified GLS scheme with linear P_1 elements. For comparison, Fig. 17 shows isodensity solution contours obtained using simplified GLS with quadratic P_2 elements on the same triangulation. The solution obtained using P_1 elements shows a

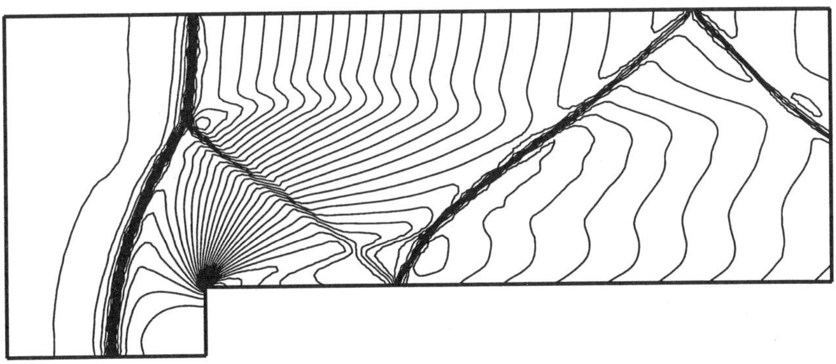

Fig. 16. Isodensity contours. Galerkin least-squares solution using P_1 (linear) elements on the adapted mesh shown in Fig. 15.

monotonic resolution of shockwaves. At supersonic Mach numbers, the GLS scheme with discontinuity capturing term and linear elements actually smears strong shocks over more cells than it does smear weak shocks. In practice, this is a beneficial numerical effect since strong shock wave are easily detected and narrowed via adaptation. The smearing effect keeps the strong shock fronts numerically smooth. Using quadratic elements, the shock front narrows to 1 or 2 cells. This causes small oscillations in the numerical solution because of the irregular triangulation. Qualitatively, the quadratic element solution provides significantly better resolution of the slip-layers emanating from the Mach triple point and the corner region. Figure 18 shows a triangulation

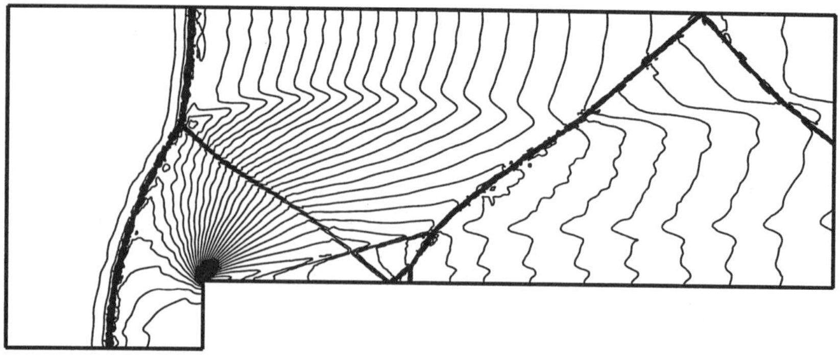

Fig. 17. Isodensity contours. Galerkin least-squares solution using P_2 (quadratic) elements on the adapted mesh shown in Fig. 15.

of the step geometry with 2 levels of conformal adaptation (22597 vertices). Isodensity contours are shown for the simplified GLS using linear elements in

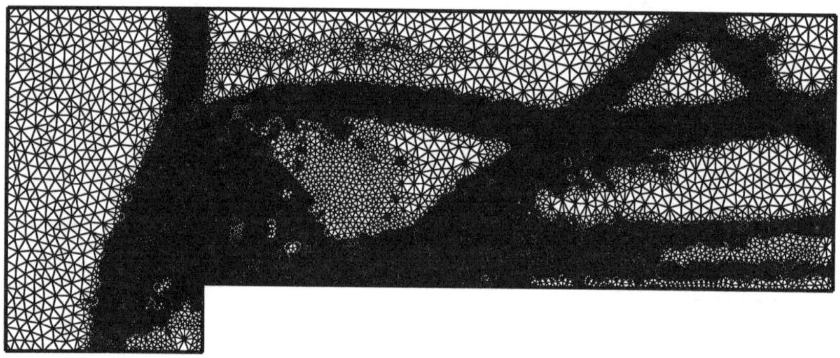

Fig. 18. Triangulation of step geometry with 2 levels of conformal adaptive refinement (22597 vertices).

Fig. 19 and quadratic elements in Fig. 20. Observe that the solution obtained using linear elements on the level 2 adapted mesh resolves roughly the same feature detail computed using quadratic elements on a level 1 adapted mesh. The solution obtained using quadratic elements on the level 2 adapted mesh resolves the various slip-layers over the entire extent of the mesh. Looking at the isodensity contours in Fig. 20, one would normally expect that the slip-layer emanating from the Mach triple point would eventually become unsteady. The present calculations fail to show the characteristic slip-layer

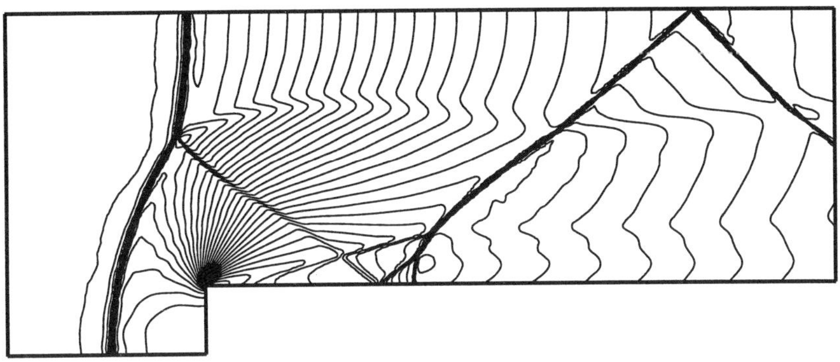

Fig. 19. Isodensity contours. Galerkin least-squares solution using P_1 (linear) elements on the partially adapted mesh in Fig. 18.

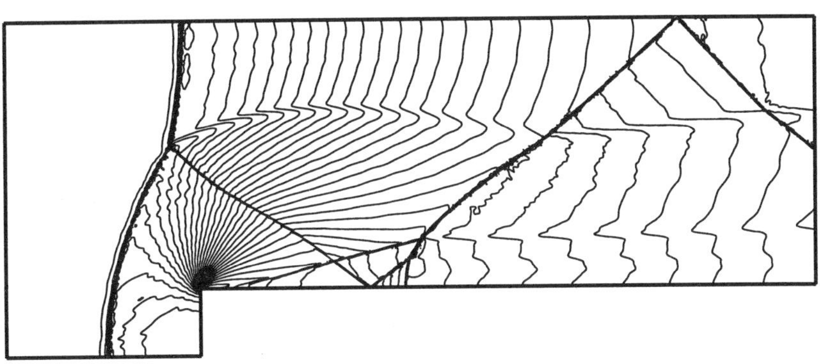

Fig. 20. Isodensity contours. Galerkin least-squares solution using P_2 (quadratic) elements on the partially adapted mesh in Fig. 18.

roll-up, see for example Cockburn and Shu [18]. This is believed to be due to the use of a backward Euler time integration with relatively large time step (corresponding to a CFL number of approximately 50-100). Apparently, the low accuracy in time suppresses the slip-layer instability.

Example 3: Simplified GLS Computation of Transonic Airfoil Flow In this test case, transonic flow is computed over the NACA 0012 airfoil geometry at Mach 0.8 with 1.25° flow angle of attack. Two meshes have been used in flow computations, see Fig. 21. The mesh in Fig. 21(a) contains approximately 8000 elements with clustering in the near airfoil region. Figure 21(b) shows a solution adapted mesh (16000 elements) obtained starting from the mesh in Fig. 21(a). Figure 22 graphs surface pressure coefficient

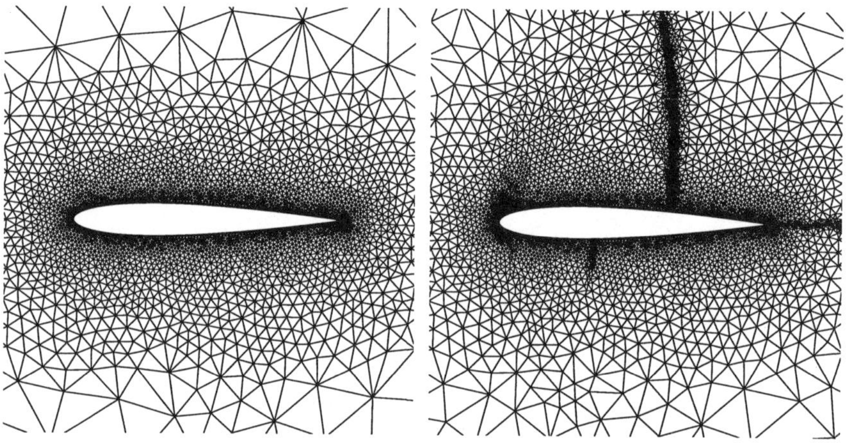

(a) Initial triangulation (8000 elements). (b) Adapted triangulation (16000 elements).

Fig. 21. NACA 0012 airfoil triangulations showing (a) initial surface clustering and (b) solution adaptive element refinement.

and surface entropy distributions obtained using the simplified GLS scheme with discontinuity capturing operator (81). The flow contains a strong upper surface shockwave and a weak lower surface shockwave. The initial mesh density is sufficiently refined so that all schemes resolve the upper and lower surface shockwaves. Consequently, the surface pressure coefficient distributions do not show significant differences between linear, quadratic, and cubic approximation. The surface entropy distributions are more revealing. In this case, a substantial improvement in the pre-shock level of entropy is observed with increasing element polynomial order. In addition, the entropy rise at the shock locations is narrowed considerably using higher order elements. Figures

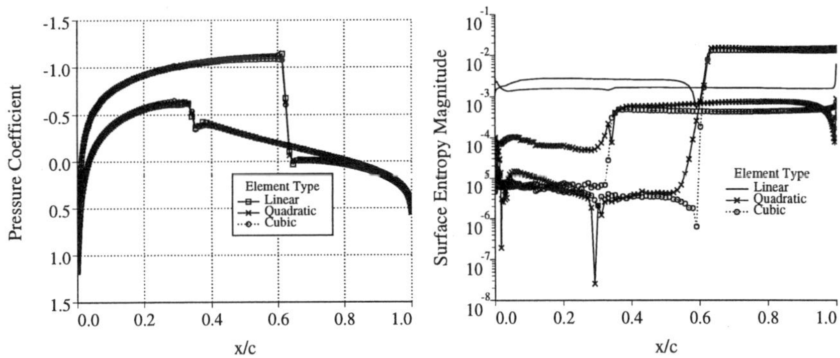

(a) Surface pressure coefficient distributions (initial triangulation).

(b) Surface entropy distributions (initial triangulation).

Fig. 22. Simplified GLS computation of transonic flow for the NACA 0012 geometry on the initial triangulation of Fig. 21(a). Comparison of (a) surface pressure coefficient and (b) surface entropy distributions using linear, quadratic, and cubic isoparametric elements.

23 and 24 show Mach number contours on the initial and adapted meshes using linear and quadratic elements. Comparing the initial and adapted mesh results, the benefits of mesh adaptation become clear. On coarse meshes, the irregular placement of elements in the shockwave transition regions results in rather noticeable solution perturbations which are then propagated downstream. In this situation, the use of high order methods only exacerbates the effect. Using mesh adaptation, these perturbations can be made arbitrarily small by sufficient local refinement. The resulting solutions obtained on adapted meshes show little or no mesh-induced solution perturbations downstream of the shockwaves. The use of high order methods with local adaptation at discontinuities appears to be a powerful combination.

Example 4: MHD Flow Computations of a Perturbed Prandtl-Meyer Fan In this example, the MHD equations are solved in the square domain $\Omega \in [-1/2, 1/2]^2$ using the simplified GLS scheme. The inflow data for this problem consists of a Mach 1.55 Prandtl-Meyer fan with origin located at $(x = -.881, y = .47147)$. A velocity aligned magnetic field, $\boldsymbol{B} = .05\boldsymbol{V}$, is introduced at the inflow boundary. The primary motivation for solving this problem is to investigate the convergence of the constraint condition $\nabla \cdot \boldsymbol{B} = 0$ as the mesh is refined and/or the element order is increased. Note that no additional stabilization, such as described in Sect. 2.7, has been added. Figure 25 shows the coarsest mesh used for the computation and Mach number contours computed from the simplified GLS scheme. Figure 26 shows contours of the x-component of the magnetic field using linear and quadratic elements

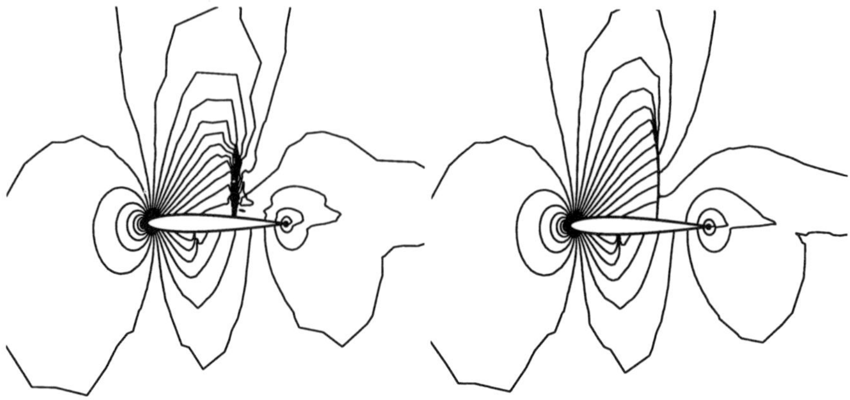

(a) Mach number contours (initial triangulation). (b) Mach number contours (adapted triangulation).

Fig. 23. Simplified GLS computation of transonic flow for the NACA 0012 geometry using linear $(X(P_2), F(P_1))$ super-isoparametric elements.

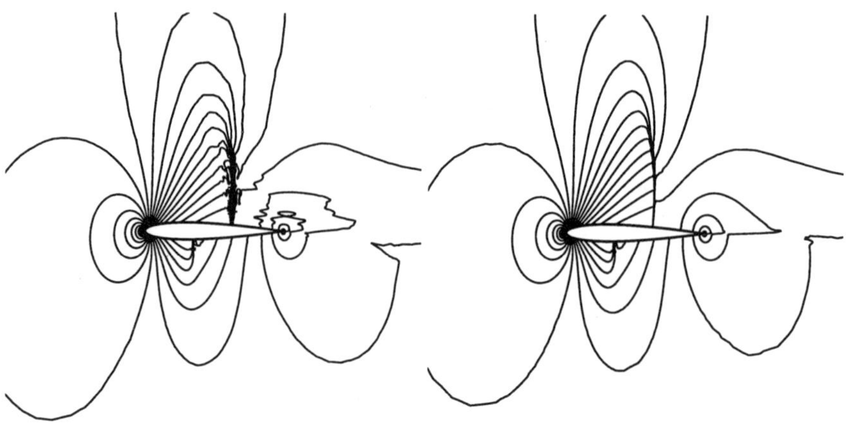

(a) Mach number contours (initial triangulation). (b) Mach number contours (adapted triangulation).

Fig. 24. Simplified GLS computation of transonic flow for the NACA 0012 geometry using quadratic $(X(P_2), F(P_2))$ isoparametric elements.

on the coarsest mesh. Solutions were then obtained on a sequence of meshes

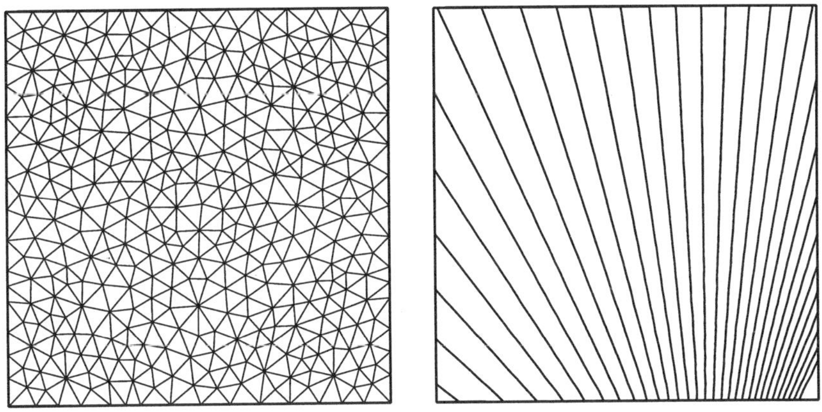

(a) Coarsest mesh (800 elements) (b) Mach number contours.

Fig. 25. MHD perturbed Prandtl-Meyer fan calculation using simplified GLS.

with decreasing mesh spacing and the L_2 norm of $\nabla \cdot \boldsymbol{B}$ measured. Figure 27 graphs the convergence of $\nabla \cdot \boldsymbol{B}$ for linear and quadratic approximations. The graphs show optimal convergence rates, $O(h)$ for linear and $O(h^2)$ for quadratic elements. Even with this result in hand, additional work is needed in this area to study the convergence properties for more general types of smooth solutions and solutions with embedded discontinuities.

Example 5: Supersonic MHD Flow Over a Circular Cylinder In this example, Mach 3 MHD flow is computed over a circular cylinder geometry. A velocity aligned magnetic field is imposed at inflow with magnitude $|\boldsymbol{B}_\infty|/(a_\infty\sqrt{\rho_\infty}) = .50$. The ratio of specific heats, γ, for the gas is fixed at 5/3. Simplified GLS calculations have been performed using the discontinuity capturing operator (81) as well as the MHD stabilization term (80) with coefficient $\epsilon_{\text{mhd2}} = .50$. Larger values of the MHD stabilization coefficient produced similar results and only served to degrade the performance of the implicit solution method. The solution adapted mesh shown in Fig. 28(a) contains approximately 18000 elements and was used in all calculations. Isodensity contours obtained from solutions computed using linear $(X(P_2), F(P_1))$ and quadratic $(X(P_2), F(P_2))$ elements are shown in Figs. 28(b-c). Even with modest local refinement, the irregular triangulation introduces small perturbations in the isodensity contours downstream of the bow shock. Additional local refinement reduces these triangulation-induced oscillations but makes the calculation rather expensive. Contours of the magnetic field magnitude

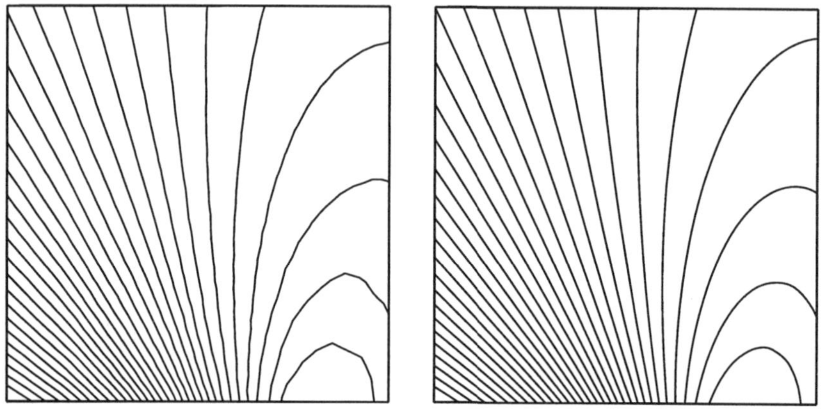

(a) B_x component of magnetic field (linear elements).

(b) B_x component of magnetic field (quadratic elements).

Fig. 26. Perturbed Prandtl-Meyer fan calculation. B_x component of magnetic field computed on coarsest mesh.

are shown in Fig. 29. Without the additional MHD stabilization term, rather large values of $\nabla \cdot \boldsymbol{B}$ are produced at the bow shock location and are advected down stream. The inclusion of the MHD stabilization term reduces this effect but does eliminate the spurious $\nabla \cdot \boldsymbol{B}$ altogether. As the magnetic field contours indicate, the magnetic field is relatively sensitive to irregularities in

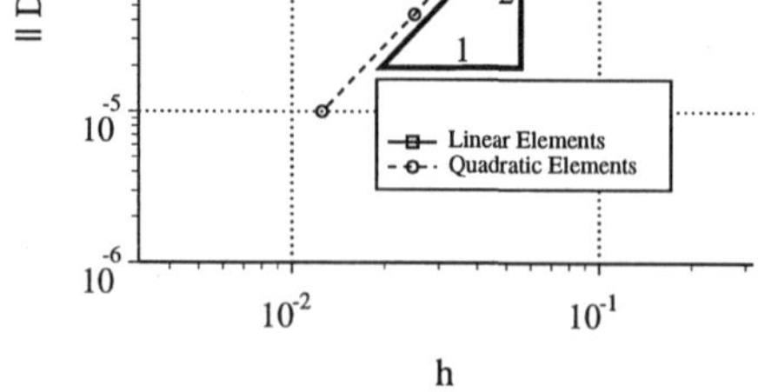

Fig. 27. Spatial convergence of $\nabla \cdot \boldsymbol{B}$ for MHD flow computed using the simplified GLS scheme with linear and quadratic elements.

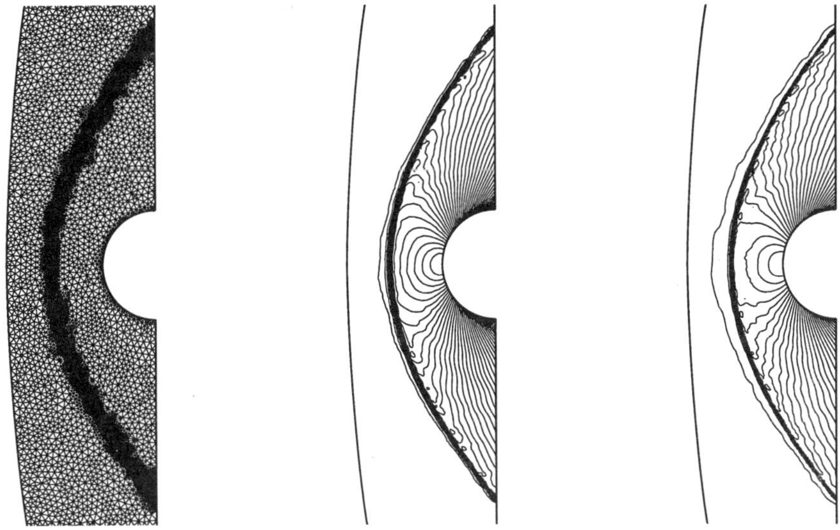

(a) Solution adapted mesh (18000 elements). (b) Isodensity contours (linear elements). (c) Isodensity contours (quadratic elements).

Fig. 28. MHD perturbed Prandtl-Meyer fan calculation using simplified GLS.

the bow shock profile. Since high order elements show increased sensitivity to mesh irregularities within the shock profile, the magnetic field contours computed using quadratic elements show only marginal improvement. Clearly, additional work in this area is needed and will be reported elsewhere.

3 Maximum Principles for Numerical Discretizations on Triangulated Domains

One of the best known tools employed in the study of differential equations is the maximum principle. Any function $f(x)$ which satisfies the inequality $f'' > 0$ on the interval $[a, b]$ attains its maximum value at one of the endpoints of the interval. Solutions of the inequality $f'' > 0$ are said to satisfy a maximum principle. Functions which satisfy a differential inequality in a domain Ω, and because of the form of the differential equation achieve a maximum value on the boundary $\partial\Omega$, are said to possess a maximum principle. Recall the maximum principle for Laplace's equation. Let $\Delta u \equiv u_{xx} + u_{yy}$ denote the Laplace operator. If a function u satisfies the strict inequality

$$\Delta u > 0 \tag{82}$$

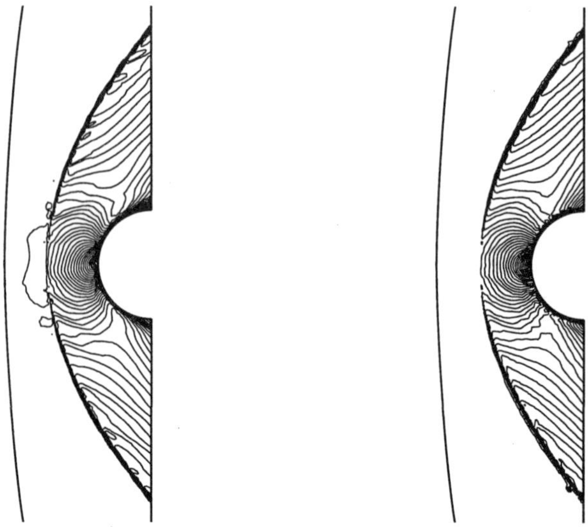

(a) Magnetic field magnitude contours (linear elements).

(b) Magnetic field magnitude contours (quadratic elements).

Fig. 29. Contours of constant magnetic field magnitude. MHD perturbed Prandtl-Meyer fan calculation using simplified GLS.

at each point in Ω, then u cannot attain its maximum at any interior point of Ω. The strict inequality can be weakened

$$\Delta u \geq 0 \tag{83}$$

so that if a maximum value M is attained in the interior of Ω then the entire function must be a constant with value M. Without any change in the above argument, if u satisfies the inequality

$$\Delta u + c_1 u_{,x} + c_2 u_{,y} > 0$$

in Ω, then u cannot attain its maximum at an interior point.

The second model equation of interest is the nonlinear conservation law equation

$$u_{,t} + (f(u))_{,x} = 0, \quad \frac{df}{du} = a(u) \ . \tag{84}$$

In the simplest setting, the initial value problem is considered in which the solution is specified along the x-axis, $u(x,0) = u_0(x)$ in a periodic or compact supported fashion. The solution can be depicted in the $x-t$ plane by a series of converging and diverging characteristic straight lines. Looking at smooth solutions of (84), Lax provides the following observation: *the total increasing and decreasing variations of a differentiable solution between any pairs of characteristics are conserved.* [42]. Moreover, in the presence of entropy satisfying discontinuities, the total variation decreases (information is destroyed) in time

$$\mathcal{I}(t+t_0) \leq I(t_0), \quad \mathcal{I}(t) = \int_{-\infty}^{+\infty} \left| \frac{\partial u(x,t)}{\partial x} \right| dx \ . \tag{85}$$

An equally important consequence of Lax's observation comes from considering monotonic solution data between two non-intersecting characteristics: *between pairs of characteristics, monotonic solutions remain monotonic*, no new extrema are created. It also follows that

1. local maxima are nonincreasing
2. local minima are nondecreasing.

When used in the design of numerical schemes, this property is sometimes referred to as the Local Extremum Diminishing (LED) condition [35].

These properties of the differential equations serve as basic design principles for numerical schemes which approximate them.

3.1 Discrete Maximum Principles for Elliptic Equations

Laplace's Equation on Structured Meshes Consider 2-D Laplace's equation with Dirichlet data

$$\Delta u = 0, \quad (x,y) \in \Omega$$

$$u = g, \quad (x,y) \in \partial\Omega \ . \tag{86}$$

From the maximum principle property, it follows that

$$\sup_{(x,y)\in\Omega} |u(x,y)| \leq \sup_{(x,y)\in\partial\Omega} |u(x,y)| \ .$$

For simplicity, consider the unit square domain

$$\Omega = \{(x,y) \in \mathbb{R}^2 : 0 \leq x,y \leq 1\}$$

with spatial grid $x_j = j\Delta x$, $y_k = k\Delta x$, and $J\Delta x = 1$. Let $U_{j,k}$ denote the numerical approximation to $u(x_j, y_k)$. It is well known that the standard second order accurate approximation

$$\mathcal{L}_\Delta U = \frac{1}{\Delta x^2}[U_{j+1,k} + U_{j-1,k} + U_{j,k+1} + U_{j,k-1} - 4U_{j,k}] \tag{87}$$

exhibits a discrete maximum principle. To see this simply solve for the value at (j,k)

$$U_{j,k} = \frac{1}{4}[U_{j+1,k} + U_{j-1,k} + U_{j,k+1} + U_{j,k-1}] \ .$$

If $U_{j,k}$ achieves a maximum value M in the interior then

$$M = \frac{1}{4}[U_{j+1,k} + U_{j-1,k} + U_{j,k+1} + U_{j,k-1}]$$

which implies that

$$M = U_{j+1,k} = U_{j-1,k} = U_{j,k+1} = U_{j,k-1} \ .$$

Repeated application of this argument for the four neighboring points yields a discrete maximum principle.

Monotone Matrices The discrete Laplacian operator $-\mathcal{L}_\Delta$ obtained from (87) is one example of a monotone matrix. A matrix $\mathcal{M}$ is a monotone matrix if and only if $\mathcal{M}^{-1} \geq 0$ (all entries are nonnegative).

Theorem 20 (Monotone Matrix). *A sufficient but not necessary condition for $\mathcal{M}$ monotone is that $\mathcal{M}$ be an M-matrix. M-matrices have the sign pattern $\mathcal{M}_{ii} > 0$ for each i, $\mathcal{M}_{ij} \leq 0$ whenever $i \neq j$. In addition $\mathcal{M}$ must either be strictly diagonally dominant*

$$\text{(strict diagonal dominance)} \qquad \mathcal{M}_{ii} > \sum_{j=1, j\neq i}^{n} |\mathcal{M}_{ij}|, \quad i = 1,2,...,n$$

or else $\mathcal{M}$ must be irreducible and

$$\text{(diagonal dominance)} \qquad \mathcal{M}_{ii} \geq \sum_{j=1, j\neq i}^{n} |\mathcal{M}_{ij}|, \quad i = 1,2,...,n$$

with strict inequality for at least one i.

Proof. The proof for strictly diagonally dominant M is straightforward. Rewrite the matrix operator in the following form

$$\begin{aligned} \mathcal{M} &= D - N, \qquad D > 0, N \geq 0 \\ &= [I - ND^{-1}]D \qquad D^{-1} > 0 \\ &= [I - P]D \qquad P \geq 0 \ . \end{aligned} \tag{88}$$

From the strict diagonal dominance of $\mathcal{M}$, eigenvalues of $P = ND^{-1}$ are less than unity. This implies that the Neumann series for $[I-P]^{-1}$ is convergent. This yields the desired result:

$$\mathcal{M}^{-1} = D^{-1}[I + P + P^2 + P^3 + ...] \geq 0 \ . \tag{89}$$

When $\mathcal{M}$ is not strictly diagonally dominant then $\mathcal{M}$ must be irreducible so that no permutation $\mathcal{P}$ exists such that

$$\mathcal{P}^T \mathcal{M} \mathcal{P} = \begin{bmatrix} \mathcal{M}_{11} & \mathcal{M}_{12} \\ 0 & \mathcal{M}_{22} \end{bmatrix} \quad \text{(reducibility)}.$$

This insures that eigenvalues of P are less than unity. Once again, the Neumann series is convergent and the final result follows immediately. □

Laplace's Equation on Triangulated Meshes
Consider solving the Laplace equation problem (86) on a planar triangulation using a Galerkin finite element approximation with linear elemental shape functions. (Results using a finite volume method are identical but are not considered here.) Next pose (86) in variational form. Let $\mathcal{S}^h \in H^1$ be the finite-dimensional space of trial functions with bounded energy which satisfy the Dirichlet boundary condition on Γ. Similarly, let $\mathcal{V}^h \in H^1_0$ denote the finite-dimensional space of functions satisfying homogeneous boundary conditions. Find $u^h \in \mathcal{S}^h$ such that for all $w^h \in \mathcal{V}^h$

$$\int_\Omega (\nabla u^h \cdot \nabla w^h)\, dx = 0 \tag{90}$$

with

$$u^h(x) = g(x), \quad x \in \Gamma.$$

From this simple equation, the following remarkable lemma is obtained:

Lemma 21. *The Galerkin finite element discretization of the 2-D Laplace equation (90) using linear elements is a monotone discretization if and only if the triangulation is a Delaunay triangulation.*

Proof. Consider a single arbitrary simplex $T = \text{Simplex}(x_1, x_2, x_3)$ and the discretization of (90) in terms of the local linear shape functions $N_i(x)$ satisfying $N_i(x_j) = \delta_{ij}$. Using these shape functions, u^h and w^h are given by $u^h(x) = \sum_{j=1}^3 N_j(x) u_j,\ x \in T$ and $w^h(x) = \sum_{j=1}^3 N_j(x) w_j,\ x \in T$. Inserting these expressions into (90) yields

$$\int_T \nabla w^h \cdot \nabla u^h \, dx = \sum_{i=1}^{3} \sum_{j=1}^{3} w_i \; u_j \; (\nabla N_i \cdot \nabla N_j) \; \text{meas}(T) \; . \tag{91}$$

These expressions can be collected pairwise for edges surrounding a vertex. After some straightforward manipulation, the following global discretization formula is obtained

$$\int_\Omega (\nabla w^h \cdot \nabla u^h)\, dx = \sum_{i=1}^{|V|} w_i \sum_{j \in \mathcal{N}_i} W_{ij} \; (u_i - u_j) = 0 \tag{92}$$

where $\mathcal{N}_i$ denotes the set of vertices adjacent to vertex v_i with weights

$$W_{ij} = (\nabla N_i \cdot \nabla N_j)\,\text{meas}(T) + (\nabla N_i' \cdot \nabla N_j')\,\text{meas}(T') \\ = \frac{1}{2}\left(\text{cotan}(\alpha_{ij}) + \text{cotan}(\alpha_{ij}')\right). \tag{93}$$

In this formula, α_{ij} and α_{ij}' are the two angles subtending the edge $e(v_i, v_j)$ as shown in Fig. 30. Since the discretization formula must hold for arbitrary

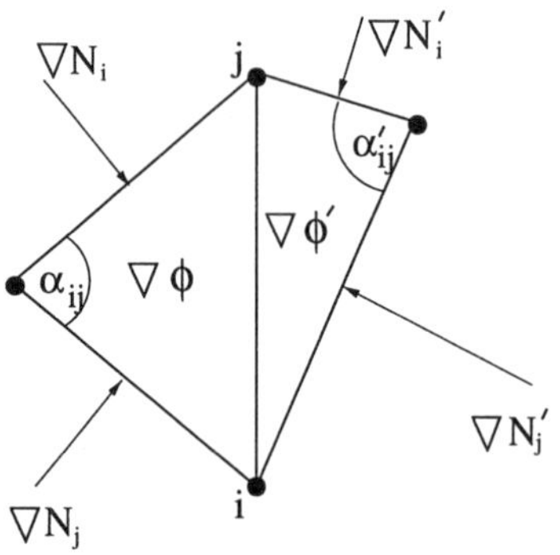

Fig. 30. Discretization weight geometry for the edge $e(v_i, v_j)$.

values of w_i at interior vertices, it can be concluded that for all interior vertices v_i

$$\sum_{j \in \mathcal{N}_i} W_{ij}\ (u_i - u_j) = 0\ . \tag{94}$$

Written in this form, the discretization is monotone if all weights are nonnegative, $W_{ij} \geq 0$. Further simplification of the edge weight formula is possible

$$W_{ij} = \frac{1}{2}\left(\text{cotan}(\alpha_{ij}) + \text{cotan}(\alpha_{ij}')\right) \\ = \frac{1}{2}\left(\frac{\cos(\alpha_{ij})}{\sin(\alpha_{ij})} + \frac{\cos(\alpha_{ij}')}{\sin(\alpha_{ij}')}\right) \\ = \frac{1}{2}\left(\frac{\sin(\alpha_{ij} + \alpha_{ij}')}{\sin(\alpha_{ij})\sin(\alpha_{ij}')}\right)\ . \tag{95}$$

Since $\alpha_{ij} < \pi$, $\alpha_{ij}' < \pi$, the denominator is always positive. Hence, nonnegativity requires that $\alpha_{ij} + \alpha_{ij}' \leq \pi$. Some trigonometry reveals that for the configuration shown in Fig. 31 with circumcircle passing through $\{v_i, v_j, v_k\}$, the sum $\alpha_{ij} + \alpha_{ij}'$ depends on the location of p with respect to the circumcircle in the following way:

$$\alpha_{ij} + \alpha_{ij}' < \pi, \quad p \text{ exterior}$$

$$\begin{aligned} \alpha_{ij} + \alpha'_{ij} &> \pi, \quad p \text{ interior} \\ \alpha_{ij} + \alpha'_{ij} &= \pi, \quad p \text{ cocircular} . \end{aligned} \tag{96}$$

Also note that by considering the circumcircle passing through $\{v_i, v_j, v_p\}$,

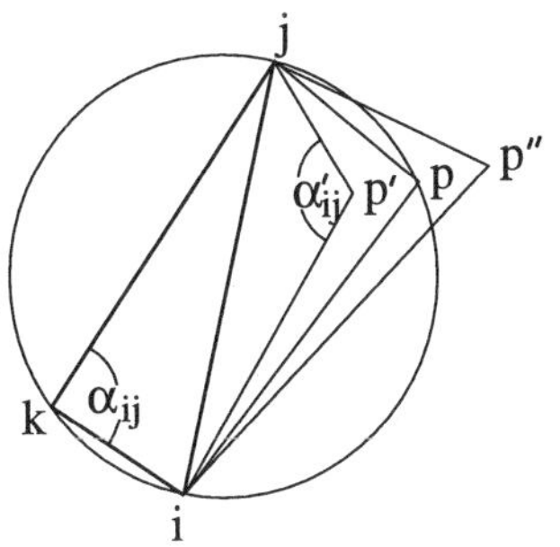

Fig. 31. Circumcircle test for adjacent triangles, p' interior, p'' exterior, p cocircular.

similar results would be obtained for v_k. The condition of nonnegativity implies a circumcircle condition for all pairs of adjacent triangles whereby the circumcircle passing through either triangle cannot contain the fourth point. This is precisely the *unique* characterization of the Delaunay triangulation [21] which completes the proof. □

Observe from equation (93) that $\text{cotan}(\alpha) \geq 0$ if $\alpha \leq \pi/2$. Therefore, a sufficient but not necessary condition for nonnegativity of the Laplacian weights is that all angles of the mesh be less than or equal to $\pi/2$. This is a standard result in finite element theory [15] and applies in two or more space dimensions.

Given lemma 21, it becomes straightforward to obtain a discrete maximum principle for Laplace's equation on Delaunay triangulations using a Galerkin finite element approximation.

Theorem 22. *The discrete Laplacian operator obtained from the Galerkin finite element discretization of (90) with linear elements exhibits a discrete maximum principle for arbitrary point sets in 2-D if the triangulation of these points is a Delaunay triangulation.*

Proof. From lemma 21, a one-to-one correspondence exists between nonnegativity of weights and Delaunay triangulation. Assume a Delaunay triangulation of the point set so that for some arbitrary interior vertex v_0 and all adjacent neighbors v_i the weights $W_{0i} \geq 0$. Next solve for u_0

$$u_0 = \frac{\sum_{i \in \mathcal{N}_0} W_{0i} u_i}{\sum_{i \in \mathcal{N}_0} W_{0i}} = \sum_{i \in \mathcal{N}_0} \sigma_i u_i$$

with

$$\sigma_i = \frac{W_{0i}}{\sum_{i \in \mathcal{N}_0} W_{0i}}$$

which satisfies $\sigma_i \geq 0$ and $\sum_{i \in \mathcal{N}_0} \sigma_i = 1$. This implies u_0 is a convex combination of the neighboring values, therefore

$$\min_{i \in \mathcal{N}_0} u_i \leq u_0 \leq \max_{i \in \mathcal{N}_0} u_i \ . \tag{97}$$

If u_0 attains a maximum value M then all $u_i = M$. Repeated application of (97) to neighboring vertices in the triangulation establishes the discrete maximum principle. □

One can ask if the result concerning Delaunay triangulation and the maximum principle extends to three space dimensions. Unfortunately, the answer is no. This can be demonstrated by counterexample. The resulting formula for the three-dimensional Laplacian edge weight is

$$W_{0i} = \frac{1}{6} \sum_{k=1}^{d(v_0,v_i)} |\Delta \boldsymbol{R}_{k+\frac{1}{2}}| \operatorname{cotan}(\alpha_{k+\frac{1}{2}}) \ . \tag{98}$$

In this formula, $\mathcal{N}_0$ is the set of indices of all adjacent neighbors of v_0

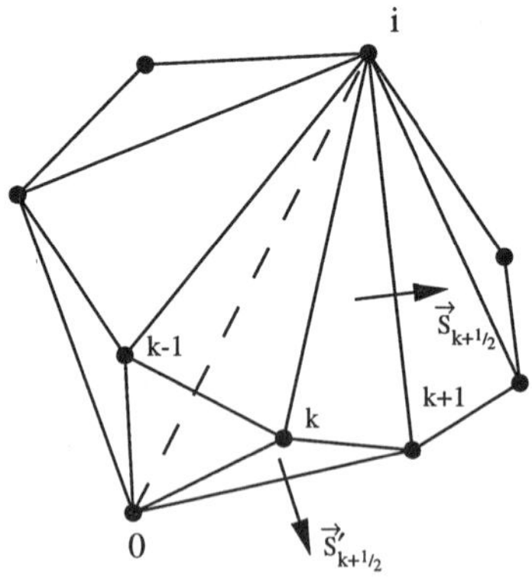

Fig. 32. Set of tetrahedra sharing interior edge $e(v_0, v_i)$ with local cyclic index k.

connected by incident edges, k a local cyclic index describing the associated vertices which form a polygon of degree $d(v_0, v_i)$ surrounding the edge $e(v_0, v_i)$, $\alpha_{k+\frac{1}{2}}$ is the face angle between the two faces associated with $\vec{S}_{k+\frac{1}{2}}$ and $\vec{S}'_{k+\frac{1}{2}}$ which share the edge $e(v_k, v_{k+1})$ and $|\Delta \boldsymbol{R}_{k+\frac{1}{2}}|$ is the magnitude of the edge, see Fig. 32. A maximum principle is guaranteed if all $W_{0i} \geq 0$. Unfortunately, it is possible to describe a valid Delaunay triangulation with one or more $W_{0i} < 0$. It will suffice to consider the Delaunay triangulation of N points in which a single point v_0 lies interior to the triangulation and the remaining $N-1$ points describe vertices of boundary faces which completely cover the convex hull of the point set. Consider a subset of the N vertices, in

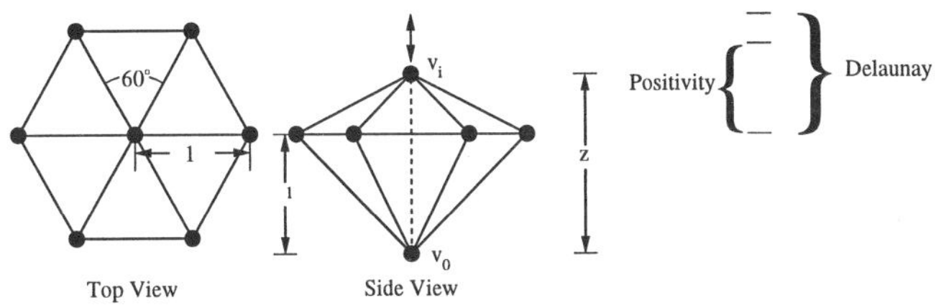

Fig. 33. Subset of 3-D Delaunay Triangulation that fails to maintain nonnegativity.

particular consider an interior edge incident to v_0 connecting to v_i as shown in Fig. 33 by the dashed line segment and all neighbors adjacent to v_i on the hull of the point set. In this experiment the height of the interior edge, z, serves as a free parameter. Although it will not be proven here, the remaining $N-8$ points can be placed without conflicting with any of the conclusions obtained for looking at the subset.

It is known that a necessary and sufficient condition for the 3-D Delaunay triangulation is that the circumsphere passing through the vertices of any tetrahedron must be point free [41]; that is to say that no other point of the triangulation can lie interior to this sphere. Furthermore, a property of locality exists so that only adjacent tetrahedra need be inspected for the satisfaction of the circumsphere test. For the configuration of points shown in Fig. 33, convexity of the point cloud constrains $z \geq 1$ and the satisfaction of the circumsphere test requires that $z \leq 2$

$$1 \leq z \leq 2. \quad \text{(Delaunay Triangulation)}$$

From (98), $W_{0i} \geq 0$ if and only if $z < 7/4$

$$1 \leq z \leq \frac{7}{4}. \quad \text{(Nonnegativity)}$$

This indicates that for $7/4 < z \leq 2$ a valid Delaunay triangulation exists which does not satisfy a discrete maximum principle. In fact, the Delaunay triangulation of 400 points randomly distributed in the unit cube revealed that approximately 25% of the interior edge weights were of the wrong sign (negative).

Keep in mind that from (98) the sufficient but not necessary condition for nonnegativity that all face angles be less than or equal to $\pi/2$. This is consistent with the known result from [15].

3.2 Discrete Total Variation and Maximum Principles for Hyperbolic Equations

This section examines discrete total variation and maximum principles for scalar conservation law equations. Begin by considering the nonlinear conservation law equation

$$u_{,t} + (f(u))_{,x} = 0, \ u(x,0) = u_0(x), \quad x,t \in \mathbb{R} \times \mathbb{R}^+$$

which is discretized in the conservation form

$$\begin{aligned} U_j^{n+1} &= U_j^n - \frac{\Delta t}{\Delta x}(h^n_{j+\frac{1}{2}} - h^n_{j-\frac{1}{2}}) \\ &= H(U^n_{j-l}, U^n_{j-l+1}, ..., U^n_{j+l}) \end{aligned} \tag{99}$$

where $h^n_{j+\frac{1}{2}} = h(U^n_{j-l+1}, ..., U^n_{j+l})$ is the numerical flux function satisfying the consistency condition

$$h(U, U, ..., U) = f(U) \ .$$

A finite-difference scheme (99) is said to be *monotone* in the sense of Harten, Hyman, and Lax [31] if H is a monotone increasing function of each of its arguments.

$$\text{(HHL monotonicity)} \qquad \frac{\partial H}{\partial U_i}(U_{-k}, ..., U_k) \geq 0 \ \ \forall \ -k \leq i \leq k \ .$$

This is a strong definition of monotonicity. In Crandall and Majda [20] it is proven that schemes on Cartesian grids satisfying this condition converge to the physically relevant, entropy satisfying solution. Kröner el al. [39] have recently proven a similar result for monotone upwind finite volume schemes on triangulated domains. Unfortunately, HHL monotone schemes in conservation form are at most first order spatially accurate. Very few results are known concerning the convergence of high order accurate approximations. Johnson and Szepessy [38] have shown convergence to entropy solutions using streamline diffusion with specialized shock capturing operators. Kröner et al. [40] have obtained measure-valued convergence of higher order upwind finite volume schemes for scalar conservation laws in several space dimensions.

To circumvent the first order accuracy of monotone schemes, Harten introduced a weaker concept of monotonicity. A grid function U is called monotone if for all i

$$\min(U_{i-1}, U_{i+1}) \leq U_i \leq \max(U_{i-1}, U_{i+1}) \ .$$

A scheme is called monotonicity preserving if monotonicity of U^{n+1} follows from monotonicity of U^n. Observe the close relationship between monotonicity preservation in time and the discrete maximum principle for Laplace's equation (97) in space. It follows immediately from the definition of monotonicity preservation that

1. local maxima are nonincreasing
2. local minima are nondecreasing

which is a known property of the conservation law equation. Using this weaker form of monotonicity, Harten [29] introduced the notion of total variation diminishing schemes. Define the total variation in one dimension as

$$TV(U) = \sum_{-\infty}^{\infty} |U_i - U_{i-1}| \ .$$

A scheme is said to be total variation diminishing (TVD) if

$$TV(U^{n+1}) \le TV(U^n)$$

This is a discrete analog of the total variation statement (85) given for the conservation law equation. Harten has proven that schemes which are HHL monotone are TVD and schemes that are TVD are monotonicity preserving. Furthermore, it can be shown that all *linear* monotonicity preserving schemes are at most first order accurate. Thus high order accurate TVD schemes must necessarily be nonlinear.

Unfortunately, two motivations suggest a further weakening of the concept of monotonicity. The first motivation concerns a negative result by Goodman and Le Veque [28] that conservative TVD schemes on Cartesian meshes in two space dimensions are first order accurate. The second motivation is the apparent difficulty in extending the TVD concept to arbitrary unstructured meshes. The first motivation inspired Spekreijse [50] to develop a new class of local extremum diminishing schemes based on the idea of positivity of coefficients. The following section considers these schemes. The generalization to unstructured meshes is then given.

3.3 Maximum Principles and Local Extremum Diminishing Schemes for Hyperbolic Equations on Multidimensional Structured Meshes

Consider the following conservation law equation in two space dimensions

$$u_{,t} + (f(u))_{,x} + (g(u))_{,y} = 0 \ . \tag{100}$$

Next, construct a discretization of (100) on a logically rectangular mesh

$$\frac{U^{n+1}_{j,k} - U^n_{j,k}}{\Delta t} = A^+_{j+\frac{1}{2},k}(U^n_{j+1,k} - U^n_{j,k}) + A^-_{j-\frac{1}{2},k}(U^n_{j-1,k} - U^n_{j,k})$$

$$+ B^{+}_{j,k+\frac{1}{2}}(U^n_{j,k+1} - U^n_{j,k}) + B^{-}_{j,k-\frac{1}{2}}(U^n_{j,k-1} - U^n_{j,k}) \quad (101)$$

with *nonlinear* coefficients

$$A^{\pm}_{j+\frac{1}{2},k} = A(..., U^n_{j-1,k}, U^n_{j,k}, U^n_{j+1,k}, ...)$$

$$B^{\pm}_{j-\frac{1}{2},k} = B(..., U^n_{j-1,k}, U^n_{j,k}, U^n_{j+1,k}, ...) \ .$$

Theorem 23. *The scheme (101) exhibits a discrete maximum principle at steady-state if all coefficients are uniformly bounded and nonnegative*

$$A^{\pm}_{j\pm\frac{1}{2},k} \geq 0 \quad B^{\pm}_{j\pm\frac{1}{2},k} \geq 0 \ .$$

Furthermore, the scheme (101) is local extremum diminishing under a CFL-like condition

$$\Delta t \leq \min_{\forall j\, k} \frac{1}{\sum_{\pm}(A^{\pm}_{j\pm\frac{1}{2},k} + B^{\pm}_{j,k\pm\frac{1}{2}})} \ .$$

Proof. The first task is to prove a discrete maximum principle at steady-state by solving for the value at (j,k)

$$\begin{aligned} U_{j,k} &= \frac{\sum_{\pm}(A_{j\pm\frac{1}{2},k} U_{j\pm1,k} + B_{j,k\pm\frac{1}{2}} U_{j,k\pm1})}{\sum_{\pm}(A_{j\pm\frac{1}{2},k} + B_{j,k\pm\frac{1}{2}})} \\ &= \sum_{\pm}(\alpha_{j\pm\frac{1}{2},k} U_{j\pm1,k} + \beta_{j,k\pm\frac{1}{2}} U_{j,k\pm1}) \end{aligned} \quad (102)$$

with the constraints

$$\alpha_{j-\frac{1}{2},k} + \alpha_{j+\frac{1}{2},k} + \beta_{j,k-\frac{1}{2}} + \beta_{j,k+\frac{1}{2},k} = 1,$$

$\alpha_{j\pm\frac{1}{2},k} \geq 0$, and $\beta_{j,k\pm\frac{1}{2}} \geq 0$. Requiring positivity of all coefficients implies convexity of (102), it then follows that

$$\min(U_{j\pm1,k}, U_{j,k\pm1}) \leq U_{j,k} \leq \max(U_{j\pm1,k}, U_{j,k\pm1}) \ . \quad (103)$$

If $U_{j,k}$ attains a maximum value M at (j,k), then

$$M = U_{j-1,k} = U_{j+1,k} = U_{j,k-1} = U_{j,k+1} \ .$$

Repeated application of (103) to neighboring mesh points establishes the maximum principle.

Next, it is straightforward to establish a CFL-like condition such that the scheme exhibits a local extremum diminishing property in time by again seeking positivity of coefficients and a convex local mapping from U^n to U^{n+1}.

$$U^{n+1}_{j,k} = \left(1 - \Delta t \sum_{\pm}(A^{\pm}_{j\pm\frac{1}{2},k} + B^{\pm}_{j,k\pm\frac{1}{2}})\right) U^n_{j,k} + \Delta t \sum_{\pm}(A_{j\pm\frac{1}{2},k} U^n_{j\pm1,k}$$

$$+ B_{j,k\pm\frac{1}{2}} U^n_{j,k\pm 1}) = \gamma_{j,k} U^n_{j,k} + \sum_{\pm} (\alpha_{j\pm\frac{1}{2},k} U^n_{j\pm 1,k} + \beta_{j,k\pm\frac{1}{2}} U^n_{j,k\pm 1}) \tag{104}$$

with the derivable constraints

$$\gamma_{j,k} + \alpha_{j-\frac{1}{2},k} + \alpha_{j+\frac{1}{2},k} + \beta_{j,k-\frac{1}{2}} + \beta_{j,k+\frac{1}{2},k} = 1$$

and $\alpha_{j\pm\frac{1}{2},k} \geq 0$, and $\beta_{j,k\pm\frac{1}{2}} \geq 0$. Equation (104) is a local convex mapping under the CFL-like condition for nonnegativity of $\gamma_{j,k}$

$$\Delta t \leq \min_{\forall j\, k} \frac{1}{\sum_{\pm}(A^{\pm}_{j\pm\frac{1}{2},k} + B^{\pm}_{j,k\pm\frac{1}{2}})} \tag{105}$$

so that the local extremum diminishing property is established

$$\min(U^n_{j\pm 1,k}, U^n_{j,k\pm 1}, U^n_{j,k}) \leq U^{n+1}_{j,k} \leq \max(U^n_{j\pm 1,k}, U^n_{j,k\pm 1}, U^n_{j,k}) \ .$$

□

3.4 Maximum Principles and Local Extremum Diminishing Schemes for Hyperbolic Equations on Triangulated Meshes

This section examines the maximum principle theory for conservation laws on unstructured meshes. Specifically, our primary attention focuses on Godunov-like upwind finite volume schemes [25] utilizing solution reconstruction and evolution. Some maximum principle results for upwind finite volume schemes can be found in [22,9,6]. Note that many of these results were subsequently used in implementations of the discontinuous Galerkin method as well, see for example Bey [11]. Also note that the present analysis differs from maximum principle theory based on the "upwind triangle" scheme developed by Jameson [35], Desideri and Dervieux [22], Arminjon and Dervieux [3].

Consider the integral conservation law form of (100) for some domain Ω comprised of nonoverlapping control volumes, Ω_i, such that $\Omega = \cup \Omega_i$ and $\Omega_i \cap \Omega_j = 0, i \neq j$. Next, impose the integral conservation law statement on each control volume

$$\frac{\partial}{\partial t} \int_{\Omega_i} u \, d\Omega + \int_{\partial \Omega_i} (F \cdot n) \, dx = 0 \tag{106}$$

where $F(u) = f(u)\hat{i} + g(u)\hat{j}$. The situation is depicted for a control volume Ω_0 in Fig. 34. For two- and three-dimensional triangulations, several control volume choices are available: the triangles themselves, Voronoi duals, median duals, etc. Although the actual choice of control volume tessellation is very important, the monotonicity analysis contained in the remainder of this section is largely independent of this choice. Consequently, a generic control

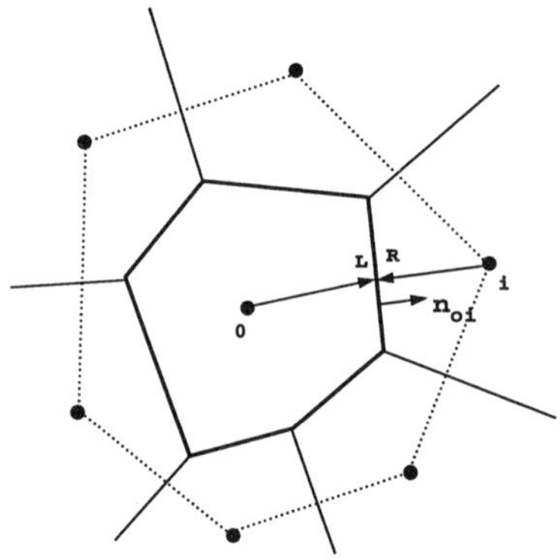

Fig. 34. Local control volume configuration for unstructured mesh.

volume Ω_0 with neighboring control volumes Ω_i, $i \in \mathcal{N}_0$, as shown in Fig. 34, is sufficient for the present analysis. In the following example, the solution data is assumed constant in each control volume. This simplifies the analysis considerably. The second example addresses the more general situation utilizing high order data reconstruction.

Example: Analysis of an Upwind Finite Volume Scheme with Piecewise Constant Reconstruction In this example, assume that the solution data $u(x,y)_i$ in each control volume Ω_i is constant with value equal to the integral average value, i.e.

$$u(x,y)_i = \overline{u}_i = \frac{1}{A_i} \int_{\Omega_i} u \, d\Omega, \ \forall \Omega_i \in \Omega$$

where A_i is the area of Ω_i. Next, define the unit exterior normal vector $\boldsymbol{n}_{0i}$ for the control volume boundary separating Ω_0 and Ω_i. It is also useful to define a normal vector $\vec{n}_{0i}$ which is scaled by the length (area in 3-D) of that portion of the control volume boundary separating Ω_0 and Ω_i. Finally, to simplify the exposition, define

$$f(u; \vec{n}) = F(u) \cdot \vec{n}$$

and assume the existence of a mean value linearization such that

$$f(v; \vec{n}) - f(u; \vec{n}) = df(u, v; \vec{n})(v - u) \ . \tag{107}$$

Using this notation, construct the following upwind scheme

$$\frac{d}{dt}(A_0 \overline{u}_0) = - \sum_{i \in \mathcal{N}_0} h(\overline{u}_0, \overline{u}_i; \vec{n}_{0i}) \tag{108}$$

with

$$h(\overline{u}_0, \overline{u}_i; \vec{n}_{0i}) = \frac{1}{2}\left(f(\overline{u}_0; \vec{n}_{0i}) + f(\overline{u}_i; \vec{n}_{0i})\right) - \frac{1}{2} |df(\overline{u}_0, \overline{u}_i; \vec{n}_{0i})| (\overline{u}_i - \overline{u}_0) \ .$$

In Barth and Jespersen [9], we proved a maximum principle and local extremum diminishing properties of the scheme (108) for scalar advection.

Theorem 24. *The upwind algorithm (108) with piecewise constant solution data exhibits a discrete maximum principle for arbitrary unstructured meshes and is local extremum diminishing under the CFL-like condition*

$$\Delta t \le \min_{\forall \Omega_j \in \Omega} \frac{-A_j}{\sum_{i \in \mathcal{N}_j} df^-(\overline{u}_i, \overline{u}_j; \vec{n}_{ji})}$$

when combined with Euler explicit time stepping.

Proof. Consider the control volume surrounding v_0 as shown in Fig. 34. Recall that the flux function was constructed using a mean value linearization such that

$$f(\overline{u}_i; \vec{n}_{0i}) - f(\overline{u}_0; \vec{n}_{0i}) = df(\overline{u}_0, \overline{u}_i; \vec{n}_{0i})\,(\overline{u}_i - \overline{u}_0) \ .$$

This permits regrouping terms into the following form:

$$\frac{d}{dt}(\overline{u}_0 A_0) = -\sum_{i \in \mathcal{N}_0} f(\overline{u}_0; \vec{n}_{0i}) - \sum_{i \in \mathcal{N}_0} df^-(\overline{u}_0, \overline{u}_i; \vec{n}_{0i})(\overline{u}_i - \overline{u}_0)$$

where $(\cdot) = (\cdot)^+ + (\cdot)^-$ and $|(\cdot)| = (\cdot)^+ - (\cdot)^-$. For any closed control volume, it follows that

$$\sum_{i \in \mathcal{N}_0} f(\overline{u}_0; \vec{n}_{0i}) = 0.$$

Combining the remaining terms yields a final form for analysis

$$\frac{d}{dt}(\overline{u}_0 A_0) = -\sum_{i \in \mathcal{N}_0} df^-(\overline{u}_0, \overline{u}_i; \vec{n}_{0i})(\overline{u}_i - \overline{u}_0) \ . \tag{109}$$

To verify a maximum principle at steady-state, set the time term to zero and solve for $\overline{u}_0$

$$\overline{u}_0 = \frac{\sum_{i \in \mathcal{N}_0} df^-(\overline{u}_0, \overline{u}_i; \vec{n}_{0i})\, \overline{u}_i}{\sum_{i \in \mathcal{N}_0} df^-(\overline{u}_0, \overline{u}_i; \vec{n}_{0i})} = \sum_{i \in \mathcal{N}_0} \alpha_i \overline{u}_i$$

with $\sum_{i \in \mathcal{N}_0} \alpha_i = 1$ and $\alpha_i \ge 0$. Since $\overline{u}_0$ is a convex combination of all neighbors

$$\min_{i \in \mathcal{N}_0} \overline{u}_i \le \overline{u}_0 \le \max_{i \in \mathcal{N}_0} \overline{u}_i \ . \tag{110}$$

If $\overline{u}_0$ takes on a maximum value M in the interior, then $\overline{u}_i = M, \forall\, i \in \mathcal{N}_0$. Repeated application of (110) to neighboring control volumes establishes the maximum principle.

Next, consider Euler explicit time advancement and rearrange terms

$$\begin{aligned} \overline{u}_0^{n+1} &= \overline{u}_0^n - \frac{\Delta t}{A_0} \sum_{i \in \mathcal{N}_0} df^-(\overline{u}_0^n, \overline{u}_i^n; \vec{n}_{0i})(\overline{u}_i^n - \overline{u}_0^n) \\ &= \alpha_0 \overline{u}_0^n + \sum_{i \in \mathcal{N}_0} \alpha_i \overline{u}_i^n \ . \end{aligned} \tag{111}$$

It should be clear that coefficients in (111) sum to unity. To show a local extremum diminishing property, it is sufficient to show nonnegativity of coefficients. Clearly, $\alpha_i \geq 0 \ \ \forall\, i > 0$ so that the local extremum diminishing property is achieved if

$$\alpha_0 = 1 + \frac{\Delta t}{A_0} \sum_{i \in \mathcal{N}_0} df^-(\overline{u}_0^n, \overline{u}_i^n; \vec{n}_{0i}) \ \geq 0 \ .$$

This establishes a local extremum diminishing property in time under the CFL-like condition

$$\Delta t \leq \min_{\forall \Omega_j \in \Omega} \frac{-A_j}{\sum_{i \in \mathcal{N}_j} df^-(\overline{u}_0^n, \overline{u}_i^n; \vec{n}_{0i})} \ .$$

□

Example: Analysis of High Order Accurate Upwind Advection Schemes Using Arbitrary Order Reconstruction [8]

In this example, high order accurate upwind schemes on unstructured meshes are considered. The technique used here is to show a maximum principle for the cell averages. The solution algorithm is a relatively standard procedure for extensions of Godunov's scheme in Eulerian coordinates, see for example [25,53,19,32]. The basic idea in Godunov's method is to treat the integral control volume averages, $\overline{u}$, as the basic unknowns. Using information from the control volume averages, $k-th$ order piecewise polynomials are *reconstructed* in each control volume Ω_i

$$U^k(x,y)_i = \sum_{m+n \leq k} \alpha_{(m,n)} P_{(m,n)}(x - x_c, y - y_c)$$

where $P_{(m,n)}(x - x_c, y - y_c) = (x - x_c)^m (y - y_c)^n$ and (x_c, y_c) is the control volume centroid. The process of reconstruction amounts to finding the polynomial coefficients, $\alpha_{(m,n)}$. Near steep gradients and discontinuities, these polynomial coefficients maybe altered based on monotonicity arguments. Because the reconstructed polynomials vary discontinuously from control volume to control volume, a unique value of the solution does not exist at control volume interfaces. This non-uniqueness is resolved via exact or approximate solutions of the Riemann problem. In practice, this is accomplished by supplanting the true flux function with a numerical flux function which produces a single unique flux given two solution states. Once the flux integral is carried out (either exactly or by numerical quadrature), the control volume average of the solution can be evolved in time. In most cases, standard techniques for integrating ODE equations are used for the time evolution, i.e. Euler implicit, Euler explicit, Runge-Kutta. The result of the evolution process is a new collection of control volume averages. The process can then be repeated. The process can be summarized in the following steps:

(1) **Reconstruction in Each Control Volume**: Given integral solution averages in all Ω_j, reconstruct a $k-th$ order piecewise polynomial $U^k(x,y)_i$ in each Ω_i for use in equation (106). In faithful implementations of Godunov's method, cell averages of the solution data

$$\int_{\Omega_i} U^k(x,y)_i \, d\Omega = (\overline{u}A)_i$$

are preserved during the reconstruction process. For solutions containing discontinuities and/or steep gradients, monotonicity enforcement may be required.

(2) **Flux Evaluation on Each Edge**: Supplant the true flux by a numerical flux function. Given two solution states the numerical flux function returns a single unique flux. Using the notation of the previous section, define $f(u;\boldsymbol{n}) = (F(u)\cdot \boldsymbol{n})$ so that

$$\int_{\partial\Omega_i} f(u;\boldsymbol{n}) \, d\Gamma \approx \int_{\partial\Omega_i} h(U^L,U^R;\boldsymbol{n}) \, d\Gamma \ .$$

Consider each control volume boundary $\partial\Omega_i$, to be a collection of polygonal edges (or dual edges) from the mesh. Along each edge (or dual edge), perform a high order accurate flux quadrature. When the reconstruction polynomial is piecewise linear, single (midpoint) quadrature is usually employed on both structured and unstructured meshes

$$\int_{\partial\Omega_i} h(U^L,U^R;\boldsymbol{n}) d\Gamma \approx \sum_{j\in\mathcal{N}_i} h(U^L,U^R;\vec{n})_{ij}$$

where U^L and U^R are solution values evaluated at the midpoint of control volume edges as shown in Fig. 34. When multi-point quadrature formulas are employed, they are assumed to be of the form

$$\int_0^1 f(s) ds = \sum_{q\in Q} w_q f(\xi_q)$$

with $w_q > 0$ and $\xi_q \in [0,1]$. Let the multi-point quadrature formulas be represented by the augmented notation

$$\int_{\partial\Omega_i} h(U^L,U^R;\boldsymbol{n}) d\Gamma \approx \sum_{j\in\mathcal{N}_i}\sum_{q\in Q} w_q h(U^L,U^R;\vec{n})_{ijq} \ .$$

(3) **Evolution in Each Control Volume**: Collect flux contributions in each control volume and evolve in time using any time stepping scheme, i.e. Euler explicit, Euler implicit, Runge-Kutta, etc. The result of this step is once again control volume averages and the process can be repeated.

In the present analysis, the reconstruction polynomials $U^k(x,y)_i$ in each Ω_i are given. The result of the analysis will be conditions or constraints on the reconstruction so that a maximum principle involving cell averages can be obtained. The topic of reconstruction and implementation of the constraints determined by this analysis will be examined in a later section. Using this notation, the following upwind scheme is constructed for the configuration in Fig. 34

$$\frac{d}{dt}(A_0\overline{u}_0) = -\sum_{i\in\mathcal{N}_0}\sum_{q\in Q} w_q h(U^L,U^R;\vec{n})_{0iq} \tag{112}$$

with a numerical flux function obtained from (107)

$$\begin{aligned} h(U^L,U^R;\vec{n}_{0i}) = {} & \frac{1}{2}\left(f(U^L;\vec{n}) + f(U^R;\vec{n})\right)_{0i} \\ - {} & \frac{1}{2}|df(U^L,U^R;\vec{n})|_{0i}(U^R - U^L)_{0i} \ . \end{aligned} \tag{113}$$

To analyze the scheme, recall that the flux function was constructed using a mean value linearization such that

$$f(U^R;\vec{n}) - f(U^L;\vec{n}) = df(U^L,U^R;\vec{n})(U^R - U^L) \ .$$

This permits regrouping terms into the following form:

$$\frac{d}{dt}(\overline{u}_0 A_0) = -\sum_{i\in\mathcal{N}_0}\sum_{q\in Q} w_q\left(f(U^L;\vec{n}) + df^-(U^L,U^R;\vec{n})(U^R - U^L)\right)_{0iq} \tag{114}$$

Rewrite the first term in the sum using a mean value construction

$$\sum_{i\in\mathcal{N}_0}\sum_{q\in Q} w_q f(\overline{u}_0;\vec{n})_{0iq} + w_q\left(df(\overline{u}_0,U^L;\vec{n})(U^L - \overline{u}_0)\right)_{0iq} \ .$$

The first term vanishes when summed over a closed volume so that (114) reduces to

$$\begin{aligned} \frac{d}{dt}(\overline{u}_0 A_0) = -\sum_{i\in\mathcal{N}_0}\sum_{q\in Q} \quad & w_q\big(df(\overline{u}_0,U^L;\vec{n})(U^L - \overline{u}_0) \\ & + df^-(U^L,U^R;\vec{n})(U^R - U^L)\big)_{0iq} \ . \end{aligned}$$

By introducing difference ratios, the scheme can be written in the following form:

$$\frac{d}{dt}(\overline{u}_0 A_0) = -\sum_{i\in\mathcal{N}_0}\sum_{q\in Q} w_q\left(df^-(\overline{u}_0,U^L;\vec{n})\Psi\right)_{0iq}(\overline{u}_i - \overline{u}_0)$$

$$
\begin{aligned}
&- \sum_{i \in \mathcal{N}_0} \sum_{q \in Q} w_q \left(df^+(\overline{u}_0, U^L; \vec{n})\Phi\right)_{0iq} (\overline{u}_0 - \overline{u}_k) \\
&- \sum_{i \in \mathcal{N}_0} \sum_{q \in Q} w_q \left(df^-(U^L, U^R; \vec{n})\Theta\right)_{0iq} (\overline{u}_i - \overline{u}_0) \qquad (115)
\end{aligned}
$$

with

$$
\Psi_{0iq} = \frac{U^L_{0iq} - \overline{u}_0}{\overline{u}_i - \overline{u}_0}, \quad \Phi_{0iq} = \frac{U^L_{0iq} - \overline{u}_0}{\overline{u}_0 - \overline{u}_k}, \quad \Theta_{0iq} = \frac{U^R_{0iq} - U^L_{0iq}}{\overline{u}_i - \overline{u}_0} .
$$

In this equation, the k subscript refers to some as yet unspecified index value, $k \in \mathcal{N}_0$.

Theorem 25. *The generalized Godunov scheme with arbitrary order reconstruction (112) exhibits a discrete maximum principle at steady-state if the following three conditions are fulfilled:*

$$
\Psi_{jiq} \geq 0, \quad \Phi_{jiq} \geq 0 \quad \Theta_{jiq} \geq 0 \quad \forall j, q, \; i \in \mathcal{N}_j \qquad (116)
$$

as defined by (115). Furthermore, the scheme with Euler explicit time advancement is local extremum diminishing in cell averages under a CFL-like condition

$$
\Delta t \leq \min_{\forall \, \Omega_j \in \Omega} \frac{-A_j}{\sum_{i \in \mathcal{N}_j} \sum_{q \in Q} \left(\overline{df}^- \Psi - \overline{df}^+ \Phi + df^- \Theta\right)_{jiq}} .
$$

Proof. Assume that (116) holds. Define $\overline{df}_{0iq} = w_q df(\overline{u}_0, U^L; \vec{n})_{0iq}$ and similarly $df_{0iq} = w_q df(U^L, U^R; \vec{n})_{0iq}$. Setting the time term to zero and solving for $\overline{u}_0$ yields

$$
\begin{aligned}
\overline{u}_0 &= \frac{\sum_{i \in \mathcal{N}_0} \sum_{q \in Q} \left(\overline{df}^+ \Phi\right)_{0iq} \overline{u}_k - \left(\overline{df}^- \Psi + df^- \Theta\right)_{0iq} \overline{u}_i}{\sum_{i \in \mathcal{N}_0} \sum_{q \in Q} \left(\overline{df}^+ \Phi - \overline{df}^- \Psi - df^- \Theta\right)_{0iq}} \\
&= \sum_{i \in \mathcal{N}_0} \alpha_i \overline{u}_i \; . \qquad (117)
\end{aligned}
$$

Examining the individual coefficients, it is clear that $\sum_{i \in \mathcal{N}_0} \alpha_i = 1$ and $\alpha_i \geq 0, \forall i$. Thus a convex local mapping exists from which it is concluded that

$$
\min_{i \in \mathcal{N}_0} \overline{u}_i \leq \overline{u}_0 \leq \max_{i \in \mathcal{N}_0} \overline{u}_i \; . \qquad (118)
$$

If $\overline{u}_0$ takes on a maximum value M in the interior, then $\overline{u}_i = M, \forall \; i \in \mathcal{N}_0$. Repeated application of (118) to neighboring control volumes establishes the maximum principle.

To establish a local extremum diminishing property in time, insert an Euler explicit time advancement scheme in (115)

$$\begin{aligned}\overline{u}_0^{n+1} &= \overline{u}_0^n - \frac{\Delta t}{A_0} \sum_{i\in\mathcal{N}_0} \sum_{q\in Q} \overline{df}^-_{0iq} \Psi_{0iq} (\overline{u}_i^n - \overline{u}_0^n) \\ &\quad - \frac{\Delta t}{A_0} \sum_{i\in\mathcal{N}_0} \sum_{q\in Q} \overline{df}^+_{0iq} \Phi_{0iq} (\overline{u}_0^n - \overline{u}_k^n) \\ &\quad - \frac{\Delta t}{A_0} \sum_{i\in\mathcal{N}_0} \sum_{q\in Q} df^-_{0iq} \Theta_{0iq} (\overline{u}_i^n - \overline{u}_0^n) \\ &= \alpha_0 \overline{u}_0^n + \sum_{i\in\mathcal{N}_0} \sum_{q\in Q} \alpha_i \overline{u}_i^n \end{aligned} \tag{119}$$

with $\alpha_0 + \sum_{i\in\mathcal{N}_0} \alpha_i = 1$ and $\alpha_i \geq 0, i > 0$. A locally convex mapping in time from U^n to U^{n+1} is achieved when $\alpha_0 \geq 0$. This assures monotonicity in time. Some algebra reveals the following formula for α_0

$$\alpha_0 = 1 + \frac{\Delta t}{A_0} \sum_{i\in\mathcal{N}_0} \sum_{q\in Q} \left(\overline{df}^- \Psi - \overline{df}^+ \Phi + df^- \Theta \right)_{0iq} .$$

From this result, the local extremum diminishing property is established under the CFL-like restriction

$$\Delta t \leq \min_{\forall\, \Omega_i \in \Omega} \frac{-A_0}{\sum_{i\in\mathcal{N}_0} \sum_{q\in Q} \left(\overline{df}^- \Psi - \overline{df}^+ \Phi + df^- \Theta \right)_{0iq}}$$

such that

$$\min_{i\in\mathcal{N}_0} (\overline{u}_i^n, \overline{u}_0^n) \leq \overline{u}_0^{n+1} \leq \max_{i\in\mathcal{N}_0} (\overline{u}_i^n, \overline{u}_0^n) .$$

□

In later sections, the topic of data reconstruction and limiting is discussed. Without specifying the actual type of reconstruction, we have the following simple lemma concerning the ratios appearing in (115):

Lemma 26. *Assume $\Psi_{0iq} \geq 0$ as defined in (115). A sufficient condition for $\Phi_{0iq} \geq 0$ is that the reconstructed polynomial reduce to the cell average value, $U^k(x,y)_0 = \overline{u}_0$, at local extrema in cell averages, i.e. whenever*

$$\max_{j\in\mathcal{N}_0} \overline{u}_j \leq \overline{u}_0 \leq \min_{j\in\mathcal{N}_0} \overline{u}_j .$$

Proof. Consider an arbitrary control volume Ω_i adjacent to Ω_0. Assume that $\overline{u}_i \geq \overline{u}_0$. The stated assumption, $\Psi_{0iq} \geq 0$ implies that $U^L_{0iq} \geq \overline{u}_0$. Consequently, $\Phi_{0iq} \leq 0$ if and only if $\overline{u}_0$ is less than all adjacent neighbors, hence a local minimum. Following a similar argument, $\overline{u}_0$ is a local maximum when $\overline{u}_i \leq \overline{u}_0$ and $\Phi_{0iq} \leq 0$. □

This lemma states the rather harsh condition that the reconstruction be reduced to a piecewise constant value at local extrema. This design condition is encountered in the construction of reconstruction limiters for structured meshes as well.

4 Upwind Finite Volume Schemes for the Gasdynamic Equations

This chapter examines upwind finite volumes schemes for scalar and system conservation laws. The basic tasks in the upwind finite volume approach have already been presented: reconstruction, flux evaluation, and evolution. By far, the most difficult task in this process is the reconstruction step.

4.1 Reconstruction Schemes for Upwind Finite Volume Schemes

In the following paragraphs, the design criteria for general reconstruction operators with fixed stencil is reviewed. The reader is referred to the papers by Abgrall [1,2], Vankeirsbilck [54] and Michell [43] for a discussion of ENO and adaptive stencil reconstruction schemes.

The reconstruction operator serves as a finite-dimensional (possibly pseudo) inverse of the cell-averaging operator $\mathbf{A}$ whose j-th component $\mathbf{A}_j$ computes the cell average of the solution in Ω_j

$$\overline{u}_j = \mathbf{A}_j u = \frac{1}{A_j} \int_{\Omega_j} u(x,y)\, d\Omega \ .$$

In addition, the following properties are usually imposed on the reconstruction:

(1) **Conservation of the mean**: Simply stated, given cell averages $\overline{u}$, we require that all polynomial reconstructions u^k have the correct cell average.

$$\text{if} \quad u^k = \boldsymbol{R}^k \overline{u} \quad \text{then} \quad \overline{u} = \mathbf{A} u^k \ .$$

This means that $\boldsymbol{R}^k$ is a right inverse of the averaging operator $\mathbf{A}$

$$\mathbf{A}\boldsymbol{R}^k = I \ .$$

Conservation of the mean has an important implication. Unlike finite-element schemes, *Godunov schemes have a diagonal mass matrix.*

(2) **k-exactness**: A reconstruction operator $\boldsymbol{R}^k$ is *k-exact* if $\boldsymbol{R}^k\mathbf{A}$ reconstructs polynomials of degree k or less exactly.

$$\text{if} \quad u \in \mathcal{P}_k \quad \text{and} \quad \overline{u} = \mathbf{A}u, \quad \text{then} \quad u^k = \boldsymbol{R}^k \overline{u} = u \ .$$

In other words, $\boldsymbol{R}^k$ is a left-inverse of $\mathbf{A}$ restricted to the space of polynomials of degree at most k.

$$\boldsymbol{R}^k \mathbf{A}\Big|_{\mathcal{P}_k} = I \ .$$

This insures that exact solutions contained in $\mathcal{P}_k$ are in fact solutions of the discrete equations. For sufficiently smooth solutions, the property of k-exactness also insures that when piecewise polynomials are evaluated at control volume boundaries, the difference between solution states diminishes with increasing k at a rate proportional to h^{k+1} were h is a maximum diameter of the two control volumes. Figure 35 shows a global quartic polynomial $u \in \mathcal{P}_4$ which has been averaged in each interval. This same figure shows

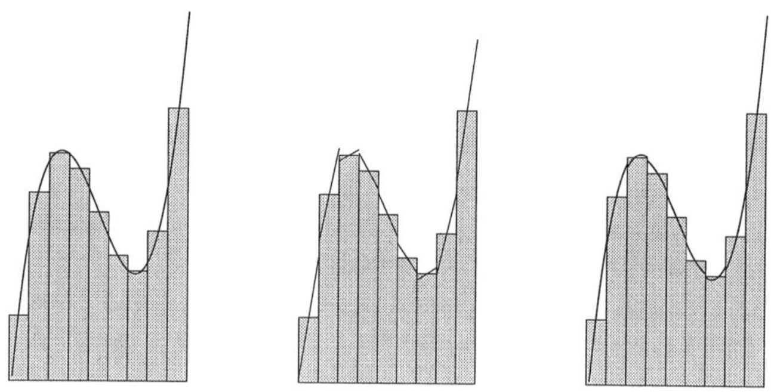

Fig. 35. Cell averaging of quartic polynomial (left), linear reconstruction (center) and quadratic reconstruction (right).

linear and quadratic reconstruction given interval averages. The small jumps in the piecewise polynomials at interval boundaries would decrease even more for cubics and vanish altogether for quartic reconstruction. Property (1) requires that the area under each piecewise polynomial is exactly equal to the cell average.

One of the most important observations concerning linear reconstruction is that one can dispense with the notion of cell averages as unknowns by reinterpreting the unknowns as pointwise values of the solution sampled at the centroid (midpoint in 1-D) of the control volume. This well known result greatly simplifies schemes based on linear reconstruction. The linear reconstruction in each interval shown in Fig. 35 was obtained by a simple central-difference formula given pointwise values of the solution at the midpoint of each interval. Note that for steady-state computations, conservation of the mean in the data reconstruction is not necessary. The implication of violating this conservation is that a *nondiagonal* mass matrix appears in the time integral. Since time derivatives vanish at steady-state, the effect of this mass matrix vanishes at steady-state. The reconstruction schemes presented below assume that solution variables are placed at the vertices of the mesh, which may not be at the precise centroid.

Green-Gauss Reconstruction Let D_0 denote the set of all triangles incident to some vertex v_0 and the exact integral relation

$$\int_{D_0} \nabla u \, d\Omega = \int_{\partial D_0} u n \, d\Gamma \ . \tag{120}$$

It is not difficult to show [9] that given function values at vertices of a triangulation, a discretization of this formula can be constructed which is exact whenever u varies linearly

$$(\nabla u)_{v_0} = \frac{1}{A_0} \sum_{i \in \mathcal{N}_0} \frac{1}{2}(u_i + u_0) \vec{n}_{0i} \ . \tag{121}$$

In this formula $\vec{n}_{0i} = \int_a^b d\vec{n}$ for any path which connects triangle centroids adjacent to the edge $e(v_0, v_i)$ and A_0 is the area of the *nonoverlapping* dual regions formed by this choice of path integration. Two typical choices are the median and centroid duals as shown below. This approximation extends

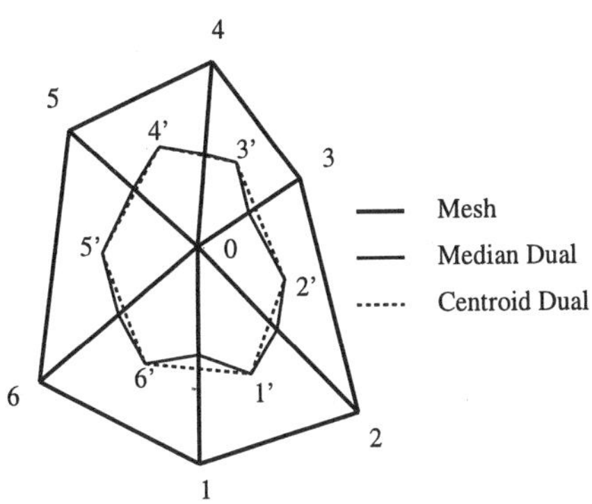

Fig. 36. Local mesh with centroid and median duals.

naturally to three dimensions. The formula (121) suggests a natural computer implementation using the edge data structure. Assume that the normals $\vec{n}_{ij}$ for all edges $e(v_i, v_j)$ have been precomputed with the convention that the normal vector points from v_i to v_j. An edge implementation of (121) can be performed in the following way:

For $k = 1, n(e)$ *! Loop through edges of mesh*

$j_1 = e^{-1}(k,1)$ *!Pointer to edge origin*
$j_2 = e^{-1}(k,2)$ *!Pointer to edge destination*
$uav = (u(j_1) + u(j_2))/2$ *!Gather*
$ux(j_1) += normx(k) \cdot uav$ *!Scatter*
$ux(j_2) -= normx(k) \cdot uav$
$uy(j_1) += normy(k) \cdot uav$
$uy(j_2) -= normy(k) \cdot uav$
Endfor

For $j = 1, n(v)$ *! Loop through vertices*
$ux(j) = ux(j)/area(j)$ *! Scale by area*
$uy(j) = uy(j)/area(j)$
Endfor

It can be shown that the use of edge formulas for the computation of vertex gradients is asymptotically optimal in terms of work done.

Linear Least-Squares Reconstruction To derive this reconstruction technique, consider a vertex v_0 and suppose that the solution varies linearly over the support of adjacent neighbors of the mesh. In this case, the change in vertex values of the solution along an edge $e(v_i, v_0)$ can be calculated by

$$(\nabla u)_0 \cdot (\boldsymbol{r}_i - \boldsymbol{r}_0) = u_i - u_0$$

where $\boldsymbol{r}$ denotes a spatial position vector. This equation represents the scaled projection of the gradient along the edge $e(v_i, v_0)$. A similar equation could be written for all incident edges subject to an arbitrary weighting factor. The result is the following matrix equation, shown here in two dimensions:

$$\begin{bmatrix} w_1 \Delta x_1 & w_1 \Delta y_1 \\ \vdots & \vdots \\ w_n \Delta x_n & w_n \Delta y_n \end{bmatrix} \begin{pmatrix} u_{,x} \\ u_{,y} \end{pmatrix} = \begin{pmatrix} w_1(u_1 - u_0) \\ \vdots \\ w_n(u_n - u_0) \end{pmatrix}$$

or in symbolic form $\mathcal{L}\ \nabla u = f$ where

$$\mathcal{L} = [\,\vec{L}_1 \quad \vec{L}_2\,]$$

in two dimensions. Exact calculation of gradients for linearly varying u is guaranteed if any two row vectors $w_i(\boldsymbol{r}_i - \boldsymbol{r}_0)$ span all of 2 space. This implies linear independence of $\vec{L}_1$ and $\vec{L}_2$. The system can then be solved via a Gram-Schmidt process, i.e.,

$$\begin{bmatrix} \vec{V}_1 \\ \vec{V}_2 \end{bmatrix} [\,\vec{L}_1 \quad \vec{L}_2\,] = \begin{bmatrix} 1 & 0 \\ 0 & 1 \end{bmatrix} .$$

The row vectors $\vec{V}_i$ are given by

$$\vec{V}_1 = \frac{l_{22}\vec{L}_1 - l_{12}\vec{L}_2}{l_{11}l_{22} - l_{12}^2}$$

$$\vec{V}_2 = \frac{l_{11}\vec{L}_2 - l_{12}\vec{L}_1}{l_{11}l_{22} - l_{12}^2} \tag{122}$$

with $l_{ij} = (\vec{L}_i \cdot \vec{L}_j)$.

Note that reconstruction of N independent variables in $\mathbb{R}^d$ implies $\binom{d+1}{2} + dN$ inner product sums. Since only dN of these sums involves the solution variables themselves, the remaining sums could be precalculated and stored in computer memory. This makes the present scheme competitive with the Green-Gauss reconstruction. Using the edge data structure, the calculation of inner product sums can be calculated for *arbitrary* combinations of polyhedral cells. In all cases linear functions are reconstructed exactly. This technique is best illustrated by example:

For $k = 1, n(e)$ *!Loop through edges of mesh*
 $j_1 = e^{-1}(k,1)$ *!Pointer to edge origin*
 $j_2 = e^{-1}(k,2)$ *!Pointer to edge destination*
 $dx = w(k) \cdot (x(j_2) - x(j_1))$ *!Weighted* Δx
 $dy = w(k) \cdot (y(j_2) - y(j_1))$ *!Weighted* Δy
 $l_{11}(j_1) = l_{11}(j_1) + dx \cdot dx$ *!* l_{11} *orig sum*
 $l_{11}(j_2) = l_{11}(j_2) + dx \cdot dx$ *!* l_{11} *dest sum*
 $l_{12}(j_1) = l_{12}(j_1) + dx \cdot dy$ *!* l_{12} *orig sum*
 $l_{12}(j_2) = l_{12}(j_2) + dx \cdot dy$ *!* l_{12} *dest sum*
 $du = w(k) \cdot (u(j_2) - u(j_1))$ *!Weighted* Δu
 $lf_1(j_1) += dx \cdot du$ *!*$\vec{L}_1 f$ *sum*
 $lf_1(j_2) += dx \cdot du$
 $lf_2(j_1) += dy \cdot du$ *!*$\vec{L}_2 f$ *sum*
 $lf_2(j_2) += dy \cdot du$
Endfor

For $j = 1, n(v)$ *! Loop through vertices*
 $det = l_{11}(j) \cdot l_{22}(j) - l_{12}^2$
 $ux(j) = (l_{22}(j) \cdot lf_1(j) - l_{12} \cdot lf_2)/det$
 $uy(j) = (l_{11}(j) \cdot lf_2(j) - l_{12} \cdot lf_1)/det$
Endfor

This formulation provides freedom in the choice of weighting coefficients, w_i. These weighting coefficients can be a function of the geometry and/or solution. Classical approximations in one dimension can be recovered by choosing geometrical weights of the form $w_i = 1./|\boldsymbol{r}_i - \boldsymbol{r}_0|^t$ for values of $t = 0, 1, 2$.

Monotonicity Enforcement When solution discontinuities and steep gradients are present, additional steps must be taken to prevent oscillations from developing in the numerical solution. One way to do this was pioneered by van Leer [53] in the late 1970's. His basic idea was to enforce strict monotonicity in the reconstruction. Monotonicity in this context means that the value of the reconstructed polynomial does not exceed the minimum and maximum of neighboring cell averages. The final reconstruction must guarantee that no new extrema have been created. When a new extremum is produced, the slope of the reconstruction in that interval is reduced until monotonicity is restored. This implies that at a local minimum or maximum in the cell-averaged data the slope in 1-D is *always* reduced to zero, see for example Fig. 37. Theorem 25 provides sufficient conditions for a discrete maximum

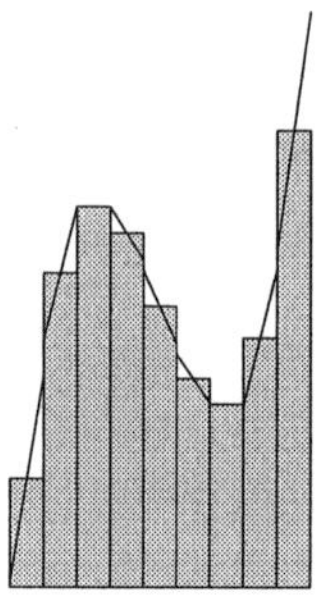

Fig. 37. Linear data reconstruction with monotone limiting.

principle in the cell averages using arbitrary order reconstruction on general control volumes. Consider the control volume interface separating Ω_0 and Ω_i as shown in Fig. 34. From Theorem 25, a maximum principle is guaranteed if for all quadrature points on the interface separating Ω_0 and Ω_i

$$\Psi_{0i} \geq 0, \quad \Phi_{0i} \geq 0 \quad \Theta_{0i} \geq 0 \ .$$

Lemma 26 states that $\Phi_{0i} \geq 0$ is always satisfied if the monotonicity enforcement algorithm reduces to piecewise constant at local extrema, i.e. when

$$\max_{j \in \mathcal{N}_0} \overline{u}_j \leq \overline{u}_0 \leq \min_{j \in \mathcal{N}_0} \overline{u}_j \ .$$

Assume that this property holds, monotonicity reduces to the following two conditions at all quadrature points:

$$\begin{aligned} &\text{(a)} \quad 0 \leq \frac{U^L - \overline{u}_0}{\overline{u}_i - \overline{u}_0} \\ &\text{(b)} \quad 0 \leq \frac{U^R - U^L}{\overline{u}_i - \overline{u}_0} \ . \end{aligned} \tag{123}$$

The second inequality appearing in (123) requires that the difference in the extrapolated states at a cell interface must be of the same sign as the difference in the cell average values. For example in Fig. 38(a) this condition is violated but can be remedied either by a symmetric reduction of slopes or by replacing the larger slope by the minimum value of the two slopes. *Observe that in one space dimension the net effect of the slope limiting in the*

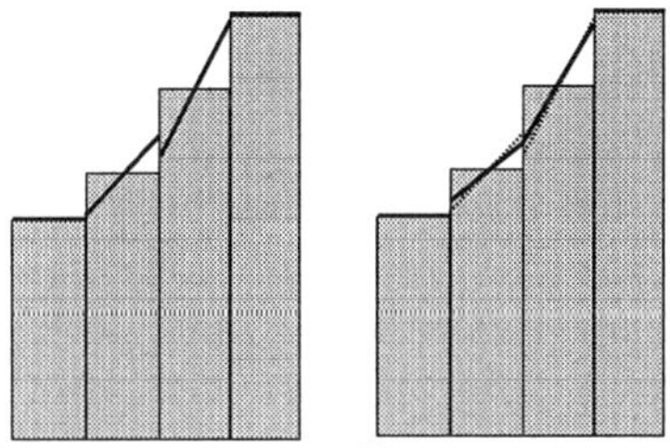

Fig. 38. (a) Reconstruction profile with increased variation violating monotonicity constraints. (b) Profile after modification to satisfy monotonicity constraints.

reconstruction process is to ensure that the total variation of the reconstructed function does not exceed the total variation of the cell averaged data.

In Barth and Jespersen [9], a simple recipe was proposed for slope limiting of linearly reconstructed data on arbitrary unstructured meshes. Consider writing the linearly reconstructed data in the following form for Ω_0:

$$U(x,y)_0 = \overline{u}_0 + \nabla u_0 \cdot (\boldsymbol{r} - \boldsymbol{r}_0).$$

Now consider a "limited" form of this piecewise linear distribution.

$$U(x,y)_0 = \overline{u}_0 + \Phi_0 \nabla u_0 \cdot (\boldsymbol{r} - \boldsymbol{r}_0)$$

$$\begin{aligned} u_0^{min} &= \min_{i \in \mathcal{N}_0} (\overline{u}_0, \overline{u}_i) \\ u_0^{max} &= \max_{i \in \mathcal{N}_0} (\overline{u}_0, \overline{u}_i) \end{aligned}$$

and require that

$$u_0^{min} \le U(x,y)_0 \le u_0^{max}$$

when evaluated at the quadrature points used in the flux integral computation. For each quadrature point location in the flux integral, compute the extrapolated state U_{0i}^L and determine the smallest Φ_0 so that

$$\Phi_0 = \begin{cases} \min(1, \frac{u_0^{max} - \overline{u}_0}{U_{0i}^L - \overline{u}_0}), & \text{if } U_{0i}^L - \overline{u}_0 > 0 \\ \min(1, \frac{u_0^{min} - \overline{u}_0}{U_{0i}^L - \overline{u}_0}), & \text{if } U_{0i}^L - \overline{u}_0 < 0 \\ 1 & \text{if } U_{0i}^L - \overline{u}_0 = 0 \end{cases} .$$

This limiter function automatically satisfies lemma 26. Condition (a) from (123) is local to the control volume and can be enforced by a further reduction in slope. In practice this step is sometimes omitted. Condition (b) is enforced using the procedure shown in Fig. 38.

Extensive numerical testing has shown that this limiter can noticeably degrade the overall accuracy of computations, especially for flows on coarse meshes. In addition the limiter behaves poorly when the solution is nearly constant unless additional heuristic parameters are added. This has prompted other researchers (c.f. Venkatakrishnan [55]) to propose alternative differentiable limiter functions, but no serious attempt is made to appeal to the rigors of maximum principle theory. The design of accurate limiters satisfying the maximum principle constraints is still an open problem in this area.

When the above procedures are combined with the flux function given earlier (113),

$$\begin{aligned} h(U^L, U^R; \boldsymbol{n}) = & \frac{1}{2}\left(f(U^L; \boldsymbol{n}) + f(U^R; \boldsymbol{n})\right) \\ & - \frac{1}{2}|df(U^L, U^R; \boldsymbol{n})|\left(U^R - U^L\right) \end{aligned} \tag{124}$$

the resulting scheme has good shock resolving characteristics. To demonstrate this, we consider the scalar nonlinear hyperbolic problem suggested by Struijs, Deconinck, et al. [51]. The equation is a multidimensional form of Burger's equation

$$u_{,t} + (u^2/2)_{,x} + u_{,y} = 0 \ .$$

This equation is solved in a square region $[0, 1.5] \times [0, 1.5]$ with boundary conditions: $u(x, 0) = 1.5 - 2x$, $x \leq 1$, $u(x, 0) = -.5$, $x > 1$, $u(0, y) = 1.5$, and $u(1.5, y) = -.5$. Figure 39 shows carpet plots and contours of the solution on regular and irregular meshes. The exact solution to this problem consists of

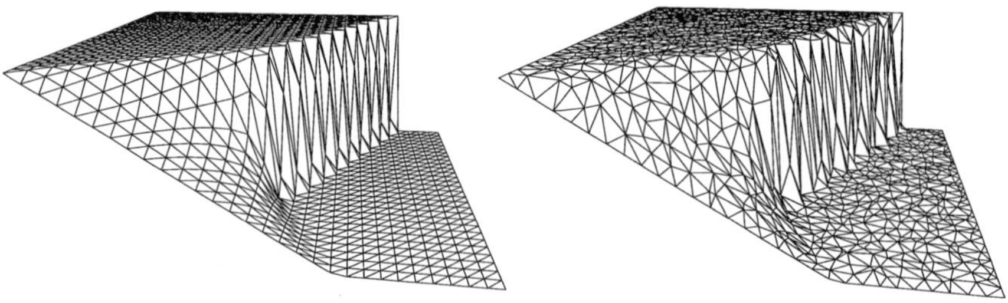

Fig. 39. Carpet plot of Burger's equation solution on (a) regular mesh, (b) irregular mesh (1800 elements).

converging straightline characteristics which eventually form a shock which propagates to the upper boundary. The carpet plots indicate that the numerical solutions on both meshes are monotone. Even so, most people would prefer the solution on the regular mesh. This is an unavoidable consequence of irregular meshes. The only remedy appears to be mesh adaptation.

Next, consider a test problem which solves the two-dimensional scalar advection equation

$$u_{,t} + (yu)_{,x} - (xu)_{,y} = 0$$

or equivalently

$$u_{,t} + \overrightarrow{\lambda} \cdot \nabla u = 0, \quad \overrightarrow{\lambda} = (y, -x)^T$$

on a mesh centered about the origin, see Fig. 40, containing 3200 elements. Discontinuous inflow data is specified along an interior cut line, $u(x,0) = 1$ for $-.6 < x < -.3$ and $u(x,0) = 0$, otherwise. The exact solution is a solid body rotation of the cut line data throughout the domain. The discontinu-

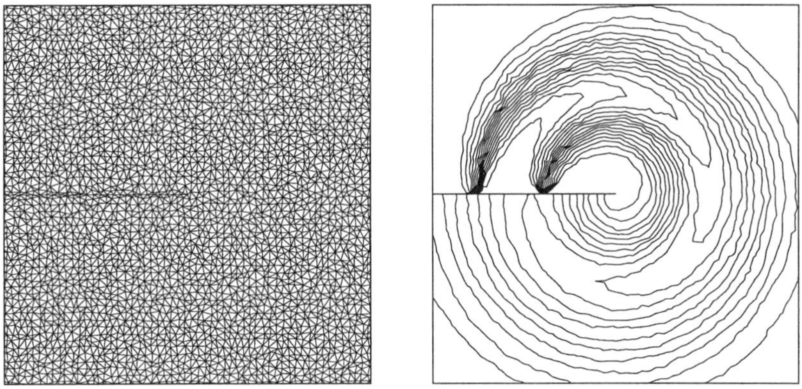

Fig. 40. Mesh for the circular advection problem (left) and solution contours obtained using piecewise constant reconstruction (right).

ities admitted by this equation are similar to the linear contact and slip line solutions admitted by the Euler equations. Figure 40 also shows solution contours obtained using piecewise constant reconstruction. The discontinuities are quickly smeared by the computation. Figure 41 displays solution contours obtained using piecewise linear and a piecewise quadratic reconstruction technique discussed in [7,6]. The improvement from piecewise constant reconstruction to piecewise linear is quite dramatic. The improvement from piecewise linear to piecewise quadratic also looks impressive. The width of the discontinuities is substantially reduced with little observable mesh dependence. Note however, that the quadratic approximation used for this computation has roughly quadrupled the number of solution unknowns because of the use of 6-noded triangles.

4.2 Numerical Solution of the Euler Equations Using Upwind Finite Volume Approximation

The section discusses the extension and application of the scalar finite volume scheme to the Euler equations. One advantage of the finite volume method is the relative ease in which this can be done.

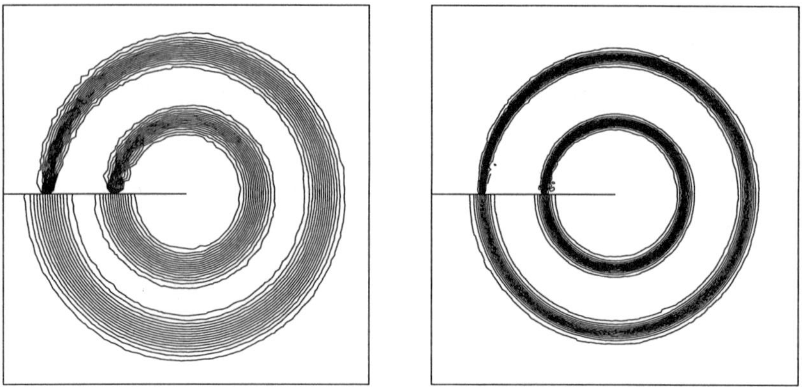

Fig. 41. Solution contours, piecewise linear reconstruction (left) and piecewise quadratic reconstruction (right).

Extension of Scalar Advection Schemes to Systems of Equations The extension of the scalar advection schemes to the Euler (and Navier-Stokes) equations requires three rather minor modifications:

1. *Vector Flux Function.* The scalar flux function is replaced by a vector flux function. In the present work, the mean value linearization due to Roe [46] is used. The form of this vector flux function is identical to the scalar flux function (113), i.e.

$$\begin{aligned} \mathbf{h}(\boldsymbol{u}^R, \boldsymbol{u}^L; \boldsymbol{n}) = & \frac{1}{2} \left(\boldsymbol{f}(\boldsymbol{u}^R; \boldsymbol{n}) + \boldsymbol{f}(\boldsymbol{u}^L; \boldsymbol{n}) \right) \\ & - \frac{1}{2} |A(\boldsymbol{u}^R, \boldsymbol{u}^L; \boldsymbol{n})| \left(\boldsymbol{u}^R - \boldsymbol{u}^L \right) \end{aligned} \tag{125}$$

   where $\boldsymbol{f}(\boldsymbol{u}; \boldsymbol{n}) = \boldsymbol{F}(\boldsymbol{u}) \cdot \boldsymbol{n}$, and $A = d\boldsymbol{f}/d\boldsymbol{u}$ is the flux Jacobian.
2. *Componentwise limiting.* The solution variables are reconstructed componentwise. In principle, any set of variables can be used in the reconstruction (primitive variables, entropy variables, etc.). Note that conservation of the mean can make certain variable combinations more difficult to implement than others because of the nonlinearities that may be introduced. The simplest choice is obviously the conserved variables themselves. When conservation of the mean is not important (steady-state calculations), we typically use primitive variables in the reconstruction step.
3. *Weak Boundary Conditions.* Boundary conditions for inviscid flow at solid surfaces are enforced weakly. For solid wall boundary edges, the flux is

calculated with $\boldsymbol{V} \cdot \boldsymbol{n}$ set identically to zero

$$\boldsymbol{f}(\boldsymbol{u};\boldsymbol{n}) = \begin{pmatrix} 0 \\ n_x p \\ n_y p \\ 0 \end{pmatrix} .$$

Boundary conditions at far field boundaries are also done weakly. Define the characteristic projectors of the flux Jacobian A in the following way:

$$P^{\pm} = \frac{1}{2}[I \pm \text{sign}(A)] .$$

At far field boundary edges the fluxes are assumed to be of the form:

$$\boldsymbol{f}(\boldsymbol{u}^n;\boldsymbol{n}) = (\boldsymbol{F}(\boldsymbol{u}^n_{proj}) \cdot \boldsymbol{n})$$

where $\boldsymbol{u}^n_{proj} = P^+ \, \boldsymbol{u}^n + P^- \, \boldsymbol{u}_\infty$ and $\boldsymbol{u}_\infty$ represents a vector of prescribed far field solution values. At first glance, prescribing the entire vector $\boldsymbol{u}_\infty$ is an overspecification of boundary conditions. Fortunately the characteristic projectors remove or ignore certain combinations of data so that the correct number of conditions are specified at inflow and outflow.

Example: Supersonic Oblique Shock Reflections In this example, two supersonic streams (M=2.50 and M=2.31) are introduced at the left boundary. These streams interact producing a pattern of supersonic shock reflections down the length of the converging channel. The mesh is a subdivided 15x52 mesh with perturbed coordinates as shown in Fig. 42. This same figure

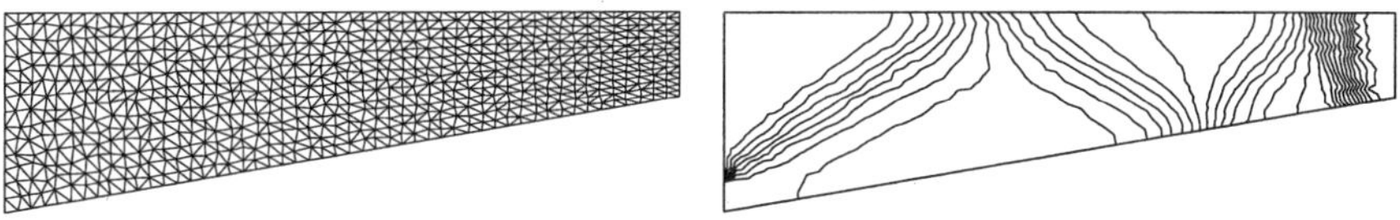

Fig. 42. (a) Channel mesh, 780 vertices (left). (b) Mach contours, piecewise constant reconstruction (right).

also shows Mach contours for the numerical solution obtained using piecewise constant reconstruction. As expected, the piecewise constant reconstruction scheme severely smears the shock system while the scheme based on a linear solution reconstruction shown in Fig. 43 performs very well. The piecewise quadratic approximation shows some improvement in shock wave thickness although the improvement is not dramatic given the increased number of unknowns used in this particular scheme [7]. The number of unknowns required for the quadratic approximation is roughly four times the number required

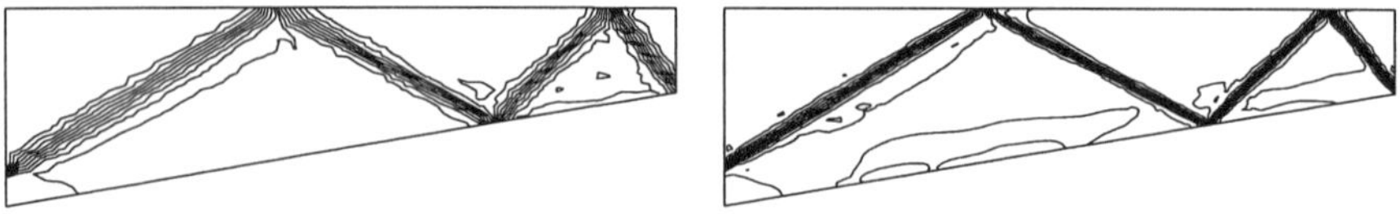

Fig. 43. Mach contours. (a) Piecewise linear reconstruction (left). (b) Piecewise quadratic reconstruction (right).

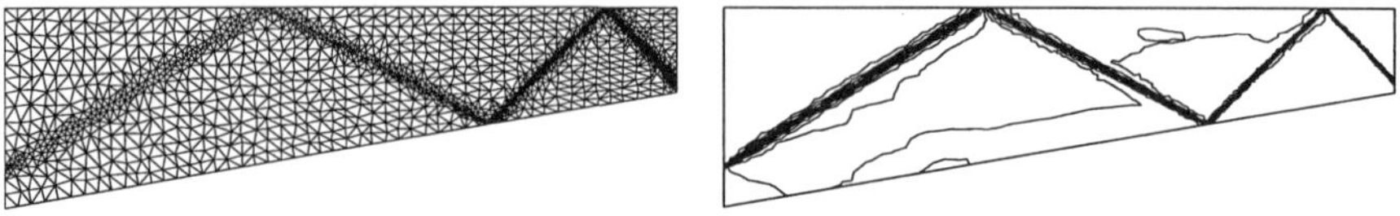

Fig. 44. (a) Adapted channel mesh, 1675 vertices. (left). (b) Mach contours, piecewise linear reconstruction (right).

for the piecewise linear scheme. This lack of dramatic improvement is not a surprising result since the solution has large regions of constant flow which do not benefit greatly from the quadratic approximation. At solution discontinuities the quadratic scheme reduces to a low order approximation which again negates the benefit of the quadratic reconstruction. To provide a fair comparison, Fig. 44 shows the same mesh adaptively refined. The number of mesh points has roughly doubled. A numerical solution was then obtained using linear reconstruction. The results are very comparable to the calculation performed using quadratic reconstruction while requiring less that 10% as much computational time.

Example: Transonic Airfoil Flow Figure 45 shows a simple Steiner triangulation and the resulting solution obtained with a linear reconstruction scheme for transonic Euler flow ($M_\infty = .80, \alpha = 1.25°$) over a NACA 0012 airfoil section. Even though the mesh is very coarse with only 3155 vertices, the upper surface shock is captured cleanly with a profile that extends over two cells of the mesh. Clearly, one benefit of using unstructured meshes is the ability to locally adapt the mesh to resolve flow features. Figure 46 shows an adaptively refined mesh and solution for the same flow. The flow features in Fig. 46 are clearly defined with a weak lower surface shock now visible. Figure 47 shows the surface pressure coefficient distribution on the airfoil. The discontinuities are monotonically captured by the scheme.

5 Conclusions

Several numerical methods for solving hyperbolic equations based on finite element and finite volume discretization have been considered. Fundamental to these methods are differing concepts such as symmetrization, energy

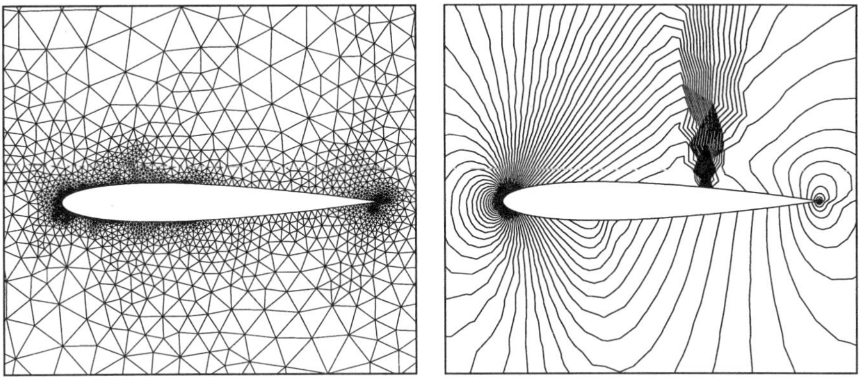

Fig. 45. (a) Initial triangulation of airfoil, 3155 vertices (left). (b) Mach number contours, $M_\infty = .80, \alpha = 1.25°$.

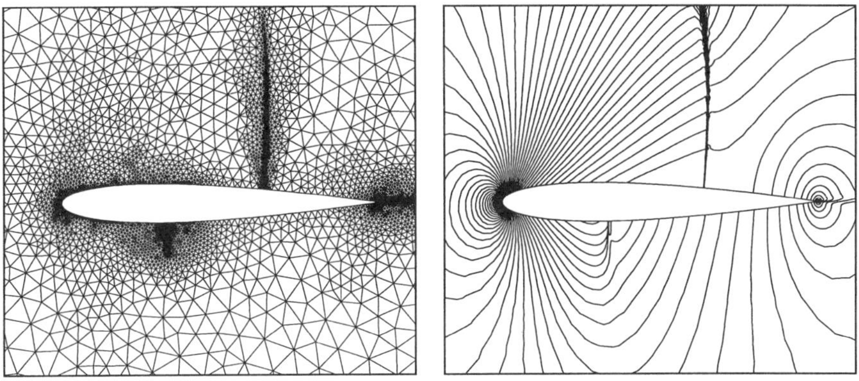

Fig. 46. (a) Solution adaptive triangulation of airfoil, 6917 vertices. (b) Mach number solution contours on adapted airfoil.

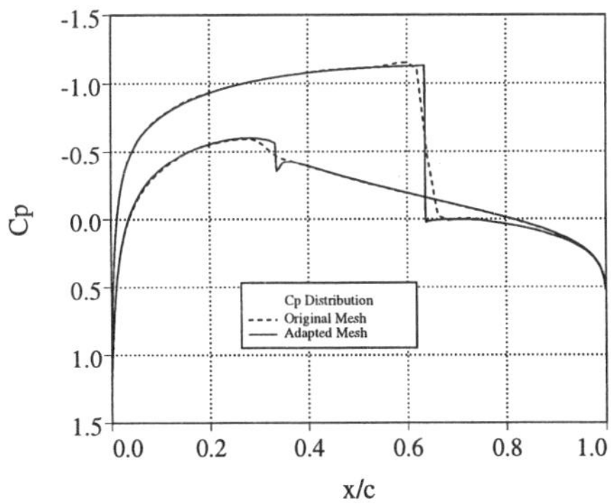

Fig. 47. Comparison of C_p distributions on initial and adapted meshes.

boundedness, and maximum principles. Ideally, a numerical method should possess all these properties. Except for very low order accurate methods, this is presently not the case. This suggests that there is still considerable room for the development of improved numerical methods.

References

1. R. Abgrall. An essentially non-oscillatory reconstruction procedure on finite-element type meshes. *Comp. Meth. Appl. Mech. Engrg.*, 116:95–101, 1994.
2. R. Abgrall. On essentially non-oscillatory schemes on unstructured meshes. *J. Comp. Phys.*, 114:45–58, 1995.
3. P. Arminjon and A. Dervieux. Construction of TVD-like artificial viscosities on two-dimensional arbitrary FEM grids. *J. Comp. Phys.*, 106(1):176–198, 1993.
4. D. Balsara. Higher-order Godunov schemes for isothermal hydrodynamics. *Astrophysical J.*, 420:197–203, 1994.
5. T. J. Barth. Some notes on shock resolving flux functions part 1: Stationary characteristics. Technical Report TM-101087, NASA Ames Research Center, Moffett Field, CA, May 1989.
6. T. J. Barth. Unstructured grids and finite-volume solvers for the Euler and Navier-Stokes equations, March 1991. von Karman Institute Lecture Series 1991-05.
7. T. J. Barth. Recent developments in high order k-exact reconstruction on unstructured meshes. Technical Report 93-0668, AIAA, Reno, NV, 1993.
8. T. J. Barth. Aspects of unstructured grids and finite-volume solvers for the Euler and Navier-Stokes equations, March 1994. von Karman Institute Lecture Series 1994-05.
9. T. J. Barth and D. C. Jespersen. The design and application of upwind schemes on unstructured meshes. Technical Report 89-0366, AIAA, Reno, NV, 1989.
10. F. Bassi and S. Rebay. High-order accurate discontinuous finite element solution of the 2D Euler equations. *J. Comp. Phys.*, 138(2):251–285, 1997.
11. K. Bey. A Runge-Kutta discontinuous finite element method for high speed flows. Technical Report 91-1575, AIAA, Honolulu, Hawaii, 1991.
12. M. Brio and C. C. Wu. An upwind differencing scheme for the equations of ideal magnetohydrodynamics. *J. Comp. Phys.*, 75:400–422, 1988.
13. G. Chiocchia. Exact solutions to transonic and supersonic flows. Technical Report AR-211, AGARD, 1985.
14. P. G. Ciarlet and P.-A. Raviart. The combined effect of curved boundaries and numerical integration in isoparametric finite element methods. In A.K. Aziz, editor, **The Mathematical Foundations of the Finite Element Method with Application to Partial Differential Equations**, pages 409–474, New York, 1972. Academic Press.
15. P. G. Ciarlet and P.-A. Raviart. Maximum principle and uniform convergence for the finite element method. *Comp. Meth. Appl. Mech. Engrg.*, 2:17–31, 1973.

16. B. Cockburn, S. Hou, and C.W. Shu. TVB Runge-Kutta local projection discontinuous Galerkin finite element method for conservation laws IV: The multidimensional case. *Math. Comp.*, 54:545–581, 1990.
17. B. Cockburn, S.Y. Lin, and C.W. Shu. TVB Runge-Kutta local projection discontinuous Galerkin finite element method for conservation laws III: One dimensional systems. *J. Comp. Phys.*, 84:90–113, 1989.
18. B. Cockburn and C.W. Shu. The Runge-Kutta discontinuous Galerkin method for conservation laws V: Multidimensional systems. Technical Report 201737, ICASE, NASA Langley R.C., 1997.
19. P. Collela and P. Woodward. The piecewise parabolic methods for gasdynamical simulations. *J. Comp. Phys.*, 54, 1984.
20. M. G. Crandall and A. Majda. Monotone difference approximations for scalar conservation laws. *Math. Comp.*, 34:1–21, 1980.
21. B. Delaunay. Sur la sphére vide. *Izvestia Akademii Nauk SSSR*, 7(6):793–800, 1934.
22. J. Desideri and A. Dervieux. Compressible flow solvers using unstructured grids, March 1988. von Karman Institute Lecture Series 1988-05.
23. A. C. Galeão and E. G. Dutra do Carmo. A consistent approximate upwind Petrov-Galerkin method for convection-dominated problems. *Comput. Meth. Appl. Mech. Engrg.*, 68, 1989.
24. F. R. Gantmacher. *Matrix Theory.* Chelsea Publishing Company, New York, N.Y., 1959.
25. S. K. Godunov. A finite difference method for the numerical computation of discontinuous solutions of the equations of fluid dynamics. *Mat. Sb.*, 47, 1959.
26. S. K. Godunov. An interesting class of quasilinear systems. *Dokl. Akad. Nauk. SSSR*, 139:521–523, 1961.
27. S. K. Godunov. The symmetric form of magnetohydrodynamics equation. *Num. Meth. Mech. Cont. Media*, 1:26–34, 1972.
28. J. D. Goodman and R. J. Le Veque. On the accuracy of stable schemes for 2D conservation laws. *Math. Comp.*, 45, 1985.
29. A. Harten. High resolution schemes for hyperbolic conservation laws. *J. Comp. Phys.*, 49:357–393, 1983.
30. A. Harten. On the symmetric form of systems of conservation laws with entropy. *J. Comp. Phys.*, 49:151–164, 1983.
31. A. Harten, J. M. Hyman, and P. D. Lax. On finite-difference approximations and entropy conditions for shocks. *Comm. Pure and Appl. Math.*, 29:297–322, 1976.
32. A. Harten, S. Osher, B. Enquist, and S. Chakravarthy. Uniformly high-order accurate essentially nonoscillatory schemes III. *J. Comp. Phys.*, 71(2):231–303, 1987.
33. T. J. R. Hughes, L. P. Franca, and M. Mallet. A new finite element formulation for CFD: I. symmetric forms of the compressible Euler and Navier-Stokes equations and the second law of thermodynamics. *Comp. Meth. Appl. Mech. Engrg.*, 54:223–234, 1986.

34. T. J. R. Hughes and M. Mallet. A new finite element formulation for CFD: III. the generalized streamline operator for multidimensional advective-diffusive systems. *Comp. Meth. Appl. Mech. Engrg.*, 58:305–328, 1986.
35. A. Jameson. Analysis and design of numerical schemes for gas dynamics. Technical Report TR 94-15, RIACS, NASA Ames R.C., Moffett Field, CA, 1995.
36. C. Johnson. *Numerical Solution of Partial Differential Equations by the Finite Element Method.* Cambridge University Press, Cambridge, 1987.
37. C. Johnson and J. Pitkäranta. An analysis of the discontinuous Galerkin method for a scalar hyperbolic equation. *Math. Comp.*, 46:1–26, 1986.
38. C. Johnson and A. Szepessy. Convergence of the shock-capturing streamline diffusion finite element methods for hyperbolic conservation laws. *Math. Comp.*, 54:107–129, 1990.
39. D. Kröner, M. Rokyta, and M. Wierse. A Lax-Wendroff type theorem for upwind finite volume schemes in 2-d. *East-West J. Numer. Math.*, 4(4):279–292, 1996.
40. D. Kröner, S. Noelle, and M. Rokyta. Convergence of higher order upwind finite volume schemes on unstructured grids for scalar conservation laws in several space dimensions. *Numer. Math.*, 71(4):527–560, 1995.
41. C. Lawson. Properties of n-dimensional triangulations. *CAGD*, 3:231–246, 1986.
42. P. D. Lax. *Hyperbolic Systems of Conservation Laws and the Mathematical Theory of Shock Waves.* SIAM, Philadelphia, Penn., 1973.
43. C. R. Michel. Improved reconstruction schemes for the navier-stokes equations on unstructured meshes. Technical Report 94-0642, AIAA, 1994.
44. M. S. Mock. Systems of conservation laws of mixed type. *J. Diff. Eqns.*, 37:70–88, 1980.
45. K. G. Powell. An approximate Riemann solver for magnetohydrodynamics (that works in more than one dimension). Technical Report 94-24, ICASE, NASA Langley R.C., 1994.
46. P. L. Roe. Approximate Riemann solvers, parameter vectors, and difference schemes. *J. Comput. Phys.*, 43, 1981.
47. P. L. Roe and D. S. Balsara. Notes on the eigensystem of magnetohydrodyamics. *SIAM J. Appl. Math.*, 56(1):57–67, 1996.
48. D. Serre. Remarks about the discrete profiles of shock waves. *Mat. Contemp.*, 11:153–170, 1996.
49. F. Shakib. *Finite Element Analysis of the Compressible Euler and Navier-Stokes Equations.* PhD thesis, Stanford University, Department of Mechanical Engineering, 1988.
50. S. Spekreijse. *Multigrid Solution of the Steady Euler-Equations.* PhD thesis, Centrum voor Wiskunde en Informatica, Amsterdam, 1987.
51. R. Struijs. An adaptive grid polygonal finite volume method for the compressible flow equations. Technical Report 89-1959-CP, AIAA, 1989.
52. R. Struijs. *A Multi-Dimensional Upwind Discretization Method for the Euler Equations on Unstructured Grids.* PhD thesis, T.U. Delf and the VKI Institute, 1994.

53. B. van Leer. Towards the ultimate conservative difference schemes V. a second order sequel to Godunov's method. *J. Comp. Phys.*, 32, 1979.
54. P. Vankeirsbilck. *Algorithmic Developments for the Solution of Hyperbolic Conservation Laws on Adaptive Unstructured Grids*. PhD thesis, Katholiek Universiteit van Leuven, 1993.
55. V. Venkatakrishnan. On the accuracy of limiters and convergence to steady state. Technical Report 93-0880, AIAA, Reno, NV, 1993.
56. P. Woodward and P. Colella. The numerical simulation of two-dimensional fluid flow with strong shocks. *J. Comp. Phys.*, 54:115–173, 1984.

Druck: Strauss Offsetdruck, Mörlenbach
Verarbeitung: Schäffer, Grünstadt